FOREST ECOLOGY

Edited by

Lindsay Patterson

Toronto Academic Press

FOREST ECOLOGY

Editor: Lindsay Patterson

Toronto Academic Press
4164 Lakeshore Road
Burlington ON L7L 1A4
Canada
www.tap-books.com

Email: orders@arclereducation.com

ISBN: 978-1-77956-328-6

ABOUT THE EDITOR

Lindsay Patterson is an ecologist, wildlife researcher, and copy editor passionate about animals, conservation, and the power of the written, and the spoken word. She has an undergraduate degree in English and Philosophy studies, an honors degree in Environmental Sciences, and a doctorate in Ecology, which has deepened her understanding of working hard for what you love. Her passion for wildlife research began with the Chacma baboons of South Africa's Table Mountain National Park and led her to the Cederberg Mountain's elusive Cape leopards, Madagascar's unique diversity of lemurs, frogs, and snakes, and KwaZulu-Natal's precocious vervet monkeys. In her 10-year career as a natural history researcher, she worked on over 200 nature documentaries for award-winning production houses, including Love Nature, National Geographic, Discovery, and Smithsonian. In her downtime, she loves camping in remote spaces and is an avid reader of everything.

Table of Contents

List of Figures

List of Tables

PREFACE

Forests occupy more than 4 billion hectares of the Earth's land surface, performing a multitude of functions both as natural formations and as products of human activity. A forest ecosystem comprises organisms, including plants, animals, and microorganisms, and their habitats such as soil and air. The functioning of these ecosystems is characterized by energy flow and nutrient cycling, biomass production, and the decomposition of dead organic matter by microbes. Microorganisms play critical roles in regulating ecosystem processes, from nutrient cycling to soil carbon storage, fluxes, the transformation of aqueous solutes, and the processing of water pollutants.

Human activities are integral to ecosystem dynamics, and perceptions of forests are evolving. Forests hold 80% of all aboveground carbon and about 40% of all belowground carbon, playing a crucial role in terrestrial functions. Furthermore, forest soils contain a large amount of mineral nutrients essential for forest growth and regeneration. In recent years, human activities have significantly altered the structure and functioning of forest ecosystems worldwide. Deforestation and conversion to agricultural land are the most visible threats, but forests are also increasingly exposed to atmospheric nutrient deposition and climate change.

The three main functions performed by a forest ecosystem are protective, productive, and social. Protective functions include soil and water conservation, climate regulation, and habitat provision for biodiversity. Productive functions encompass timber, non-timber forest products, and ecosystem services like carbon sequestration. Social functions involve recreational, cultural, and educational benefits provided by forests.

Forest structure is both a product of and a driver of ecosystem processes and biological diversity. Recent studies have shown that changes in forest structure due to management practices or disturbances can have undesirable consequences for other components of forest ecosystems and humankind. For instance, deforestation can lead to loss of biodiversity, altered water cycles, and increased greenhouse gas emissions. Conversely, sustainable forest management practices can enhance forest resilience, promote biodiversity, and support ecosystem services.

Given the crucial role forests play in maintaining ecological balance and supporting human well-being, this book is devoted to the structure and functioning of forest ecosystems. It aims to deepen our understanding of how forest structure influences ecosystem processes and biodiversity, the impact of human activities on forests, and strategies for sustainable forest management. By addressing these topics, we can better appreciate the complex dynamics of forest ecosystems and develop informed practices to preserve and enhance their functions for future generations.

Content/Structure of the Book

The content coverage in this book is organized into nine chapters, which are:

Chapter 1 provides foundational knowledge about forests and forestry, starting with defining these terms. It offers a detailed classification of forests, emphasizing tropical forests and their various types, such as tropical rainforests, lowland evergreen rainforests, and mangrove forests. Each forest type is described by its geographic delimitation and unique characteristics. The chapter also explores the different branches of forestry and their interrelations, providing a comprehensive overview of silviculture, including its definition, objectives, and scope.

Chapter 2 examines forest ecosystems, discussing their types and characteristic features. It examines the structure of forest ecosystems and concepts related to plant communities, such as vertical stratification and community development. The chapter covers nutrient cycling and water relations, focusing on the water, oxygen, carbon, and nitrogen cycles. It also addresses plant physiology in stress environments, discussing drought stress, water logging, salinity, and alkalinity. The conservation of forest ecosystems is highlighted, emphasizing the importance and methods of conservation.

Chapter 3 explores the historical development and evolution of city forests, tracing their roots from the nineteenth century and examining changes in the twentieth century. It discusses the sustainability of forests in urban ecosystems, the tasks involved in managing urban forests, and the challenges such as water scarcity, urban sprawl, and pest and disease concerns. The chapter also addresses community resistance to change and strategies for convening expectations.

Chapter 4 focuses on the components of forest soils, including soil minerals, water, organic matter, microorganisms, and gases. It discusses nutrient cycling dynamics and the decomposition process, highlighting its regulation and control. The chapter examines forest soil productivity, potential impacts on soil properties, and measures for protection. Enhancement methods for forest soil productivity, such as fertilization, genetic improvement, and species selection, are also covered.

Chapter 5 examines the role of forests in the hydrological cycle, focusing on forest roots, plant and animal life, and sheltering. It discusses the effects of forests on annual runoff, floods, erosion, sedimentation, and water quality. The chapter also evaluates land treatment methods, such as rapid infiltration and infiltration procedures.

Chapter 6 explores the components and importance of the environment, including the atmosphere, hydrosphere, lithosphere, and biosphere. It discusses principles of environmental management and conservation, such as the Polluter Pays Principle and the Precaution Principle. The impact of deforestation on climate change and other environmental aspects is also examined.

Chapter 7 discusses forest regeneration objectives and ecology, covering both natural and artificial regeneration methods. It explores the processes, requirements, advantages, and disadvantages of natural regeneration. The chapter also covers artificial regeneration methods, their scope, objectives, merits, and demerits, and factors affecting the choice of regeneration methods.

Chapter 8 addresses important diseases and insect pests of nurseries, plantations, and natural forests. It discusses the occurrence of pests and diseases, their management, and the assessment of losses. The chapter also covers biological control methods and molecular tools for developing disease-resistant trees.

Chapter 9 examines the impact of global climate change and warming on forests, discussing major and lesser greenhouse gases. It explores the effects and causes of deforestation, modern deforestation trends, and the severe consequences of deforestation on global climate and biodiversity.

Audience

This book is intended for a diverse audience that includes students, researchers, and professionals in the fields of forestry, ecology, environmental science, and natural resource management. It serves as a comprehensive resource for undergraduate and graduate students studying these disciplines, providing them with foundational knowledge and advanced insights into forest ecosystems. Researchers and academics will find the book valuable for its detailed analysis of forest structure, function, and dynamics, as well as its exploration of current trends and challenges in forest ecology. Environmental scientists and natural resource managers can use the book as a practical guide to understanding the complex interactions within forest ecosystems and applying this knowledge to conservation and sustainable management practices. Additionally, policymakers and environmental advocates will benefit from the book's thorough examination of the impact of human activities on forests and the importance of forests in global environmental health. By addressing a broad spectrum of topics, from the basics of forest classification to the latest research on climate change and forest regeneration, this book aims to equip its readers with the knowledge and tools necessary to contribute to the preservation and sustainable management of forests worldwide.

Learning Features

- Start-of-chapter learning objectives.
- Key terms in each chapter.
- Summary points at the end of each chapter.
- Review questions at the end of each chapter.
- Case studies and activities in each chapter for solving real problems.
- References at the end of each chapter.

CHAPTER 1

General Aspects of Forests

LEARNING OBJECTIVES

After studying this chapter, you will be able to:

- Define the terms "forest" and "forestry" and understand their significance.
- Classify forests based on geographic and ecological criteria.
- Explore the branches of forestry and their interrelationships.
- Explain the concepts of silviculture.
- Discuss the roles of forests in biodiversity preservation, nutrition-sensitive food systems, and landscape management.
- Analyze historical developments in forestry and human interactions with forests.

KEY TERMS FROM THIS CHAPTER

Biodiversity
Forestry
Landslide prevention
Tropical rainforests
Forest
Human interactions with forests
Silviculture
demonstrate

INTRODUCTION

Forest use and management have been around as long as man. Forests were originally primarily used for food, fuelwood, and construction materials. The early management of forests consisted primarily of burning, clearing, and other land use, particularly for agriculture, settlements, and infrastructure. Early industrialization increased the pressure on forests. Conversion and overuse have resulted in a significant reduction in forest areas in Europe, the Middle East, India, China, and, more recently, parts of North America. Currently, forests cover approximately one-fourth of the earth's surface. We rely on forests for survival, from the air to the wood we use. Furthermore, watersheds are protected, soil erosion is reduced, and climate change is mitigated. They provide habitat for animals as well as sources of income for people. Even though we rely so heavily on forests, we remain silent and allow them to vanish. Forests provide us with shelter, livelihood opportunities, water, and food and fuel security.

All of these activities, whether direct or indirect, involve forests. Some are easy to identify, such as trees, fruits, paper, and wood. Others are less obvious, such as byproducts found in common household items such as medicines, cosmetics, and detergents.

How important is it to sit in the shade of a forest tree and drink cool water or eat lunch in the morning? What is the worth of a view of a pristine grove of conifers or a stand of hardwood trees in full autumn bloom? And how do we value wildlife habitats, flood protection, landslides, and avalanches, as well as clean water, air, and, perhaps most importantly, climate? Humanity relies on forests for a variety of essential services.

For a long time, we valued the forest for things with a high financial value, such as wood and wood products, but were unable to demonstrate or capture the values of its services, which are difficult to quantify or even invaluable. It is not because our forests do not recognize these services. We've long recognized the value of our forests' key ecosystem services, but we've been slow to realize that paying for them is worthwhile, especially when the costs and responsibilities of forestry management are not in the public sector. We must understand how forests function to better understand what it takes to manage forest resources efficiently.

1.1. DEFINITION OF FOREST AND FORESTRY

The theories and practices of everything which consists of creating, preserving and scientifically managing, and using their resources is called as forestry (Anon., 1966). It covers everything related to the development and management of forests, including the harvest, marketing and use of all forest goods and services. It not only includes forestry management but also new forest management. Some legal definitions of' forest' are grounded on actual vegetation, while others are based on a defined area, which has no vegetation but really is legally within the competence of the national forest and natural resources management agency.

According to India's Forest Conservation Act of 1980, any land is forest land in any land record, whether or not there is any vegetation on the land.

The Philippines has provided a definition on accordance with the slope of the land -any untitled land is considered to be forestland that has a slope greater than 18%. "For example, not all countries classify shrubs or bamboos, and some countries count trees planted along road sides or fruit trees, as forests, each EU member State has its own specific forestry definition (and, in Belgium, each one of the three provinces has its own definition of forests).

Bamboos in some countries are considered grass, but bamboos are regarded as trees in the majority of Asian countries. Shrubs and bushes can be the most common woodland vegetation in the arid and semi-arid regions, yet shrubs are not classified as trees in all nations. In China, windbreaks, and certain orchards are included in the facts of the Ministry of Forests, whereas in Korea, forests are specifically excluded from its definition.

In terms of the FAO report of 1990; the forests in the developed countries are considered as land areas covered by 20% of the trees. The report of 2000 changed this definition to align it with the definition of a 10% tree cover for developing countries. For instance, the forest cover recorded increased from 40,000 km^2 to 158,000 km^2 (from 1990 to 2000) in Australia.

1.2. CLASSIFICATION OF FORESTS

Forests are sometimes classified as biomes, which are forest types that are closely related to the earth's climate regions. There are three main forest biomes: tropical, temperate, and boreal. Tropical temperate forests emerged during the world's cooling period, which began at the beginning of the Cenozoic period (65.5 million years ago).

Tropical forests are frequently mistaken for evergreen and moist forests, but only 60% of tropical forests are rainforests. Tropical forests, which account for more than half of the world's species and cover only one-tenth of the world's land area, are extremely important in terms of global biogeochemical cycles.

1.2.1. Geographic Delimitation and Varieties of Tropical Forests

Some key geographical and climate system facts must be considered when determining why and where tropical forests exist. The tropics are defined as the region between the Tropic of Cancer in the Northern Hemisphere at 23°26⊠09.9⊠ (or 23.43608°) N and the Tropic of Capricorn in the Southern Hemisphere at 23°26⊠09.9⊠ (or 23.43608°) S; these latitudes correspond to the axial tilt of the Earth.

Because of the high flow of solar energy in that region, sunlight is projected at a 908–influence angle onto the planet's surface. This high solar power flow causes a high rate of water evaporation in tropical oceans and evapotranspiration on tropical surfaces. As a result, warm, moist air rises in tropical latitudes.

As warm, moist air rises near the equator, it expands and cools due to the decrease in atmospheric pressure with altitude. This cooling causes the water vapor in the air to condense, forming clouds and leading to precipitation. The air mass, now cooler and drier, descends in subtropical regions, roughly between 23.5° N and 23.5° S latitude. This descent creates high-pressure areas and is associated with the formation of deserts. This pattern is part of the global wind circulation known as Hadley cells.

In contrast, the equatorial region, where the warm, rising air is prevalent, remains consistently warm and wet. This is due to the high levels of precipitation associated with the condensation of moist air, supporting lush tropical forests along the equatorial axis.

1.2.2. Tropical Rainforests

The primary types of tropical forests differ in their annual rainfall distribution, elevation, and soil type. Tropical forests under constant wets with no months of precipitation less than 100 mm are generally known as "tropical rainforests," though some also distinguish tropical 'moist woods' from 'rainforests' in a strict sense, which have annual precipitation of more than 4,000 mm. Dry periods are severe in tropical and dry forests, also known as 'monsoon forests,' during which most trees shed their leaves to reduce water loss (Figure 1.1).

Seasonal drought in half-green rainforests is less severe, and leafless periods are not comparable. Tropical forests are classified by elevation, soil type, and differences in forestry structure within these broad moisture regimes. The distinctive structural features include canopy height, crown layering, and the presence (or absence) of various climbers and epiphytes.

Other important physiognomic descriptors of tropical forests are tree support, crown shape, leaf structure, and flower/fruit formation position. Globally, the most important types of tropical humid forests are lowland evergreen rainforests, upper and low mountain forests, peat swamp forests, heath forests, cooler wild swamp forests, and mangrove forests.

Figure 1.1. *Tropical rainforests are vital for our planet, providing oxygen, absorbing carbon dioxide, and supporting incredible biodiversity.*

Source: Wikipedia, creative commons license.

1.2.3. Lowland Evergreen Rainforests

The tall and lush evergreen forests that most people associate with tropical forests are known as lowland evergreen rainforests. Large, canopy-like emergent trees with multiple layers of vegetation characterize the forests. Lowland evergreen rainforests are generally very diverse, with over 1,000 trees per square kilometer in the richest Amazonian and Southeast Asian forests.

Canopy and emerging trees frequently have large scattered crowns that reach 420 m when mature, are more than 1 m wide in lowland evergreen forests, and frequently have planks that provide physical support. The upper canopy layer contains lesser-known trees, treelets, and an herbaceous layer of ground vegetation.

Cauliflora and ramiflora are among the most common trees in lowland tropical rainforests. In general, there are a lot of lianas, which are wooden climbers that germinate in the winery but have climbing mechanisms (such as tendrils or hooks) that allow them to use independent trees for support. Lianas that reach the canopy remain anchored to the forest floor.

We also know about epiphytes, which are plants that live in other plants (usually trees) but are never rooted in the ground.

Many tropical epiphytes, such as orchids, ferns, and bromeliads, contribute significantly to tropical biodiversity.

Another group of plants found in tropical forests are hemi-epiphytes, which germinate in the canopy and produce roots that grow down the host's trunks to become rooted in the ground. The largest tropical semi-epiphyte group is figs (many Ficus species), the fruit of which is a valuable commodity for several vertebrate species.

1.2.4. Semi-Evergreen Rainforests

Semi-evergreen rainforests replace evergreen lowland rainforests in areas where moderate drought occurs annually. As the name indicates, certain trees in these forests have a seasonal period of unleavened leaf and up to a third of the top canopy of the leafy component. Laub and evergreen trees clump, so that the certain parts of the forest may be completely dominated by lawns. The variety of species is usually slightly lower in semi-green tropical forests than in the evergreen rainforest.

1.2.5. Tropical Dry Deciduous Forests

Many canopy trees lose their leaves in the time of dry season in tropical dry leaf forests, also known as monsoon forests. The dry season for trees is 2–6 months without leaves. For much of the year, high levels of light penetrate the understanding of these forests and common grass species. Savannah ecosystems are classified in many tropical dry forests.

1.2.6. Montane Rainforests

An observer in the tropics will notice strong changes in altitude vegetation. The highest mountain forests near the summit are short (1.5–1.8 m), with a canopy overlooking the gnarring branches with small thick leaves. The forests are rich in bryophytes and epiphytes, but there are virtually no butters, woody grape-climbers and vines (Figure 1.2).

The upper and lower mountain forests differ significantly in structure, with the latter resembling the lowland evergreen rainforest. Low mountain forests share many characteristics with lowland forests. Although adiabatic cooling ultimately drives changes in vegetation structure at altitude, wind exposure and soil characteristics may also play an important role.

Figure 1.2. *Montane rainforests.*

Source: Wikimedia Commons.

1.2.7. Heath Forest

Heath forests, also known as Kerangas in some parts of Asia, have a low, uniform canopy with low stratification. The canopy's relatively small leaves are brownish and reddish in color. The canopy is relatively open, and the level of light in the understanding is high.

Many plants, such as the pitcher plants from Borne (Nepenthes) found in Asian Kerangas, are unusual for heat forests. Podzolic soils grow in highly acidic and well-drained heat forests. In blackish or tea, draining water is generally colored and highly acidic.

1.2.8. Mangrove Forests

Mangrove, also known as "mangal," is associated with salt or brackish water and is thus found along tropical coasts. In terms of floristic diversity, the other forest types are severely underdeveloped. The Rhizosphoraceae are the dominant family in the system. Because the trees grow in salt or brackish water, special adjustments are required to tolerate high levels of salt and anaerobic conditions in their root areas (Figure 1.3).

Figure 1.3. *Illustration of mangrove forests.*

Source: https://blogionik.org/blog/2017/06/20/filter-salt-mangroves/.

Stilt roots and pneumatic species allow them to keep a portion of their rooting system above water, avoiding the problems associated with anaerobic habitat. Mangroves can also retain extremely high levels of salt in their tissue, allowing them to absorb saline water. Swamp forest formations are distinguished primarily by their habitat rather than by the structure and physiognomy of the forest.

Peat accumulates, but plant material does not decompose. The latter glacial maximum has caused rising water levels in certain tropical peat swamp forests, as well as in other areas with limited drainage along river paths, such as oxbow lakes. The forests are most common in Asia and the neotropics.

Sumptuous freshwater forests connect low-lying areas near river banks and are frequently flooded by water. These frequent floods result in high soil nutrients and rapid treetop development, particularly when flooding causes physical disturbances to the forest.

1.2.9. Tropical Land Masses

Tropical forests exist on all three major tropical landmasses: America, Africa, and Asia.

Differences in geologic history, climate, topography, and extent among these three areas result in distinct biota. The United States, or Neotropical region, has the world's largest tropical forest area, accounting for roughly half (4,106 km^2).

1.2.10. Tropical Forest Characteristics

The vegetation here is luxuriant, given the low seasonal difference, given the forest's geographical location and its high humidity.

- High animal and plant biodiversity.

- Evergreen trees.
- Dark and sparse undergrowth interspersed with clearings.
- Small litters (organic matter settled to the ground).
- Presence of 'strangler' creepers, such as *Ficus* spp..
- Presence of buttresses, (i.e., winged waves at the base of the timber), and of spindle roots in living trees and trees.

The forest is a highly productive and dynamic ecosystem that constantly changes in response to climate, disease, and human activity. Temperate forests grow in moderate climates between the tropics and boreal regions in both the northern and southern hemispheres. They are also known as "forests of four seasons," because they frequently experience four seasons in midlatitude climates.

1.2.10.1. Temperate Forest

This broad range includes a wide variety of forest types, including broad-ranging leafy forests, pine forests, and temperate forests that are relatively regionally constrained. The temperate forest biome is one of the most profitable in terms of net primary production while also being the most densely populated in terms of human occupation. However, the temperate forest biome is one of the most recent in terms of geology.

The majority of the trees in this biome evolved between 90 and 70 million years ago (late cretaceous), and only around 60 million years ago (in the tertiary) did a composition of trees related to what is now known as the "temperate forest biome" emerge. Temperate forests cover vast areas of North America, Eurasia, and a small portion of the southern hemisphere.

Eastern America and Canada, as well as Europe, China, Japan, and western Russia, have the highest proportion of "signature temperate forest types." Temperate forests have relatively long growing seasons and reasonable rainfall, which can spread fairly evenly throughout the year or concentrate in a specific season, with the most temperate forests dominating the feeding stocks that lose their leaves in the winter.

Drier climate conditions in western North America, for example, may be widespread, with evergreen pines and other drought-tolerant conifers. Temperate rainforests, which make up two-thirds of North America's Pacific Northwest, have a milder, more moist climate that is frequently maritime in nature than other temperate forests. The temperate deciduous forest appears scary in the winter because most of the leaves have fallen off the trees.

Wildlife can survive the winter or migrate to warmer climates in these forests. Spring is a period of regeneration, with hardwoods blown away and flowering shrubs and forbs proliferating. As the days shorten and the temperatures drop in the autumn, the leaves of the trees change color, and the animals begin to store food for the winter and/or the energetic migration needs.

1.2.10.1.1. The Flora of Temperate Forests

Many temperate forests have productive soils that can support a wide variety of trees. Temperate forests contain maples, oaks, elms, and birches. Conifers such as pines and hemlocks may play a minor role in hard wood-dominated communities, but these needle-leaved trees have the potential to re-establish themselves as dominant species in some temperate ecosystems, such as the temperate North American rainforest and Southeast American pine forests.

Evergreen leafy trees such as "living oaks" in California and southern Europe, as well as eucalypts in Australia, are typical of so-called Mediterranean climates. Moose, ferns, and understory shrubs are all common in temperate forests.

Temperate forests support a diverse range of wildlife due to their temperate climate and plentiful food resources. Koalas, possums, wombs, and other marsupials live in Australian temperate forests, while deer, bears, squirrels, rabbits, foxes, and wolves live in North American and Eurasian ecosystems. Giant and Red Bamboo Pandas live in China's temperate forests. Many migratory songbirds nest in temperate forests and eat flowers, berries, seeds, and insects in the spring and summer.

1.2.10.2. Boreal Forest

The boreal forest is considered the world's largest land-based biome. The boreal forest, which spans continents and many countries, makes a significant contribution to global biodiversity and climate. The boreal eco-region in North America extends from Alaska's tundra to Newfoundland.

Taiga, a boreal forest in Russia spanning approximately 12 million cubic kilometers, is one of the world's largest biomes. The forest is a mosaic of successor and sub-climax plant communities that are vulnerable to changing environmental conditions on both continents, including formerly glaciated regions and patchy glacier areas.

It has a small number of species, primarily sprouts, firs, and conifers that grow along waterways, with a few caduceus trees. The boreal forest appears to be associated with the location of the Arctic Summer airmass, which typically begins where it spreads its southern limit and extends to the southernmost in winter. As a result, it lies between the summer and winter positions of the Arctic Front.

The boreal forest, also known as the taiga, spans the subarctic and cold continental climate zones. It experiences significant temperature variations between winter and summer, with long, harsh winters that can last up to six months with average temperatures below freezing, and short, warm summers that typically last 50–100 days. For instance, Verkhoyansk, Russia, has experienced temperature extremes ranging from below -90° F (-68° C) to above 90° F (32° C).

Despite receiving an average annual rainfall of 15–20 inches (380–510 mm), the climate remains relatively humid due to low evaporation rates. The boreal forest is home to a variety of water bodies, including bogs, fens, ponds, shallow lakes, rivers, and wetlands. The forest covers approximately 19.2 million square miles (49.8 million square kilometers), making up about 33% of the world's land surface.

In terms of water and energy balance, the boreal ecosystem often behaves like an arid landscape during the growing season. This is because, while the moss layer remains moist throughout the summer, the poor soils and harsh climate conditions limit photosynthetic activity and result in low evapotranspiration rates. Much of the precipitation is retained by the moss and only slowly reaches the semi-impermeable layers below. Additionally, the dense vegetation canopy reduces the amount of solar radiation reaching the forest floor, further limiting photosynthetic processes.

As a result, much of the available surface energy dissipates as sensitive heat, causing the formation of deep (3000 m) and turbulent boundary layers in the atmosphere. These findings on surface energy partitioning are expected to have a significant impact on the development of weather and climate models, which typically depict the boreal environment as a freely evaporating surface.

Most importantly, moisture levels in the moss and humus layer never fall low enough to cause moisture stress in subsurface vegetation. If this finding is examined, it appears that root area moisture, a difficult variable to quantify on large spatial scales, has little effect on the surface energy balance.

Instead, important photosynthesis and evaporation control variables appear to be spring soil temperature, summer and autumn relative humidity, and air temperature in the atmosphere. This novel method of controlling regional evaporation rates is relevant to the question of whether the boreal ecosystem is a carbon sink or source.

Boreal forest expansion takes place at high elevations along mountain spines. This occurs in Eastern North America up to New Jersey, then in West Virginia, and finally in the Appalachians of the South. The trees in the north are spruce and balsam fir, while those in the south are Fraser firs. Fir typically grows at the highest altitude.

Yellow birch and eastern hemlock are also prevalent. These forests begin at 4,500 feet in the southern Appalachian Mountains and cool to sea level in the north (Maine and Canada). Because of its small land coverage, the Southern Appalachian boreal forest is unjustifiably considered a highly endangered ecosystem.

1.2.10.2.1. Boreal Forest Soils

The soils in this forest are known as podzols, which is a Russian word for ash (the color of the soils), and their development is known as podzolization. Podzolization happens when soil solutions form beneath needle-leaf trees. This means that iron and aluminum from the A horizon leach into the B horizon. Clays and other minerals migrate into the lower strata, leaving the upper layer sandy.

Because of the low temperatures, decomposition occurs slowly, and soil microorganism activity is limited. The hot needles of the dominant trees gradually degrade, forming a mat over the soil. Tannins and other acids cause high acidity in the upper soil, while evergreen trees' permanent shade reduces evaporation and wet soils.

In some cases, it is nearly year-round. In comparison to more southerly forests, this tends to restrict nutrient cycling.

1.2.10.2.2. Major Plant Species

Conifers, which adapt well to the harsh climate and thin, acidic soils, are by far the most common tree species. This area's typical species include black and white spruce, Tamarack, balsam fir, and Jack pine.

The four main generations include evergreen spruce (Picea), fir (Abies), and pine (Pinus), as well as deciduous larch or tamarack (Larix), the boreal biome's dominant vegetation. In North America, one or two spruce species are the most common. The Scot pine is a common feature of taiga in Scandinavia and Western Russia.

In the early succession, low leaf trees and shrubs belong to both the primary and secondary successions. Alder, birch (Betula), and aspen (Populus) are more prevalent. The boreal region's so-called culmination groups are now known to have a 200-year cycle of nitrogen-applying spruce-fir forests and nitrogen-accumulating aspen forests.

The conical or spike-shaped needle-leaf trees that are commonly found in boreal areas are adapted to the cold and physiological droughts of winter, as well as the short growing season. Needle-Leaf reduces the surface area over which water can be lost (transpired), particularly in the winter when frozen soil prevents plants from refilling.

Thick wax-coatings-an aquatic cuticle in which stomachs are sunk and protected against drying winds can also be found on boreal conifer needles. Evergreen habit— foliage retention enables plants to photosynthesize in the spring rather than wasting time growing leaves in the short growing season. (Note: Deciduous larks predominate in areas with nearly continuous permafrost and where the weather remains too cold and dry for waxy spruce and fir needles.) The dark green of the spruce and fir needles allows the leaf to absorb the most heat from the sun, resulting in photosynthesis at an earlier stage.

In European and Asian boreal forests, two other species, such as Norway and Siberia, influence spruce growth. The lynx is common throughout Russia's vast Siberian region and in wetlands. Larks are feces and are more common at the northern end. Evergreen species thrive in severe winters and have short growing seasons.

These trees can also snow in the winter, preventing them from breaking up and allowing them to begin photosynthesizing earlier in the spring, when the weather is favorable. Muskegs are common in the whole boreal forest-low lies, water-filled depressions or bogs-and occur in poorly drained glacial depressions. The spongy mat over ponded water is forming the spongy moss. Tundra species such as cotton grass and heat family shrubs are growing on this mat.

Black spruce and larch surround the edge. The sphinx moose may enhance water logging-it can keep up to 4,000% of its dry weight in water once it has been established. Species are often restricted when they are found. Some trees may reproduce by lying, as the likelihood of germination is low.

Pine forests are located in the sandy outwash plains and in former dune regions, and are dominated by jack pines in North America. These substrates are low in nutrients and drought and do not accept spruce and fir. In level areas underlain by permafrost, larch forests claim the thin waterlogged substratum. These forests are opened with shrubbery, moose and lichens understories. Larixlarichina's stalls are located phenomena in Alaska, but vast areas dominated by Larixdihurica are the result of extreme continentality and almost constant permafrost in Siberia east of the Yenisei River.

The boreal forest of North America is home to over 200 species of birds, as well as to species such as caribou, lynx, coyote and timber wolf. Since most trees have cones, animals evolved to make seeds from the cones, and the trees are adapted to prevent them, which often have spines on the cones. Since the spines are mostly cones. Crossbills (crossed by beaks) are seed extractors that are highly efficient. Highly lignified food is hard to digest for herbivores. Moose in the boreal are usually large herbivores. In sometimes of the winter, caribou use the forest as a refuge.

Moose (*Alces alces*), in Europe known as elk, tend to prefer deciduous browse and herbaceous plants, whereas lichens are used to caribou and needles can be eaten. The two large herbivores therefore have various food requirements–orange peel is an early (young) forest species, while caribou are a late (older) wood species. The beaver (*Castor canadensis*), which was the foundation for the first time in North America's trade in fur, is also a cattle of early succession communities.

Bears and wolves (where they were not exterminated) are abundant in the boreal. Snowshoe hares and lynx are usual throughout the environment, with unusually large feet to cross the snow. The boreal forest itself may be most characteristic of the lynx (Felis Lynx) and of various Weasel families (e.g., wolverine, fishing, pine martin, mink, hermit, sand). Snowshoe or different hares, red squirrels, lemmings and voles are the mammalian herbivores that they feed. Insect eaters such as wood warblers migrate among birds and leave after their breeding season. Seed eaters (e.g., sparrows) and omnivores (e.g., raven) are generally residents throughout the year.

In poor cone years, the taiga can be saved by the residents such as the grosbeak in the morning, pine siskin and red crossbill. The situation in the Boreal regions is alarming with these facts at hand. There are all problems that can be overlooked because the Boreal areas are not popular to fret. All these are problems.

Remember that forests, when cut down, take longer than forests that are logged in the tropical areas of the earth to regrow at these extreme, polar latitudes. Air pollution from smelter and nuclear power plants as well as weapons testing and water pollution & habitat destruction are particular problems that boreal regions face if marketing of Northern shipping routes becomes an adverse effect of new minerals and extraction of oil/gas new threats to endangered species Conservation and environmental groups believe. Large reserves able to sustain their ecological dignity must be adequately dedicated and careful environmental assessments must be simply carried out before any large-scale industrialized activity is decided by governments. Boreal forests are a huge amount of carbon, their biomass being so huge and so essential that the world's carbon dioxide decrease and the world's oxygen increase can occur in their maximum growth stage in northern summer and spring. For the world's ecosystem, the Boreal forests are equally important as the tropical forests.

All concerned with forestry and the environment should pay equal attention. Global change in the environment and the social, economic, and political globalization processes that drive concerns now influence local wildlife and management practices. At the same time, political changes and alliances facilitate the evolution of new institutions and the interaction between governmental institutions.

Some aim at facilitating further forest resource utilization and promoting economic development, while others are designed more to monitor or mitigate some of the impacts of these transformations on the environment and society. A series of environmental regimes, such as the Kyoto Protocol and the Biological Diversity Convention, are evolving at the international level in ways that could influence the development strategies of nations in the forest.

Decentralization at a really local level facilitates a return to more local forms of forestry management rather than a state-focused one.

1.3. BRANCHES OF FORESTRY AND THEIR RELATIONS

Mainly, the forestry has been grouped as follows:

1. **Basic Forestry:** Basic Forestry addresses the theory, practice and use of the production and production of forests. In India, the geographic area is around 32,80,500 km^2. Agriculture is the scientific and artistic research of plants and animals used by human beings. The forests cover 7,50,000 square kilometers. Agriculture is the largest enterprise in India (22.9% approximately), while the agricultural (cultivated) area is around 46%. In this profession, almost 70% of the population is employed.

   However, it is unfortunately incapable of meeting the ever-increasing population needs despite this. Like agriculture, forestry is also a good profession with a long-life differential. Silviculture covers definite aspects of theory and practice of forest rearing, planting, tree growing methods, growth and care until final harvesting.

   A. **Silviculture:** It is a significant component of the basic forest sector. But it refers to the cultivation of forest trees in simple words.

   B. **Forest Mensuration:** The extent of forest products is in simple language. However, the measurements from, voltage, age and increase of single trees, stands or whole woods, either standing or after felling are determined (e.g., height, diameter, volume, etc.). The subject is linear measurements of area, volume and weight.

   C. **Silvicultural Systems:** The silvicultural systems can be described as a process by which forestry crops are cultivated and harvested and supplanted by new crops of different forms are produced in accordance with the accepted sets of silvicultural Principles.

   D. **Forest Management:** The use of science, technique and economics in a forest estate to produce some desired outcomes. In reality, business methods are used for the process of a forest plantation. The Society of American Foresters (SAF) has portrayed it as the implementation of business methods and technical principles of forestry to the deployment of forest property.

   E. **Forest Use:** Forestry branch dedicated to the harvest, conversion, disposal as well as the use of forest products (British Common Wealth Forest Terminology, 1953).

The forestry branch deals in harvesting, processing, and delivery for the consumer of the forest products in accordance with SAF (1983).

F. **Law on Forests:** Law covers every action rule. The rules and laws imposed by the state on its citizens' actions because they have been penalized for breaching them.

Forest Law is classified as:

a. Constitutional Legislation:

Establishes the foundational principles and responsibilities for forest management and protection at the highest level of government authority.

b. Public Legislation:

Encompasses laws and regulations that govern the sustainable management, conservation, and use of forest resources for the benefit of the general public.

c. Private Legislation:

Includes regulations affecting individual landowners and entities, crucial for implementing sustainable practices and protecting forest resources on private lands.

d. Forest Wildlife Law:

Specifically protects wildlife within forest ecosystems, such as the Wildlife Protection Act of 1972, which safeguards animals and birds from poaching and habitat loss.

e. Forest Policy:

Guides the strategic framework for forestry management, focusing on balancing economic, social, and environmental objectives, as exemplified by policies like SAF (1983).

2. **Applied Forestry:** This contains those subjects which reference other subjects but make up the essence of forestry:

a. Dendrology.
b. Forest ecology.
c. Forest economy.
d. Forest entomology.
e. Forest fire.
f. Forest genetics.
g. Forest pathology.
h. Forest seed technology.
i. Forest soils.
j. Forest statistics.
k. Forests surveying.
l. Remote sensing.

Social forestry-social resent:

1. Agroforestry.
2. Forestry extension.
3. Afforestation.
4. Recreation forestry, etc.

Some other forestry branches:

1. Aesthetic forestry.
2. Commercial forestry.
3. Community forestry.
4. Extensive forestry.
5. Farm forestry
6. Extensive forestry.
7. Multiple use forestry.

1.4. DEFINITION, OBJECTIVES, AND SCOPE OF SILVICULTURE

Forestry is the art and science of controlling the forest and its establishment, development, structure, health, and quality to meet the various needs and values of landowners and society, such as wildlife habitat, timber, resource water, and sustainable restoration and recreation.

1.4.1. Definition

Forest treatment options include thinning, harvesting, seeding, pruning, prescribed burning, and planting. Intermediate treatments (thinning) are used after establishment or regeneration but before final harvesting to improve the growth, quality, vigor, and composition of the stand. The regeneration (harvesting) treatments are used to establish the new age class of trees on mature platforms.

Regeneration methods are grouped into four categories: coppice, even-aged, two-aged, and uneven-aged. In various ways, it has been defined. The definitions of forestry are given below:

- The art and science of forest crops is Silviculture. —Indian Forest and forest products Terminology (1957)
- Generally, the art and science of the cultivation of forest crops, which is based on the knowledge of silvios. More specifically, forest theory and practice of creating, composing, forming and growing. —Society of American Foresters (1983)
- Silviculture relates to certain elements of forest crop theory and practice, methods of tree growing, growth and after-care until the last harvest. — *Rao (1987)*

In simple words, Silviculture is the producing and nurturing stands of trees. Silva is the Latin term for forest and culture for cultivation. Silviculture is therefore the real art of forestry without exaggeration (Figure 1.4).

The silvicultural prescription shall apply to all vegetation activity, including prescribed fire, wildlife habitat improvement, timber harvesting, and tree removal in human security campgrounds. A forestry prescription is a document that contains a planned series of therapies aimed at changing a stand's current structure and achieving one of its management goals. The prescription usually considers ecological, financial, and social goals and limitations.

Figure 1.4. *Silviculture process.*

Source: https://www.forestryimages.org/browse/detail.cfm?imgnum=3948010.

Before the Forest Service implements a project or treatment, a certified forestry scientist must prepare or revise forestry prescriptions. Essentially, forestry is a discipline that addresses human needs through forest management. Because forests typically take a long time to grow, forestry relies on ecosystem management that is similar to natural ecosystems.

As a result, forestry is commonly known as applied ecology for forests. Agricultural Objectives A Silvicultural object is a collection of techniques that can be used to help achieve specific forest management objectives. It aims to achieve its objectives by manipulating the forest's composition and structure.

In most cases, forestry aims to improve the growth and quality of potential plant trees in a timber-producing forest. In the wild, we can find many species in one location. Some species avoid competition, while others remain at lower levels. Some of them die due to competition. Economic considerations typically govern silvicultural factors.

If there are many species, a forester may select those that are economically more important. The forester can also remove trees that are at risk of destruction. Because we are not intimately familiar with the natural and economic factors, the deviation from the purely natural path is not always predictable.

In nature, succession is the process by which one species or species group is replaced by another, and a stage emerges when a more stable species appears. Forestry studies allow foresters to understand the entire course of natural primogeniture on a given site, as well as the manner and rate at which existing crops are replaced or changed. This experience assists foresters in determining where and how to control succession.

1.4.2. Objectives

The main goals of Silviculture can be summarized as:

1. **Control of cropping composition and the invention of more economic value species:** A large number of species may flourish under natural conditions at the expense of desirable species, either below their value or less valuable. There are two ways of controlling:
   i. Through removal or removal of lower species.
   ii. The creation of better conditions to regenerate and grow desired species.
2. **Stand Density control, maximum volume production:** Trees will most likely grow too dense or too open in natural forests. When the trees are too dense, the production of wood is divided over a great number of people and none of them is as large as possible. If the trees are too small, production would be smaller, but individual trees with higher dimensions may grow enough.

   If the trees are too small, then they cannot efficiently and even insufficiently use the site to refurbish the region. Neither are good conditions for maximum production of wood. Forestry helps keep or retain sufficient trees per unit area so that maximum timber production is guaranteed through the optimal use of soil.

   The dense prop can provide a substantial increase in production by saving the trees otherwise these trees would have died.
3. **Afforestation of blank and underground areas:** A large area of forests is stored or blank because of fire, intrusion, illegal filling or certain natural causes. However, these areas are suitable for growing trees. Farming helps us by planting or seedling, to afford these areas with appropriate trees. Silviculture guidelines are intended to know the best collection period, kindergarten technique, and planting details, etc.
4. **Quality timber production:** Quality timber is not produced in unmanaged forests, due to strong competition or low competition. Many trees are malformed, defective and occasionally ill. Proper controls can boost the production of harmful agencies.

   The appropriate Silvicultural techniques and methods are used to control insects, fungi, fire, wind, pasture, sanding, etc., that affect wood quality.
5. **Rotation control Period:** Rotation is calculated from regeneration to harvest for years. When the number of trees is greater, the growth of each tree is therefore slow, in unmanaged forests it takes longer to reach harvestable size.

   The science of forestry helps to regulate crop density at different sizes and ages, helping to reach exploitable dimensions far more quickly. Thus, crop rotation can be reduced by adjusting the crop density. It helps also to identify short crops for rotation.
6. **Facilitate forest management and use:** Good forests exist in non-managed forests in difficult areas where wood management and harvesting are difficult.

The growth and distribution of forests in managed forests can be easily planted to make efficient and cost-efficient use of the products. In different places of this age group and species composition, it is possible to arrange the forest so that management is easy.

7. **Creation and introduction of exotic man-made woodlands:** forestry technologies help us to replace natural forests wholly or partially with mad human forests of the same or other species. Where there is, there are no valuable and desirable species in the forest. This important species can be cultivated.

   The forest can artificially be harvested and regenerated if the forest consists of desirable species but is not properly regenerating. Identifying appropriate exotics according to the geography, trail raising, plantations, selection of appropriate exotic species, nursery perfection and exotic planting techniques are some of the forestry methods that help in the introduction in large-scale exotic species.

8. **The purpose:** To provide maximum protection of the site and to ensure intangible returns from the forests. Protection of site and intangible returns: Significant intangible returns include climate moderation, increased precipitation, reduction of soil erosion, and soil and water conservation floods which enable a large number of wild animals to shelter. Silviculture enables the tree needs and its impact on the site to be understood.

   The species that may damage the site are discarded. Only those species are preferred, which provide the site full protection and guarantee the continued flow of intangible benefits.

1.4.3. Scope of Silviculture

Forestry is expanding beyond traditional timber management objectives. Foresters are working with professional fellows and farmers throughout the forest system to develop innovative forestry prescriptions that meet a variety of resource management goals on other public lands and, gradually, on private lands. Some of these innovations include therapies, schedules, or intensities that are not supported by published literature or ongoing research. The management of forests changed paradigms at the turn of the 21st century. Standard processes such as rainforest deforestation were critical in light of the new forest conceptual approaches widely recommended over three decades ago (Franklin et al., 2002). Furthermore, the tactics used to make strategic decisions have changed as a result of these reviews of strategic forest management thinking. For example, three decades ago, forest management was primarily used to produce wood, so the terms were widely regarded as synonymous. Some professionals still believe that forestry cannot be used in any context other than wood production, and that a different term should be used for manipulation in a forest without an objective.

1.5. STATUS OF FORESTS AND THEIR ROLE

Indian forests range from tropical wet and evergreen forests in the northwest and southwest to tropical rainforests, as well as tropical and dry thorny forests in central and western India. India's forests can be divided into 16 major types and 221 subtypes. The western Himalayas, east Deccan, northeast, and Western Ghats contain the vast majority of these forests. They are also an important biodiversity repository, serving more than just the needs for timber, fuelwood, and forests. They also provide a diverse range of essential products and services to ecosystems, as well as support for millions of forest-dependent people. Forests, on the other hand, are almost completely ignored in the Indian economy, and their economic and social value is underestimated. In recent years, India has successfully reduced deforestation rates and overall forestry cover for several years, with an increase in treed plants (such as agroforestry plantations, tea and coffee estates with trees, fruit garden plants, cocoon plants, and more) treated as forests since 2001 by the Indian Forest survey. The quality of existing forest resources, however, is deteriorating due to a variety of factors. While India covers only 2.4% of the world's land area, it is home to 17% of the world's population and 18% of the world's animals. India's per capita forest cover is also 0.064 ha, while the global average is 0.64 ha. As a result, India's forests are under biotic pressure, causing a decline in forest cover quality and productivity.

1.5.1. The Role of Forests, Trees, and Wild Biodiversity for Nutrition-Sensitive Food Systems and Landscapes

The importance of forests in the country's social, cultural, economic, and industrial development, as well as the preservation of their ecological balance, is becoming more widely recognized. Forests help to promote fair, inclusive, and sustainable economic development while also protecting natural heritage by providing a variety of socioeconomic and environmental benefits. Forests also provide opportunities for maximizing broad development and poverty reduction in forest-dependent communities, as well as addressing global issues such as climate change and biodiversity.

Food safety and nutrition provide numerous benefits to forests, trees, and natural biodiversity in nutritionally sensitive food systems and landscapes. Foods (such as fruits, vegetables, brown meats, fish, and insects) are essential for the diet, diversity, and nutrition of people living in a diverse landscape. Forests and trees provide fuelwood, which is an important component of the food system that is frequently overlooked in rural areas around the globe. Forest and tree products contribute significantly to the incomes of those who live in and around them; purchasing them is often the only way to gain access to the financial economy and thus nutritious foods.

Forestry is also proving to be resilient: during times of food scarcity, forest products become more common and can serve as livelihood safety nets. Forest and tree products may contribute to the nutritional sensitivity of global food systems once they reach the market, especially if markets are encouraged and developed in a nutritionally sensitive manner. Biodiversity, forests, and trees outside of forests all benefit from ecosystem services, which are necessary for sustainable and nutritious agricultural systems.

A better understanding of the significance of these relations and their spatial levels is required to ensure that they are not overlooked in politics and practices. Forests and trees are important in agricultural systems because of their role in food sustainability and nutrition sensitivity. To achieve the goal of reducing global malnutrition, nutrition must become a cross-cutting issue not only in health, development, education, and agriculture, but also in forestry and conservation.

Forestry, agricultural trees, and biodiversity are widely acknowledged as critical to the sustainability of food and agricultural systems; even large-scale monoculture processes in developed countries rely on remote forests and biodiversity to serve ecosystems, such as water management and the development of genetic reproduction and innovations. In many rural development countries, biodiversity agriculture habits are constantly required for long-term improvement and play an important role in biodiversity preservation. Health experts have also long observed that a single approach to human and ecosystem health is neither practical nor effective. Forests play an important role in larger landscapes with many farming systems, as they cover approximately 30% of the Earth's surface (FAO, 2011).

1.5.2. The Role of Forests in the Preservation of Biodiversity

Wildlife roles contribute to the Earth's ecosystem, genetic diversity, and species. Many forest woods have particularly high genetic diversity, which can have far-reaching consequences, despite being underappreciated due to their loss. Temperatures and rainfall generally rise, and three major forest areas are defined based on climatic conditions: tropical, temperate, and boreal. Each of these primary forest ecosystem types contains many finer ecosystems. Tropical forests have fantastic species levels and ecosystem diversity, but many of these species are now under threat from harvesting. Forest fragmentation, exotic species introduction, and climate change all have an impact on the patterns and quantities of forest biodiversity. Small intact remains of forest and forest-dependent species were or still exist in a number of areas, particularly temperate and tropical regions, but only in isolated fragments and heavily modified landscapes (Figure 1.5).

Figure 1.5. *Forests role in the preservation of biodiversity.*

Source: https://courses.lumenlearning.com/wm-nmbiology2/chapter/the-diversity-of-life/.

Montane forests are particularly interesting because they are home to a variety of endemic species that cannot migrate to other suitable habitats, such as warmer climates. It is the world's largest and most diverse land ecosystem, containing the greatest number of plants and animals. Even the most primitive forests are uncoordinated forms of life, ranging from literal soil microbes and arthropods to lichen and Lepidoptera that adorn the forest canopy's high branches.

Tropical forests alone account for more than 50% of global biodiversity. Many uncovered species continue to live in tropical areas. Forests, particularly in the tropics, are also under significant threat. In 1995, Kimmins, a British Columbia forest researcher, estimated the projected annual increase in global demand for forest products. So, even if current logging practices are consistent, a new area the size of British Columbia will be required for future harvests. Globally, many of the current logging processes are not sustainable. There is also a significantly higher risk in regions with the highest biodiversity.

Forests are estimated to contain two-thirds of the world's terrestrial biota, with tropical forests housing the greatest proportion of the terrestrial biota. Forests are constantly converted to other uses, and collection activities disrupt them so severely that they become scrub or semi-desert. Some parts of the tropics are estimated to lose 2% of their forests each year. As a result, forests' role in biodiversity conservation diminishes by the day.

Forest ecosystems are constantly changing, often due to interruptions that completely alter the ecosystem structure, resulting in the formation of a new ecosystem and the emergence of a large number of new species. Other wood-based ecosystems remain largely stable in natural conditions over many centuries, with only a small number of trees affected by localized insects or diseases. Some life forms are essential for the survival of a specific ecosystem. Others are opportunistic due to temporary circumstances. People have used forests all over the world for thousands of years, and as population and technology have grown, their impact on forest ecosystems has accelerated. Forestry residents have frequently used fire to improve gambling conditions and clear farmers' land. Forest cultivation has existed for thousands of years in Asia, Africa, and South America. People were one of several species that manipulated the forest to benefit other species the most.

1.5.3. The Role of Forests and Forestry in the Prevention and Rehabilitation of Landslides

Landfall is expected to have an impact on population growth, infrastructure development, forest and agriculture activities in sloping areas, and an increase in rainfall incidents. Climate change in temperate and tropical Asia also tends to increase forest dieback and fire events, which impair the ability of trees to stabilize soil layers and contribute to a disproportionate death rate in poor and vulnerable countries. The role of forests in reducing landslide risks in and outside of Asia has been mitigated. These measures can be modified to further alleviate the hardships and suffering of the most vulnerable. Deep-rooted trees and bushes can reduce fast-moving slides by strengthening surface soil layers and improving drainage (Sidle et al., 2006). Forests can also help reduce and block waste streams and rocks by forming a physical barrier to sliding and falling materials (Guthrie et al., 2010). Roots can pierce the entire soil shell by providing anchors in stabilization substrates, and thick lateral systems also protect the soil's surface layers from tears. Large sweating canopies may also lower soil water levels and reduce the risk of landslides (Sidle et al., 2006; Dolidon et al., 2009).

Although woody vegetation can prevent landslides, its cover has a limited impact on deepening landslides due to evapotranspiration. Rainforests, tables, geographic convergence points, and steep slopes that rise rapidly are typically more vulnerable to terrestrial hazards. In cases of continuous heavy rainfall or earthquakes, the trees' increased stability is frequently denied. Activities in these high-risk areas should be avoided or limited, particularly when forest clearing occurs. Logging, road and path construction, and forest conversion are all examples of land use practices that contribute to slope instability. Roads contribute two orders of magnitude more than untouched and steep forests, compared to other land uses.

The majority of rural Asia has built highways with insufficient engineering attention and, in many cases, inadequate drainage. The greatest risk of sloping from central highways is due to surface water interception, overload, and slopes. The removal of wood from inclining soil reduces rooting strength for up to two decades, even after regeneration (Sidle et al., 2006). In temperate climates, the risk of landslides increases as roots degrade three years after timber harvest, and their susceptibility remains high until roots regenerate after about 15 years. Crop growth changes due to soil depletion can reduce regeneration and regeneration in tropical climates.

Deep terrains are also dried to a lower level, increasing the risk of slides and permanently reducing root depth and strength as trees are converted to cropland. Placement and terracing, low ground cover, reduced rooting depth and strength, and decreased perspiration are all potential risks when compared to forests. Some risks, particularly those with a high tree component, can be mitigated by using land for agricultural forestry.

1.6. HUMAN INTERACTIONS WITH FORESTS

Previous sections discussed non-living habitats, forest plants, and the interactions of forest animals. This section discusses the human component of the forest web of life and the use and impact of forests. Human-forest interactions can have a variety of effects on both forests and people. We have a long history of forests and their wealth, but they are not limitless and are becoming increasingly vulnerable to extinction. For a long time, human history and forestry have been inextricably linked. For thousands of years, forests have been seen as enchanting, spiritual destinations. They have inspired people's traditions, representing knowledge, fertility, and life (Figure 1.6).

The forest's value extends far beyond aesthetic and cultural considerations; it connects many areas critical to human and environmental health. Forests have long been valued as sources of feed, fuel, materials, and land, but scientific advancements have demonstrated that they also provide important, less visible benefits. They act as a massive carbon sink, absorbing and storing carbon dioxide from the atmosphere in their woody biomass. Their extensive underground root networks connect soils, prevent erosion, and protect against flooding and landslides. They have an impact on local climate patterns, lower tropical sparkling temperatures, and activate local cloud systems, bringing precipitation to life. In short, forests, which are rich in both tangible and intangible resources, are critical to the planet's health.

Your appetite for wealth has become unsustainable over the centuries. Intensive logging, land development, and urban sprawl have resulted in the clearing of 46% of the world's forest since civilization began. Forestry products now power a multibillion-dollar, millions-dependent industry. However, the sustainable management of these valuable resources remains a challenge with a yearly loss of 7.6 million hectares of forests, mostly from the tropics, despite the commendable efforts of international agencies and national governments. Protecting environmental forests has become a "terrible issue," while still reaping long-term benefits from their numerous assets. Forests are interlinked systems with intricate dependencies: forestation programs that attempt to replace losing ecosystems may result in excessive demands of water and society's deficiencies.

Small-scale selective logging may fragment forests and reveal carbon and biodiversity bruising. Forest systems represent disciplines. A multidisciplinary approach and a thorough understanding of their complexities are required for successful management. However, more and more of us no longer live beside forests. Cities will be increasingly attractive by 2050 from 3.5 billion to 6 billion. As cities are increasingly congested, water scarce, flood-exposed and polluted, they are constantly under pressure for a high quality of life while at the same time minimizing the environmental consequences, a challenge to be addressed by the United Nations Sustainable Development Goals.

To address this problem, urban projects need to combine a broad range of knowledge and expertise, including forestry, in urban planning programs and have clear objectives. In an increasingly urbanized world, with all of its challenges, we cannot treat people and forests separately. We come closer by revisiting the value of forests to strike a balance and achieve a sustainable way of life. There is always reason to hope that these great mutants will last and encourage future generations.

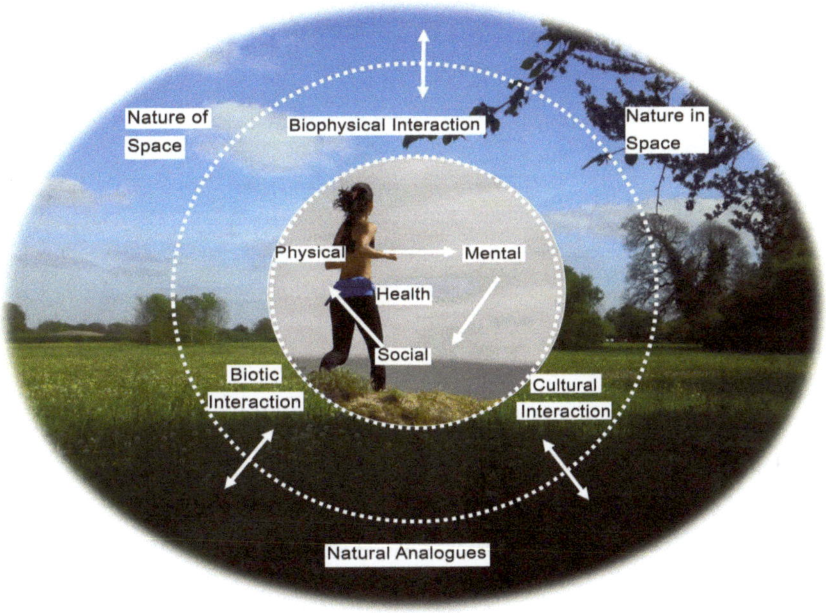

Figure 1.6. *Human-forest interaction.*

Source: https://www.frontiersin.org/journals/public-health/articles/10.3389/fpubh.2016.00260/full.

1.7. HISTORY OF FORESTRY DEVELOPMENT

Forest conservation and sustainable management are critical components of achieving long-term global development objectives in terms of the environment, economy, and society. In many areas of natural resource management and environmental protection, the term "sustainability" is used as a replacement or successor to the earlier concept of conservation. Sustainability has always been a major concern in the science and practice of forestry. In fact, the systematic forestry survey and scientific approach to forest management as renewable resources began in the first place. It was concerned that forest resources would be depleted unintentionally, resulting in unacceptable social and financial consequences. The area and diversity of resources represented by forests, as well as the social values associated with their protection and maintenance, have largely changed in recent years, laying the groundwork for "sustainable management of forests," as we now understand it.

This was simply intended to prevent local forest scarcity and was developed as a highly technological method for modeling growth, mortality, and risk to achieve a theoretical level of permanent wood removal. Changes in scientific understanding of the ecological function of forest ecosystems have called into question the notion that wood yield sustainability implies the preservation of all the components and natural processes required to maintain long-term health and productivity.

The ongoing uncertainty about what is sustainable from a social, economic, and ecological standpoint makes forestry management extremely difficult to optimize. In the United States and around the world, there is ongoing public debate about what constitutes sustainable forest management practices. This final section will provide a brief history of sustainable forestry, examine how natural resource and forest preservation policies have evolved, and investigate additional policies to create local and global opportunities for sustainable forest management.

Forest management practices (FMPs) are intended to manage forests in such a way that current forestry resources, products, services, and values, as well as societal wishes, can be met without jeopardizing future generations. The concept of forest sustainable development originated in Central Europe during the 17th and 90th centuries, when timber shortages were caused by wars, fuel consumption, land clearing for agriculture, and game cover; all forest demand was high. It was critical to produce high yields of forest products from the limited human timber fields (Bavarian State Food, Africans and Forestry Ministry, 2000). Stands have been harvested, resulting in a significant and likely reduction in mature stand output. To ensure the long-term maintenance of a non-declining yield, sustainability was intended to control the composition, stock, and growth of stand species. Related issues arose as a result of the long-standing production-centric approach to forest management and politics in the United States National Forest System (Floyd, 2002; Helms, 1998). For many urban resorts, particularly those in high-developed countries, sustainable production of high timber yields against other forest value appears to be less important than it was in the mid-1800s and earlier periods, and the concept of sustainability has recently expanded to include many of the more complex dimensions (i.e., the UNEP, 2011).

A new approach to forest maintenance has emerged as a result of forest management societies' new understanding of forests as ecosystems and changes in forest values; concepts that concern values outside of specific forestry management units, as we consider resource-dependent, landscape– and migratory-bird communities, and biodiversity. In forestry, however, these modern concepts of sustainability in the United States have been subsumed for at least a century, and much longer if the origins of sustainable forestry management from Late Medieval Europe are considered. Indeed, local forest resource use exceeds local supply, which is the primary reason for forest management science and practice. It acknowledged the potentially catastrophic consequences of forest resource depletion on the community level, in addition to providing an initial glimpse into the green understanding of forest biotechnology as key fundamental concepts of forestry sustainability. The concept of sustainable forestry has evolved significantly over the last century.

The intricate workings and diversity of forest resources, values, and ecological services have introduced new challenges and opportunities to our comprehensive understanding of forest ecosystems and their diverse nature. Forest management has always been a major focus of international trade and communication, whether in Oregon or Congo. The large and small forests of one of the world's largest natural biomes are now regarded as locals capable of significantly influencing global climate.

This increased understanding of the importance and vulnerability of forests to deliberate and unintentional destruction has given new urgency to ensure their survival. In the heated debate over the use of forest resources, governments have pitted their energies and faith against their own people, while the forest industry has opposed its communities.

The expansion has resulted in renewed efforts to find ways to collaborate in pursuit of common sustainable forest management objectives. Countless local meetings and nearly every international summit have brought together perspectives on certain fundamental principles of sustainable forest management. These generally accepted guidelines for the sustainable management of forests have begun to provide a common basis for evaluating forest conditions and trends, identifying the actions required to bring forest management closer to the sustainable objectives, and ensuring the financial and political resources required for the successful implementation of actions from forest to natural resource levels.

Each nation and forest owner will draw their own lessons and carry out their own work. The vision of forests as a source of sustainable renewable value sheds new light on individual responsibilities and opportunities for protection and management. And seeing forests in accordance with widely accepted principles of sustainable forest management – achieved locally, nationally, and internationally through broadly democratic processes – provides new opportunities for even the smallest timber or community forest managers to directly contribute to achieving global joint sustainability goals.

Many environmental challenges exist today that overwhelm our ability to "do anything about them" as individuals. However, regardless of size, those who play a role in global forest management are fortunate. They have the power and resources to achieve a noble goal: to preserve a critical component of the earth's living systems for current and future generations (Shirong et al., 2018).

CASE STUDY

AFRICAN FORESTS

African forests are critical ecosystems that play a vital role in biodiversity conservation, climate regulation, and the livelihoods of millions of people. Covering about 20% of the continent's land area, these forests are home to a wealth of plant and animal species, many of which are endemic and threatened by habitat loss. The Congo Basin, the second-largest tropical forest area in the world, is a significant carbon sink and a key player in mitigating climate change. However, African forests face significant pressures from deforestation due to logging, agriculture, mining, and urbanization. The intersection of economic development and environmental preservation poses a complex challenge for policymakers and communities.

In recent years, initiatives aimed at sustainable forest management have gained traction across Africa. Programs such as REDD+ (Reducing Emissions from Deforestation and Forest Degradation) seek to incentivize forest conservation while promoting the rights and needs of indigenous communities. Successful examples include community-based forestry projects in countries like Cameroon and Ghana, where local populations are empowered to manage their resources sustainably. These projects have demonstrated that involving communities in conservation efforts not only protects biodiversity but also improves their economic resilience.

Despite these positive developments, significant challenges remain. The increasing demand for land for agriculture, particularly palm oil and cocoa production, threatens forest integrity. Additionally, illegal logging and weak enforcement of environmental regulations exacerbate deforestation. Balancing the need for economic growth with the imperative of conserving African forests will require innovative solutions, including enhanced policy frameworks, better governance, and increased funding for conservation initiatives.

Discussion Questions

- How can African nations balance economic development and forest conservation while ensuring the rights and livelihoods of local communities?
- What role do international organizations and agreements play in supporting sustainable forest management in Africa, and how can their effectiveness be improved?

ROLE MODEL

WANGARI MAATHAI: A CHAMPION OF FOREST CONSERVATION

Early Life and Education

Wangari Maathai was born on April 1, 1940, in Nyeri, Kenya. Growing up in a rural village, she developed a deep appreciation for nature and the environment. She excelled in her studies and earned a scholarship to the United States in the early 1960s, where she attended Mount St. Scholastica College in Kansas. Maathai later transferred to the University of Pittsburgh, where she earned her bachelor's degree in biological science, becoming one of the first women in East and Central Africa to hold such a degree. She continued her education, earning a master's degree in veterinary anatomy from the University of Nairobi.

Career and Activism

Maathai's career began in academia, but she soon became involved in environmental activism. In 1977, she founded the Green Belt Movement, a grassroots organization focused on tree planting, environmental conservation, and women's empowerment. The movement aimed to combat deforestation in Kenya, restore the environment, and provide economic opportunities for women by promoting sustainable land management practices. Under her leadership, the Green Belt Movement mobilized thousands of women, planting over 51 million trees across Kenya and raising awareness about the importance of environmental conservation.

Legacy and Recognition

Wangari Maathai's tireless advocacy for the environment earned her global recognition. In 2004, she became the first African woman to receive the Nobel Peace Prize for her contributions to sustainable development, democracy, and peace. Her work transcended national boundaries, inspiring a global movement for environmental justice. Maathai authored several books and served as a member of various international organizations, including the United Nations Environment Programme. She continued her activism until her death on September 25, 2011, leaving behind a legacy of environmental stewardship and empowerment.

Wangari Maathai is remembered not only as a champion of forest conservation but also as a symbol of resilience, demonstrating that grassroots movements can lead to significant environmental and social change. Her life continues to inspire individuals and organizations worldwide to prioritize the protection of our planet.

SUMMARY

This chapter provides a comprehensive introduction to the fundamental aspects of forestry. It begins by defining what constitutes a forest and the broader field of forestry, emphasizing their importance in ecological and economic contexts. The chapter then goes to the classification of forests, detailing various types such as tropical rainforests, lowland evergreen rainforests, and mangrove forests among others, each characterized by unique geographic and ecological features.

Branches of forestry are explored next, highlighting their interconnections and roles in forest management. Silviculture, defined as the art and science of forest cultivation, is examined in terms of its objectives and scope, emphasizing its crucial role in sustaining forest ecosystems and biodiversity. The chapter also addresses the status of forests globally and their multifaceted roles, including their contribution to nutrition-sensitive food systems, biodiversity conservation, and efforts in natural disaster mitigation like landslides.

Human interactions with forests and the historical development of forestry practices conclude the chapter, illustrating how societal needs and environmental awareness have shaped the management and conservation of forests over time.

REVIEW QUESTIONS

1. What is the difference between a forest and forestry?
2. Describe the classification of tropical forests mentioned in the chapter.
3. What are the main objectives of silviculture?
4. How do forests contribute to biodiversity preservation?
5. Discuss the historical development of forestry and its impact on modern forest management.

REFERENCES

1. Aerts, R., Boot, R. G. A., & van der Aart, P. J. M. (1991). The relation between above- and belowground biomass allocation patterns and competitive ability. *Oecologia, 87*(4), 551–559. https://doi.org/10.1007/BF00317525.
2. Blair, B. (2001). Effect of soil nutrient heterogeneity on the symmetry of belowground competition. *Plant Ecology, 156*(2), 199–203. https://doi.org/10.1023/A:1013838212818.
3. Bläsing, O. E., Gibon, Y., Günther, M., Höhne, M., Morcuende, R., Osuna, D., Thimm, O., Usadel, B., Scheible, W. R., & Stitt, M. (2005). Sugars and circadian regulation make major contributions to the global regulation of diurnal gene expression in Arabidopsis. *Plant Cell, 17*(11), 3257–3281. https://doi.org/10.1105/tpc.105.034439.
4. Børja, I., De Wit, H. A., Steffenrem, A., & Majdi, H. (2008). Stand age and fine

root biomass, distribution and morphology in a Norway spruce chrono sequence in southeast Norway. *Tree Physiology, 28*(5), 773–784. https://doi.org/10.1093/treephys/28.5.773.

5. Cahill, J. F., McNickle, G. G., Haag, J. J., Lamb, E. G., Nyanumba, S. M., & St Clair, C. C. (2010). Plants integrate information about nutrients and neighbors. *Science, 328*(5982), 1657–1659. https://doi.org/10.1126/science.1187971.
6. Casper, B. B., & Jackson, R. B. (1997). Plant competition underground. *Annual Review of Ecology and Systematics, 28*, 545–570. https://doi.org/10.1146/annurev.ecolsys.28.1.545.
7. Chen, J. H. (2007). Study on the distribution of root density of *Robinia pseudoacacia* L. and *Pinus tabulaeformis* Carr. *Arid Zone Research, 24*(6), 647–651.
8. Curtis, E. J., Gorrod, E. J., Ellis, M. V., & Chisholm, L. A. (2019). A spatio-temporal analysis of canopy dynamics and intra-stand competition in a riparian forest, south-eastern Australia. *Forest Ecology and Management, 432*, 189–199. https://doi.org/10.1016/j.foreco.2018.09.029.
9. Day, K. J., Hutchings, M. J., & John, E. A. (2003). The effects of spatial pattern of nutrient supply on yield, structure and mortality in plant populations. *Journal of Ecology, 91*(3), 541–553. https://doi.org/10.1046/j.1365-2745.2003.00813.x.
10. de Kroon, H., Huber, H., Stuefer, J. F., & van Groenendael, J. M. (2005). A modular concept of phenotypic plasticity in plants. *New Phytologist, 166*(1), 73–82. https://doi.org/10.1111/j.1469-8137.2005.01351.x.
11. Deng, J. M., Wang, G. X., Morris, E., Wei, X. P., Li, D. X., Chen, B. M., Zhao, C. M., Liu, J., & Wang, Y. U. N. (2006). Plant mass-density relationship along a moisture gradient in north-west China. *Journal of Ecology, 94*(5), 953–958. https://doi.org/10.1111/j.1365-2745.2006.01159.x.
12. Ding, J., Ding, J., Kong, D., Zhang, Z., Cai, Q., Xiao, J., Liu, Q., & Yin, H. (2020). Climate and soil nutrients differentially drive multidimensional fine root traits in ectomycorrhizal-dominated alpine coniferous forests. *Journal of Ecology, 108*(6), 2544–2556. https://doi.org/10.1111/1365-2745.13407.
13. Feng, D. D., Xue, Q. Q., Qi, J. Y., Wang, L. J., Li, H., Men, L., & Zhang, Z. W. (2020). Community structure and occurrence regularity of Pyralidae in artificial seabuckthorn plantation. *Journal of Shanxi Agricultural Sciences, 48*(5), 793–798.
14. Gao, Y. H. (2023). Effects of different thinning methods on the growth of *Pinus sylvestris* plantation. *Science and Technology of Qinghai Agriculture and Forestry, 4*, 82–85.
15. Glass, N. T., Yun, K., Dias de Oliveira, E. A., Zare, A., Matamala, R., Kim, S. H., & Gonzalez-Meler, M. (2023). Perennial grass root system specializes for multiple resource acquisitions with differential elongation and branching patterns. *Frontiers in Plant Science, 14*, 1146681. https://doi.org/10.3389/fpls.2023.1146681.
16. Grime, J. P. (2007). The scale–precision trade-off in spatial resource foraging by plants: Restoring perspective. *Annals of Botany, 99*(6), 1017–1021. https://doi.org/10.1093/aob/mcm052.

17. Guo, Y., Yao, J. F., Dong, Y., Yan, J., Yang, N., Feng, Y. H., Wei, X., & Liang, W. J. (2023). Root distribution characteristics of monoculture and mixture of *Pinus tabulaeformis* and *Robinia pseudoacacia* plantation. *Ying Yong Sheng Tai Xue Bao, 34*(12), 2881–2888.

18. Gustafsson, L., Baker, S. C., Bauhus, J., Beese, W. J., Brodie, A., Kouki, J., Lindenmayer, D. B., Lõhmus, A., Pastur, G. M., & Messier, C. (2012). Retention forestry to maintain multifunctional forests: A world perspective. *BioScience, 62*(7), 633–645. https://doi.org/10.1525/bio.2012.62.7.6.

19. Hermans, C., Hammond, J. P., White, P. J., & Verbruggen, N. (2006). How do plants respond to nutrient shortage by biomass allocation? *Trends in Plant Science, 11*(12), 610–617. https://doi.org/10.1016/j.tplants.2006.10.007.

20. Hodge, A. (2009). Root decisions. *Plant Cell & Environment, 32*(6), 628–640. https://doi.org/10.1111/j.1365-3040.2009.01940.x.

21. Huang, F. X. (2023). Effects of different fertilization methods on seedling growth of *Pinus sylvestris*. *Agricultural Technology and Equipment, 12*, 129–130, 133.

22. Huang, X. M., Liu, S. R., Wang, H., Hu, Z. D., Li, Z. G., & You, Y. M. (2014). Changes of soil microbial biomass carbon and community composition through mixing nitrogen-fixing species with *Eucalyptus urophylla* in subtropical China. *Soil Biology and Biochemistry, 73*, 42–48. https://doi.org/10.1016/j.soilbio.2014.02.020.

23. Jackson, R. B., Canadell, J. G., Ehleringer, J. R., Mooney, H. A., Sala, O. E., & Schulze, E. D. (1996). A global analysis of root distributions for terrestrial biomes. *Oecologia, 108*(3), 389–411. https://doi.org/10.1007/s004420050089.

24. Jiali, S., Hongbo, L., & Aiping, Z. (2022). Effects of nutrient heterogeneity on shoot and root growth of *Zea mays* and intraspecific competition. *Journal of China Agricultural University, 27*(2), 11. https://doi.org/10.11861/jcau20220918.

25. Kang, X. Y. (2016). Thinking and practices for strategy on a new round genetic improvement of *Populus tomentosa*. *Journal of Beijing Forestry University, 38*(1), 1–8. https://doi.org/10.13332/j.1000-1522.2016.01.001.

26. Kim, B. M., Horita, J., Suzuki, J. I., & Tachiki, Y. (2021). Resource allocation in tragedy of the commons game in plants for belowground competition. *Journal of Theoretical Biology, 529*, 110858. https://doi.org/10.1016/j.jtbi.2021.110858.

27. Li, C. B., Zheng, Z., Peng, Y., Nie, X. Q., Yang, L. C., Xiao, Y., & Zhou, G. Y. (2019). Precipitation and nitrogen addition enhance biomass allocation to aboveground in an alpine steppe. *Ecology and Evolution, 9*(22), 12193–12201. https://doi.org/10.1002/ece3.5797.

28. Li, D. D., Jiang, Y. C., Luo, C. L., Feng, Y., Wang, A. P., & Dong, Q. (2022). Effects of nitrogen and phosphorus separate application and intercropping on soil nutrients and enzyme activities at flowering stage of spring wheat. *Journal of Shanxi Agricultural Sciences, 50*(7), 1414–1421.

29. Li, Y., Mei, B., & Linhares-Juvenal, T. (2019). The economic contribution of the world's forest sector. *Forest Policy and Economics, 100*, 236–253. https://doi.org/10.1016/j.forpol.2018.12.002.

30. Lin, Y., Berger, U., Grimm, V., & Ji, Q. U. (2012). Differences between symmetric and asymmetric facilitation matter: Exploring the interplay between modes of positive and negative plant interactions. *Journal of Ecology, 100*(6), 1482–1491. https://doi.org/10.1111/j.1365-2745.2012.02006.x.
31. Lobet, G., Couvreur, V., Meunier, F., Javaux, M., & Draye, X. (2014). Plant water uptake in drying soils. *Plant Physiology, 164*(4), 1619–1627. https://doi.org/10.1104/pp.113.231704.
32. Lu, B., Qian, J., Hu, J., Wang, P. H., Jin, W., Tang, S., He, Y., & Zhang, C. (2022). The role of fine root morphology in nitrogen uptake by riparian plants. *Plant and Soil, 472*(1), 527–542. https://doi.org/10.1007/s11104-021-05257-5.
33. Lu, Y. C., Zhang, S. G., Lei, X. D., Ning, J. K., & Wang, Y. X. (2009). Theoretical basis and implementation techniques on close-to-nature transformation of plantations. *World Forestry Research, 22*(1), 20–27.
34. McCormack, M. L., Guo, D., Iversen, C. M., Chen, W., Eissenstat, D. M., Fernandez, C. W., Li, L., Ma, C., Ma, Z., Poorter, H., et al. (2017). Building a better foundation: Improving root-trait measurements to understand and model plant and ecosystem processes. *New Phytologist, 215*(1), 27–37. https://doi.org/10.1111/nph.14488.
35. McNickle, G. G., Cahill, J. F., & Karban, R. (2009). Plant root growth and the marginal value theorem. *Proceedings of the National Academy of Sciences of the United States of America, 106*(12), 4747–4751. https://doi.org/10.1073/pnas.0900655106.
36. Ratcliffe, S., Holzwarth, F., Nadrowski, K., Levick, S., & Wirth, C. (2015). Tree neighborhood matters: Tree species composition drives diversity-productivity patterns in a near-natural beech forest. *Forest Ecology and Management, 335*, 225–234. https://doi.org/10.1016/j.foreco.2014.10.016.
37. Robinson, D. (2002). Root proliferation, nitrate inflow and their carbon costs during nitrogen capture by competing plants in patchy soil. In *Interactions in the Root Environment: An Integrated Approach: Proceedings of the Millennium Conference on Rhizosphere Interactions, IACR-Rothamsted, United Kingdom 10–12 April 2001* (pp. 41–50). Springer.
38. Schenk, H. J. (2006). Root competition: Beyond resource depletion. *Journal of Ecology, 94*(4), 725–739. https://doi.org/10.1111/j.1365-2745.2006.01126.x.
39. Shirong, L., Yujing, Y., & Hui, W. (2018). Development strategy and management countermeasures of planted forests in China: Transforming from timber-centered single objective management towards multi-purpose management for enhancing quality and benefits of ecosystem services. *Acta Ecologica Sinica, 38*(1), 1–10. https://doi.org/10.5846/stxb201710301929.
40. Song, S. H., Zhu, B. B., Jin, L., Chang, J. C., Hu, L., Song, L. P., Chang, Q., & Song, Y. F. (2022). Effects of forest gaps disturbance on regeneration pattern of *Pinus sylvestris* var. *mongolica* plantations. *Journal of Northeast Forestry University, 50*(4), 6–10. https://doi.org/10.13927/j.issn.1000-5382.2022.03.007.
41. Sun, D. J., Wen, Y., Luo, Y. H., Li, X. Q., Zhang, W. X., & Ming, A. G. (2015). Effect of close-to-nature management on species diversity in a *Cunninghamia*

lanceolata plantation. *Forest Research, 28*(2), 202–208. https://doi.org/10.13624/j.cnki.forestres.2015.02.004.

42. Sun, Y. B., Bi, H. X., Duan, H. Q., Peng, R. D., & Wang, J. J. (2019). Variation of fine root distribution and belowground competition in apple-soybean intercropping system. *Chinese Journal of Ecology, 38*(2), 459–466. https://doi.org/10.13292/j.1000-4890.201902.013.
43. Tilman, D. (1982). *Resource Competition and Community Structure* (MPB-17, Vol. 17). Princeton University Press.
44. Von Wettberg, E. J., & Weiner, J. (2003). Larger *Triticum aestivum* plants do not preempt nutrient-rich patches in a glasshouse experiment. *Plant Ecology, 169*(1), 85–92. https://doi.org/10.1023/A:1022354223033.
45. Wang, H. Y., Liu, J. G., Yuan, H. C., Duan, W. B., Zhu, S. W., & Mu, M. X. (2024). Effects of three extracts on physiological indexes of 2-year-old *Pinus sylvestris* var. *mongolica* seedlings. *Forest Engineering, 40*(1), 44–54. https://doi.org/10.13930/j.cnki.1003-8819.2024.01.007.
46. Wang, M. L., Jiang, Y. S., Wei, X., Cai, S. F., Qi, X. X., & Li, F. (2010). Effects of planting density and fertilization on growth characters and artemisinin of *Artemisia annua* L. *Journal of Plant Nutrition and Fertilizers, 16*(2), 185–190. https://doi.org/10.11674/zwyf.2010.02022.
47. Wang, Q., & Cheng, Y. (2004). Response of fine roots to soil nutrient spatial heterogeneity. *Chinese Journal of Applied Ecology, 15*(8), 1063–1068. https://doi.org/10.13287/j.1001-9332.2004.0125.
48. Warton, D. I., Wright, I. J., Falster, D. S., & Westoby, M. (2006). Bivariate line-fitting methods for allometry. *Biological Reviews, 81*(2), 259–291. https://doi.org/10.1017/S1464793105006957.
49. Wei, S. P., Liang, W. J., Wei, X., Bu, R. Y., & Yao, J. F. (2022). Natural regeneration of *Larix principis-rupprechtii* plantations with different densities and its influencing factors. *Chinese Journal of Applied Ecology, 33*(9), 2687–2694. https://doi.org/10.13287/j.1001-9332.2022.0234.
50. Weiner, J. (2004). Allocation, plasticity and allometry in plants. *Perspectives in Plant Ecology, Evolution and Systematics, 6*(3), 207–215. https://doi.org/10.1078/1433-8319-00019.
51. Wen, L., Lei, P. F., Xiang, W. H., Yan, W. D., & Liu, S. G. (2014). Soil microbial biomass carbon and nitrogen in pure and mixed stands of *Pinus massoniana* and *Cinnamomum camphora* differing in stand age. *Forest Ecology and Management, 328*, 150–158. https://doi.org/10.1016/j.foreco.2014.05.025.
52. Wu, F. Z., Wang, K. Y., Yang, W. Q., & Lu, Y. J. (2005). Effects of *Fargesia denudata* density on its biomass distribution pattern. *Chinese Journal of Applied Ecology, 16*(5), 991–995. https://doi.org/10.13287/j.1001-9332.2005.0157.
53. Xiao, W. J., & Ghen, Y. X. (2022). Application of Grey Correlation Analysis to the evaluation of HDS sediment in different phytoremediation copper sulfide. *Jiangxi Science, 40*(6), 1108–1113. https://doi.org/10.3969/j.issn.1000-5606.2022.06.026.

54. Yan, Z., Eziz, A., Tian, D., Li, X., Hou, X. H., Peng, H., Han, W., Guo, Y. L., & Fang, J. (2019). Biomass allocation in response to nitrogen and phosphorus availability: Insight from experimental manipulations of *Arabidopsis thaliana*. *Frontiers in Plant Science, 10*, 598. https://doi.org/10.3389/fpls.2019.00598.

55. Yang, F., Nan, H., Gao, S., Wang, X., Zhang, M., & Liu, W. (2022). Effects of underground competition on leaf and root development of *Pinus tabulaeformis* and *Picea asperata* seedlings. *Guangxi Forestry Science, 51*(4), 307–316.

56. Yang, Y. H., Li, W. L., Huang, P., Zong, J. W., & Li, Z. Z. (2011). Effect of potassium fertilizer and plastic film mulching on growth and size inequality in *Lilium davidii* var. *unicolor*. *Acta Prataculturae Sinica, 20*(2), 217–222. https://doi.org/10.3969/j.issn.1003-1881.2011.02.003.

57. Zhang, H., Ye, R. H., & Wang, J. (2024). Effects of irrigation intensity on seedling growth of *Pinus sylvestris* var. *mongolica* in Mu Us Sandy Land. *Protection Forest Science and Technology, 2*, 1–3.

58. Zhang, L., He, M., & Zhang, K. (2023). Response of biomass allocation and allometric growth of *Caragana korshinskii* to nitrogen and phosphorus addition. *Acta Ecologica Sinica, 43*(6), 6627–6636. https://doi.org/10.5846/stxb202208213433.

59. Zhang, Y., Li, Y., Zhao, H., Cheng, Y., Wang, W., Jiang, Z., & Cai, J. (2022). Relationship between hydraulic efficiency and gas exchange and growth of six poplar clones. *Science of Silvae Sinicae, 58*(2), 118–126. https://doi.org/10.11707/j.1001-7488.20220217.

60. Zhao, X. M., Sun, X. Y., Kang, X. Y., & Wang, H. Y. (2012). Dynamics of soil organic carbon and total nitrogen contents in short-rotation triploid *Populus tomentosa* plantations. *Acta Ecologica Sinica, 32*(15), 4714–4721. https://doi.org/10.5846/stxb201203100479.

CHAPTER 2

Forest Ecosystems

LEARNING OBJECTIVES

After studying this chapter, you will be able to:

- Understand the types and characteristic features of forest ecosystems.
- Describe the structure of forest ecosystems and their components.
- Explain the concepts of plant communities, including vertical stratification and community development.
- Discuss nutrient cycling and water relations in forest ecosystems, focusing on the water, oxygen, carbon, and nitrogen cycles.
- Analyze the physiological adaptations of plants to stress environments.
- Evaluate the importance of conservation efforts in maintaining forest ecosystems.

KEY TERMS FROM THIS CHAPTER

Alkalinity	Carbon cycle	Conservation
Drought stress	Forest ecosystems	Nitrogen cycle
Nutrient cycling	Oxygen cycle	Plant community
Salinity	Structure	Vertical stratification
Water cycle	Water relations	Waterlogging

INTRODUCTION

The scientific study of the relationship between organisms and the environment in which they live is known as ecology. Both biotic and abiotic components are present in it. As opposed to a classification that is based on organizational level or complexity, such as population or community ecology, forest ecology is one of the branches that falls under the biotically oriented classification of different types of ecological study.

Due to the fact that trees are capable of growing to greater heights than other types of plant life, there is the possibility of a diverse array of forest structures. The fact that there are an infinite number of possible spatial arrangements of trees of different species and sizes results in a microenvironment that is extremely complex and diverse. This microenvironment is characterized by significant variations in the environment over both large and small distances, including solar radiation, temperature, relative humidity, and wind velocity.

Furthermore, the bio-structure, quality, and quantity of land, in addition to the levels of various soil nutrients, frequently differ significantly from underground for a forest ecosystem. This is the case from the surface. When compared to other types of terrestrial plants, forests are typically characterized by a substantial degree of heterogeneity. Consequently, this heterogeneity can make it possible for a high level of biodiversity to exist among plant and animal species.

The study of forests encompasses a wide range of organizational levels, beginning with the individual organism and progressing all the way up to the ecosystem. The term "forest" refers to a region home to several different organisms; consequently, the study of forest ecology typically concentrates on the population, community, and ecosystem levels.

In a logical sense, tree species are an important component of forest research; however, other components, such as wildlife and soil nutrients, frequently play a significant role in various other forest life forms and abiotic components. Because of this, the ecology of forests is a very diverse and significant subfield of the study of ecology.

2.1. FOREST ECO-SYSTEMS

World forests are essential not only for the well-being of all the people who live on the planet but also for the planet's overall health. For the sake of the benefits of society and the variety of life, it is of the utmost importance to protect forests from deforestation and other potential adverse effects of civilization. Additionally, forestry contributes to climate control, which is just one of the many ecological benefits it offers. The reduction in the force of rainfall on the surface of the soil helps to prevent erosion, as well as the absorption of water and the prevention of water from running off directly and removing topsoil.

In addition to their functionality as water filters, forests also collect and store water, as well as collect water underground. The perspiration that occurs in forests causes an increase in humidity in the atmosphere, which in turn affects both temperature and precipitation. There will be a movement of seeds, and plants and trees will begin to grow wherever there is sufficient land and water.

The establishment of a forest will take place in due time if the conditions are favorable. A forest once covered almost all of the land in the tropics, just as it did the majority of the land in the more humid regions of the continents. There are a many different kinds of forests. When it comes to structure, the temperate and boreal forests of colder lands can be quite simple provided that they contain only a few common trees. In most cases, they have recovered since the end of the most recent ice age. Warmer tropical regions that receive an adequate amount of precipitation typically consist of rainforests that contain a wide variety of tree species, some of which are extremely tall and others of which are shorter. The clearing of agriculture, villages, and other applications typically comes first as these forests develop on the soil that is the best and most accessible. In mountainous regions, different kinds of forests may come into existence higher on the mountainside, with trees that are shorter and soils that are more subterranean.

If the clouds continue to keep the tops and ridges wet, there may be a forest of clouds with a large number of moose and plants growing on the trees. It is common for bamboo, swamp forests, and riparian forests to inundate the riverbanks of riparian areas. It is possible for there to be forests that lose their leaves during the dry season while living in regions that experience seasonal rainfall.

There is a species of atoll or beach forest that grows on coral rocks and sand behind beaches. The majority of the trees in this forest are resistant to salt or have seeds that may float on the water (Gunderson & Holling, 2002; Gann et al., 2019).

2.1.1. Types and Characteristic Features

Types and Characteristics of forest ecosystem are explained below:

A. *Temperate Forest Ecosystem:* The ecosystem of a temperate forest is of great significance. During the transition from summer to winter, some regions experience significant climate change, and these regions contain temperate forests.

Regions that have a climate that is consistent throughout the year are home to tropical rainforests. The evergreen and caducous trees that make up the temperate forests almost always go hand in hand with one another. The leaves of trees that are known as dolphin trees fall off during the winter months. Similarly to pine trees, evergreen trees are able to maintain their leaves throughout the entire year. It is possible for either the woods or both to be one. The fourth category of the forest is known as a moderate rainforest.

In the United States of America, you can find these in the states of California, Oregon, and Washington. The redwood and sequoia trees that grow in these forests are by far the tallest in the world. The amount of precipitation that falls in a region is what determines whether or not there is a forest there. In situations where there is an abundance of rain for trees, a forest will typically be growing.

B. *Tropical Rainforest Ecosystem:* The tropical rainforest ecosystem derives its name from the abundant rainfall these regions receive, averaging around 80 centimeters or more annually. This high level of precipitation sustains the lush and diverse vegetation characteristic of rainforests. Unlike other ecosystems, tropical rainforests experience relatively stable temperatures throughout the year, with little seasonal variation. One of the most well-known tropical rainforests is the Amazon, located in Brazil, South America. This vast forest is part of the Neotropics, which refers to the tropical regions of the Americas. In addition to the Amazon, large tropical rainforest blocks can be found in Central and West Africa, such as the Congo Basin, which is the second-largest tropical rainforest in the world. These regions play a critical role in global biodiversity, carbon storage, and the regulation of the Earth's climate.

- *Tropical Rainforest Insects:* Tropical rainforest insects play essential roles in pollination, decomposition, and nutrient cycling within the ecosystem. They exhibit high biodiversity, with species ranging from ants and beetles to butterflies and mosquitoes. These insects contribute significantly to the complex food web, serving as prey for many larger animals. Some of them pose a risk to people's lives, while others are completely harmless. Of the many species of ants that live in the rainforest, army ants are just one. The name "army ants" comes from their characteristic behavior of moving in large, coordinated columns through the jungle. After reaching the pupal stage, the young larvae develop and mature, while the colony temporarily halts its movement. Once the queen lays her eggs, the ants resume their march. The rainforest is also home to beautiful butterflies, which start as caterpillars—creatures with a less appealing appearance. Through metamorphosis, caterpillars undergo a complete transformation into butterflies.

Centipedes, although not related to butterflies, also roam the forest in search of food, with some species using venom to incapacitate their prey.

- *Tropical Rainforest Birds:* The birds that inhabit rainforests are among the most stunning in the world, bringing vibrant colors to the forest canopy. The natural beauty of many tropical bird species makes them highly valued and often protected. Among the hundreds of bird species found in rainforests, parrots are particularly prominent. One such species is the scarlet macaw, known for its bright red, blue, and yellow feathers.

  The scarlet macaw measures about three feet in length from the top of its head to the tip of its tail, making it one of the longest parrot species. If macaws consume poisonous fruit, they neutralize the toxins by ingesting a special type of clay found in the rainforest. Another fascinating bird species is the toucan. Researchers estimate that there are around 33 different species of toucans in tropical rainforests around the world.

  The Hoatzin, with its awkward appearance and scruffy tail, is a unique bird often seen clumsily landing as if recovering from an accident. Despite their somewhat intimidating appearance in groups, Hoatzins are harmless. On the other hand, the brown kiwi is a flightless bird with a long beak, resembling a small mammal rather than a typical bird. Unlike most birds, kiwis live on the ground, using their specialized claws for running, digging, and self-defense.

- *Tropical Rainforest Mammals:* When it comes to flying through the rainforest, birds are not the only creatures that do so. The rainforest is home to a wide variety of mammalian species. In reality, the tropical rainforests are teeming with surprises, which can range from a harmless bat to a flying squirrel.

- *Tropical Rainforest Reptiles:* Tropical rainforests are home to a wide variety of reptiles. As cold-blooded animals, reptiles rely on external warmth to regulate their body temperature, which is why they thrive in warm environments. One notable group of rainforest reptiles includes snakes, such as the mambas. These snakes use their sharp fangs to inject venom into their prey. Another famous rainforest snake is the anaconda, one of the world's longest, reaching lengths of up to 30 feet. Unlike venomous snakes, anacondas constrict their prey, wrapping around and squeezing until the prey suffocates before swallowing it whole.

  The chameleon, a fascinating lizard species, can change its color to blend into its surroundings, helping it catch prey and hide from predators. Chameleons primarily feed on insects. Geckos, known for their cleanliness, have remarkable adhesive abilities. The flying gecko, in particular, can glide between trees to escape predators, and their grip is so strong that attempting to remove them can cause injury.

e. *Tropical Rainforest Primates:* Tropical rainforests are home to a rich diversity of primate species, each uniquely adapted to thrive in these lush, complex environments. These primates play essential roles in the ecosystem, such as seed dispersal and maintaining the balance of the forest's biodiversity. One of the most notable primates found in tropical rainforests is the howler monkey, known for its loud vocalizations, which can be heard from miles away. Howler monkeys are commonly found in the rainforests of Central and South America. Another well-known species is the spider monkey, which has long limbs and a prehensile tail that allows it to navigate the forest canopy with agility.

In the rainforests of Africa, gorillas and chimpanzees are among the most recognizable primates. Gorillas, particularly the lowland and mountain species, inhabit the dense forests of Central Africa, while chimpanzees are found in both Central and West African rainforests. These primates are highly social and exhibit complex behaviors, including tool use and sophisticated communication.

In Southeast Asia, the orangutan, native to the rainforests of Borneo and Sumatra, is a critically endangered species. Orangutans are primarily tree-dwelling and are known for their intelligence, using tools to gather food and build nests high in the trees.

These primates are not only essential to the ecological health of the rainforest but also face significant threats due to habitat destruction, primarily from deforestation and human activities. Conservation efforts are crucial to protect these species and the ecosystems they depend on.

C. *Boreal or Tai Forests:* It is the contiguous green coniferous belt of lacy trees that surrounds a significant portion of the Northern Hemisphere that constitutes the ecosystem of the boreal forest.

Within the continent of North America, it encompasses the vast majority of northern Alaska and Canada, considered for a long time to be among the largest forest ecosystems in the world. Due to the fact that it contains approximately 35% of Canada's land population, it is the largest land-based ecosystem in North America. In addition to this, it is a significant contributor to the biodiversity of Canada and acts as a major carbon sink.

The "respiration" of carbon dioxide and the "exhalation" of oxygen into the atmosphere in North America are essential processes, although the boreal is largely unknown. To put it succinctly, the Amazon rainforest is home to only a small portion of the region's full animal and plant life. Across the North American continent, the forest ecosystem is home to both the largest and the smallest species of mammals.

The boreal forest, also known as the taiga, is characterized by a wide variety of features, including large lakes, rivers, extensive wetlands, and peat bogs. This region supports a diverse array of wildlife, with older forests being particularly rich in biodiversity. Boreal forest ecosystems play a critical role in sustaining numerous species, providing essential habitats for both plant and animal life. These forests are vital for the survival of many species adapted to the harsh climate of the northern hemisphere.

2.1.2. Structure of Forest Ecosystems

The layers of the forest are home to a wide variety of organisms. These living things communicate with one another as well as with the planet. When it comes to the upkeep of the ecosystem, each organism plays a significant role or perhaps occupies a niche.

Others provide shelter for other organisms, and some may even control populations through predation. Some organisms provide food for other organisms.

- *Producers:* For their continued existence, all living things require energy. The sunlight that penetrates a forest ecosystem is the source of the trees and plants that grow there. Using carbohydrates as their primary source of nutrition, plants produce their own food.

  Within the context of the food chain and the web, plants are considered to be the primary producers because they are responsible for the production of the fundamental components that are required by other organisms. Photosynthesis is the chemical reaction that enables plants to produce their own food when they are able to do so.
- *Consumers:* It is impossible for animals to produce their own food on their own. For the sake of their survival, they are required to consume dietary sources of energy. For all intents and purposes, all animals, including birds, insects, and mammals, are considered to be consumers.

  To obtain food, consumers are dependent on the products of other animals and plants. There was previously provided information regarding the forest ecosystem that these animals inhabit. The term "herbivore" refers to primary consumers, who consume only plant organisms. It is more accurate to refer to secondary consumers as carnivores, while herbivores are the ones that feed. Tertiary consumers would be defined as carnivores that consume other carnivores as their food source. Animals and plants are both consumed by omnivores.
- *Decomposers:* Needles, leaves, and branches that have been around for a long time fall to the ground as the forest floors continue to grow. Possibly, every single plant and animal will perish. It begs the question: what exactly is going on with all of this plant and animal material? Does it tend to remain on the ground in the forest at all times? Worms, microbes, mushrooms, other insects, and ants are the organisms that are responsible for the breakdown of these materials. It is possible to reuse these items by disassembling them into their most fundamental components. Due to the fact that they maintain the ecosystem's nutrient cycle, decomposers are extremely important.

2.2. PLANT COMMUNITY CONCEPTS

The term "plant community" refers to a group of recurrent species that share a typical habitat, collectively construct a distinctive physiognomy, achieve a typical rich species portfolio and annual productivity, and make use of nutrients and energy to generate predictable levels and predictable efficiency.

The continent of North America is home to thousands of different plant species. It is possible for the successive plant communities to experience annual fluctuations, but the climax plant communities must remain constant. Plants are the primary producers of the biosphere, which means that they supply all terrestrial and aquatic animals, including humans, with food either directly or indirectly. As a result, plants are an integral part of the biosphere. The term "plant community" refers to the social groups that are comprised of different plant species, with the plant itself serving as the primary component.

The plants are able to directly absorb and store solar energy, also known as light energy, and then use this energy to produce food through the process of photosynthesis. Through the various trophic levels of the food chain, solar energy is converted into food or possibly chemical energy, and this energy is then transferred to a wide variety of animals and microorganisms. From this perspective, plants serve as a bridge between the biotic and abiotic components of the environment, ecosystem, and biosphere.

The investigation of plants takes on a much greater significance when one considers the dominant role that plants play in the biosphere as well as the significance of plants.

2.2.1. Characteristics of Plant Communities

The term "plant community" refers to the collection of plants that develop in close proximity to one another within a particular environment.

According to Whittaker (1975), plant communities are defined by the species composition, structure, and the interactions within a shared habitat, reflecting the ecological processes that shape them."

The term "ecological community" refers to a collection of populations of distinct species that coexist in the same neighborhoods and engage in social interactions with one another.

There are three fundamental characteristics that define the plant community that exists within a particular habitat. These basic characteristics include plant communities that contain two or more distinct types of plants, plant communities that contain ecologically linked species that are able to coexist and cultivate together in a particular habitat, and plant communities that have well-organized compositions and structures that have been developed to a high degree.

The term "vegetation" refers to the collection of plants that are found in a particular region. Essentially, this means that "all plants rising together in every area form its vegetation, the nature of which depends not only on the different present species but also on their relative proportions."

To put it another way, two habitats might have the same flora, but their vegetation might be different. Similarly, two habitats might have the same vegetation, but their flora might be different. In the first habitat, as opposed to the second habitat, which is characterized by dense salt trees and a sparse distribution of grass, there are two habitats that are very similar to each other. Both of these habitats contain grassland and salt trees, but the prevalence of grasses and the sparse distribution of salt trees are overwhelming.

2.2.2. Vertical Stratification of Plant Communities

An environment in which different plant communities are grown together with different life forms is conducive to the development of a wide variety of plant species. Utilized are the processes of growth and adaptation, as well as competition and natural selection, which occur among the various species of plant communities that exist within a particular region. As a consequence of this, the surface of the floor or the tree canopy develops into a selection of different strata or layers.

As a result of competition between various plant species for sunlight, which is the primary source of energy for the production of their food through photosynthesis, this pattern of vertical layering or vertical stratification has developed (Figure 2.1). The availability of sunlight is the primary factor that is responsible for the development of vertically stratified crops, which are characterized by significant differences in the height of various plant species.

Generally speaking, there are four vertical layers of plant community in a given area, and the majority of these layers are found in the deciduous forests of template regions: With regard to the community of plants, the dominant layer represents the highest level. A coronation or canopy is another name for this top stratum, which is responsible for reproducing the highest plant community limit that exists within a particular region.

Located just below the crown or perhaps the overwhelming layer, which is frequently referred to as the co-dominant layer, are the larger trees, which are significantly shorter than the largest trees.

Following the dominant or crown layer is the next layer, which is distinguished by plants that have a shrubby life form. This layer is located below the latter. The term "shrub layer" is another name for this. Herbaceous plants are responsible for the formation of the third layer, which is also referred to as the herb layer.

The fourth layer is also referred to as the moss layer or the ground layer, and it is the layer that represents the mosses that are found on the ground surface.

Figure 2.1. The illustration of stratification in a terrestrial ecosystem.

Source: By Elke Freese - Own work (selbst gezeichnet/drawn by myself), Public Domain, https://commons.wikimedia.org/w/index.php?curid=947575.

2.2.3. Development of Plant Communities

The process by which one vegetation community is replaced by another within a particular habitat is referred to as biotic succession within that habitat. The term "succession" refers to the progression of a plant community or even the entire ecological system over the course of time that encompasses the entire process of direction and sequence.

The process of sequential ecological community development is one way to define ecological succession. This definition is straightforward and easy to understand. The vegetation sequence, also known as biotic succession, is referred to as the "sere." This means that the transitional phases of the sequence are referred to as the "sere" when the succession of the vagueness community reaches a point of equilibrium or maturity because of the various phases of modifications.

This phase is characterized by plant growth, which is the maximum growth and development that vegetation can experience during this phase. Because vegetation control is dominant, a developed vegetation group is named "climax vegetation," "climax," and "climate" at the end point of the sequence. This is because the vegetation control is dominant. It should come as no surprise that the final succession of successive biotic changes will culminate in succession within the community of vegetation.

According to F. E. Clements (1916), the biotic succession of a vegetation community can be broken down into two distinct categories: primary biotic succession and secondary biotic succession.

1. *Primary Biotic Succession:* In open areas where there is no vegetation and animals, primary succession refers to the developmental sequence of vegetation.

   These may be newly formed areas or sites, with cooled basaltic surfaces and consolidated due to recent fluxes of lava, exposed lakes due to the drying of the water, newly formed sand dunes, flood plains that formed a recent alluvium, waste heaps accumulated by man, areas of the rocks which have been exposed due to the melting of ice from the glacial zones, etc.

   The first succession of vegetation on a bare rock surface, where previous vegetation and animals do not exist, begins and is completed by the following steps:

   i. The preliminary plant-free site has comparatively dry

environment. This does not mean that the climate on this site was dry because the rocks on the site are naked and without any plant, and so the environment is relatively dry, although the climate may be even moist, as a consequence of excessive evaporation. On the pure rocks of the first plant-free site are established pioneering plants.

The pioneer plants primarily comprise algae and lichens as they easily adhere to the bare rock and can easily be adapted to hot, dry or cold conditions on the initial sites.

ii. Wind-blown dust particles settled in the concerned habitat. These particles are deposited gradually around the algae and lichens. Several of the lichens discharge acids, which react in the dissolution of minerals with the rocks of this habitat.

This process begins with the pedogenesis (soil formation processes) process and a thin soil veneer is formed at the appropriate time. Although the soil zone is very thin at first, microorganisms colonize it.

iii. Soil formation continues by rock and soil organisms and with time increases the soil thickness. A few animals such as mites, ants, spiders, etc., have therefore evolved. More soil-living animals, sporadic plants and wide-open areas without plants distinguish this "sere" of successional plant community development. This kind of plant community is known as an open community or a community of pioneers.

iv. In due course, the secondary moss community will substitute the pioneering algae and lichens community.

The moss spread like thin sheets over the soils, and the soil is covered with the moss. As a consequence of the delays in evaporation, the soil moisture particles and content are increased. New dense matting of mosses also gives the soils organic matter and so organic nutrients are added to the soils.

The seasonal and perennial grasses are gradually and step by step developing together with new animal groups such as nematodes, spring tails, etc., that can produce their food from the seasonal and perennial grasses. The gradual development of grass covers the entire area and thus develops a dense vegetation cover that changes and alters the microclimates of the habitat.

Dense vegetation degrades the soil temperature and sunshine at the ground surface while increasing the soil's moisture content since dense coverage of vegetation effectively reduces moisture evaporation through the soil surface. The open community is now transformed into a shut-down community (this means that no part of the habitat concerned is vegetation-free, that is, the whole habitat area is covered by vegetation).

This is only possible when the surroundings of the habitat are wet but when the environment is dry and the area is sandy desert, and therefore, a large amount of territory remains open and without vegetation.

v. The competition for space, sunlight, water, and nutrients starts after the development of the sere of a closed community. There may be two different routes of competition among the plants such as:

In situations where the plants that inhabit the area belong to the same species, there is competition between the various members of the same species for the aforementioned elements. In this kind of competition, only the plants that are the most suitable for survival are able to thrive.

The principle of survival of the fittest is, therefore, applicable to the process of community development in plant communities. This kind of competitiveness is also referred to as intraspecific competition, which demonstrates that the plants that are the most robust survive, while the plants that are the least robust are eliminated, but species conservation continues to take place.

When the habitat of concern contains more than one species, there is competition among individuals of different species to obtain space, sunlight, water, and nutrients. In this competition, the species that are the most powerful and aggressive tend to dominate the entire vegetation community. If, on the other hand, all of the species that inhabit this habitat are of comparable power, then the equilibrium of power between the various species will be preserved, despite the fact that there will be competition between them. This will result in the preservation and maintenance of all of the species. This competition is known as interspecific competition. Herbal communities are the source of this (serious) phase of community development, which has its roots in herbal plants.

vi. In the spring season, large shrubs emerge, and the herbaceous community is gradually replaced by a scrub community. Currently, seeds from flowering plants (phanerogams) have been dispersed through airborne transportation from nearby areas, establishing themselves in a previously shrub-dominated habitat. This is a significant development, as the canopy has expanded and now surpasses the height of the scattered shrubs and trees. At this stage, the vegetation consists of a mix of mosses, grasses, lichens, shrubs, and trees. This is referred to as a forest community, with the "preclimax" stage marking the ongoing progression toward the final, stable ecosystem (climax community).

vii. The final phase or "sere" of the succession of vegetation is marked by the growth of giant and very tall trees; their density is increasing, with large trees covering the entire habitat. The roots of these large trees penetrate the earth much deeper.

The soil zone reaches the highest depth and soil profile horizons are degraded. Different micro-organisms colonize soil zones, which decompose organic matter and aid in transferring energy. It is well developed to vertically stratify the plant community.

This final phase, or "sere" of the successive development of the vegetation community represents a mature ecosystem and is known as a climax community, climax succession, climax, etc. (Gunderson & Holling, 2002; Gann et al., 2019).

2. *Secondary Biotic Succession:* Secondary succession refers to the development sequence of plant life in regions where the plant cover has been reproduced naked or bare (either partially or completely) due to vegetation damage (e.g., lava flow, prolonged drought, glaciation, natural, broad-based forestry fires, lightning, severe storms, catastrophic flood, etc.). It should be noted that these troubled ecosystems or habitats still have mature soils and original vegetation, which distinguishes them from an early stage or sere of secondary succession of plants, which begins on a bare rocky surface with no previous plants or animals.

This is the first step in the sequence, or the "sere" of primary succession. The total time required for climax or climax sequence development in secondary succession is significantly less than that required for primary succession development. The northeastern Indian hill areas where jhum cultivation or shifting is common may serve as an example of secondary succession.

The first forest to be affected by this cultivation is burned off in small areas, and the soil is then grown for farming for several years. When the soil loses its fertility, the area is abandoned, and new cultivation areas are cleared of all vegetation. The area or the old clearance has been colonized by vegetation through various phases and reaches the climax of vegetation in a short period of time, such as a few years, because the secondary succession sequence is faster than the primary succession due to the proliferation of mature soils.

When a vegetation community in a given area is disrupted by human activities before it reaches its climax stage (through slow, long-term actions like deforestation or burning), the resulting vegetation is referred to as subclimax vegetation. If disturbances persist over time, the normal progression of vegetation development does not occur. Instead, the natural stages of succession, or "seres", are diverted by these repeated disturbances, preventing the ecosystem from reaching its stable climax community.

During the deflected sere, the vegetation continues to grow as long as the troublesome factors are active. Plagioclimax is a deformed culmination, and its various stages are known as plagiosere. If factors that disrupt the succession development of the vegetation community cease or become ineffective over time, the environmental conditions at the site or habitat in question will change, resulting in new habitat conditions that cannot support and maintain plagioclimax veg formation.

Instead of plagioclimax vegetation, which results from human disturbances, new vegetation may emerge under changing environmental conditions, allowing the succession of vegetation communities to proceed under more typical circumstances. In forest vegetation ecology, plant communities play a crucial role in controlling diversity, productivity, and the overall functioning of the ecosystem, especially in response to biotic, abiotic, and extrinsic disturbance factors.

The overarching goal of ecological research on forest vegetation is to gain a better understanding of how plants interact with the other biotic and abiotic elements of the forest ecosystem. This includes understanding the interdependence of plants, animals, and microbes, as well as plant-soil feedback and plant responses to external disturbances such as fire. This section also discusses the role of understory vegetation in forest community structure and ecosystem processes.

Forest ecosystems' trees and soil vegetation, as well as the movement of slow water, help to stabilize soils. Preventing or reducing soil erosion helps to reduce pollution from sedimentation in rivers, lakes, and streams. Water pollution is typically classified into two broad categories: point source pollution, which can be attributed to an original point or a non-point source.

Increased nutrients and sedimentation would encompass non-point sources of pollution. The primary cause of non-point source pollution is sedimentation or soil movement in our waterways. Even if some soil movement occurs naturally as water flows through the ground, poorly planned human activity can significantly increase soil erosion rates.

This increased erosion may cause our water resources to become more sedimentary. Vegetation slows the movement of water and reduces soil erosion, resulting in lower pollution levels in our waterways. This invaluable assistance is delivered in a variety of ways. Raindrops that fall on leaves, stalks, and other plant parts are disrupted and redistributed, reducing the soil's direct impact. Well-established vegetation slows the movement of water across the soil's surface, reducing erosion while also allowing more water in. Finally, plant roots help to hold or lock in soil.

The removal of vegetation will expose soil, increasing the risk of soil erosion and sedimentation from streams. The higher the earth's vertical slope, or the more vulnerable it is to erosion, the greater the damage potential. When exposed to these disturbances, the finer particle clay soil is the most easily eroded.

2.3. ECOLOGICAL SUCCESSION AND CLIMAX COMMUNITY

Ecologists are interested in how communities evolve and change over time. Indeed, they have spent a lot of time studying how complex communities like forests emerge from bare rock and empty land. They investigate areas affected by volcanic eruptions, glacier retreats, wildfires, land clearing, or rock exposure.

Ecologists have observed gradual processes of change in ecological communities while studying these sites over time. In many cases, a community that emerges from a disturbed area experiences a number of composition changes over time. This change is known as ecological succession. Ecological succession refers to the gradual transformation of an ecological community's species structure over time.

The time scale can be decades or even millions of years after a mass extinction (such as a wildfire). The community begins with a small number of pioneering plants and animals and grows in complexity until it reaches autonomy or stability as a climax community. The "engine" of succession, which causes ecosystem changes, has an impact on established species' habitats. Life has the effect of altering our environment, both subtly and overtly. It is a phenomenon or process that results in more or less orderly and predictable changes in an ecological community after initial colonization or disturbance of a new habitat.

A succession may begin with the establishment or disturbance of a new, unoccupied habitat, such as a lava flow or a severe landslide caused by fire, strong winds, or logging. Primary succession begins in new habitats and is influenced by previous communities; secondary succession occurs after the disruption of a pre-existing community. Succession was one of the first ecological theories.

First documented in the northwest Indiana, Indiana dunes, ecological succession continues to be central to the ecological sciences and ecologists usually identify two types of succession that differ at their starting points:

a. In the primary succession of newly exposed or newly formed rock, living things are colonized for the first time.

b. In a secondary succession, the pre-occupied area is disturbed and subsequently recolonized after the disturbance.

Succession is usually a move from less stable, species diversity communities to higher species diversity communities which may be a lot more stable but this doesn't constitute a universal rule.

- *Primary Succession and the Pioneer Species:* When new land is formed or bare rock is exposed, a primary succession occurs, creating a habitat that can be colonized for the first time. For example, primary successions may occur after volcanoes such as those on Hawaii's Big Island.

  When lava flows into the ocean, it creates new rock. The Big Island adds approximately 32 acres of land each year. What is the primary succession of this land? Initially, the substratum, rock, degrades due to weathering and other natural forces, producing some hardy plants and lichens with few soil requirements, known as pioneer species (Figure 2.2).

Figure 2.2. Succulent plants colonizing lava in the course of primary succession on Maui.

Source: CC BY 4.0; Kim Starr.

  These species contribute to the breakdown of mineral-rich lava into soil, allowing less hardy species to grow and replace the pioneer species. Additionally, as these early species die and grow, they contribute to an ever-increasing layer of organic decomposition and soil formation. This procedure is repeated several times in a row.

  Every step of the way, new species enter an area, usually as a result of environmental changes caused by previous species. At some point, the community's composition may stabilize and stop changing.

- *Secondary Succession and Pioneer Species:* In a secondary succession, a previously occupied area is recolonized following a disruption that kills several or perhaps all of its communities. In the wildfire cleared oak and hickory forests, a standard example of secondary succession can be observed. Most vegetation is burned by wildfires, killing animals not in the position to flee the region. The nutrients are, nonetheless, brought back to earth ash.

Since a disrupted area already has a soil rich in nutrients, it could be settled a lot faster compared to primary rock. Before a fire, the vegetation of an oak and hickory forest was dominated by tall trees. Its elevation would've helped them to get solar energy, to shade the earth and others. Nevertheless, after the fire, these trees do not emerge.

Rather than the first crops, the annual crops are typically one year old and the grass grows quickly in a couple of years. By their primary succession, the first colonizers can be classified as a pioneer. Shrubs will be developed over many years, followed by small pines, oaks and hickory trees as a result of the environmental changes brought on by the growth of other species and grasses. Finally, oaks and hickory trees will dominate and form a thick canopy that will restore the community to its original state and the composition of pre-fire (Figure 2.3).

Figure 2.3. Forest succession.

Source: by Lucas Martin Frey, CC BY 3.0.

This procedure takes around 150 years to complete. The psychologist who first investigated succession saw it as a predictable process in which a community had to go through the same set of stages. They believed that the best, most stable state known as a climax community was the result of a succession, largely determined by a region's climate.

In the preceding example, the older oak and hickory forest would serve as the climax. He questioned the idea of a defined path of succession and a stable climax community today. It appears that succession may follow a predetermined path based on the circumstances.

However, stable climax communities can form, but this is uncommon in many environments. Ecosystems frequently experience disturbances that prevent a community from reaching equilibrium, or they can quickly escape if it does.

Biomass accumulation of matter and energy is the primary productivity. Green plants are the only organisms capable of collecting and converting electronic magnetism into chemical energy from reduced carbon compounds. Green plants represent the first step in this process (known as primary productivity). Secondary productivity occurs when heterotrophic organisms consume plant tissue, converting some of the matter and energy into their own biomass.

Secondary producers associated with detrital pathways and energy transfer routes account for a small portion of total forest productivity, but ecosystem process regulators, particularly nutrient cycling, are critical.

2.4. NUTRIENT CYCLING AND WATER RELATIONS

Life on the planet depends on the cycle of the elements of the biosphere. Atmospheric carbon dioxide will expire within a year or so if the breathing and fire-generating in the atmosphere do not constantly refill the green plants. We are also aware that it is necessary for life to continue to cycle nitrogen, oxygen, and water. These cycles include a gas phase and mechanisms to provide feedback, making them relatively perfect.

Any increase on one path is compensated for quickly by adjustments on other paths. However, there was concern recently that the carbon cycle may be disrupted by fossil fuels and the nitrogen cycle by the thoughtless introduction into the biosphere of pesticides and other substances. No less important to life are the elements with sedimentary cycles, such as phosphorus, calcium and magnesium.

These cycles cause continuous loss of biological systems with ultimate marine deposition in response to erosion. The substitution or return of an element with a sedimentary cycle into land biological systems depends on processes such as weathering of rocks, adding volcanic gasses or biological transfer from the sea to the soil. Sedimentary cycles are less perfect and easier to disrupt by men than carbon and nitrogen cycles.

In particular, the acceleration of losses or disruption of local cycling patterns through human activities may reduce the existing' pools' of an element in local ecosystems, limit the productivity, and thus limit the population in human beings. Many farmers, food scientists and ecologists, for example, consider that people accelerate the loss of phosphorus and this is a key resource for reducing the functioning of the biosphere. The importance to the benefit of mankind of these biogeochemical processes was recognized in an intensive study of these cycles.

The biogeochemical aspects of certain ecosystems have been focused on this interest among ecologists and foresters employed with nature-terrestrial ecosystems. The dissemination of chemical elements, as well as the absorption, retention and release rates in different ecosystems has therefore been accumulated. There has been little emphasis on the role of weathering and erosion in these systems.

However, the three main determinants of the structure and function of land ecosystems are the rate of weathering nutrients released from minerals, erosion levels and erosion loss. This information also enables the development, and linkage with the broader biogeochemical cycles, of the overall chemical budgets for ecosystems. Much of these relationships could not be determined because of the complex naturally occurring interaction between the hydrological cycle and the nutrient cycles.

This interaction complicates almost hopelessly the weathering or erosion in many ecosystems. However, under certain conditions, the integrated biogeochemical cycling study in small watershed ecosystems could benefit from such obvious barriers. The movement into the creation of matter and conversation of organic and inorganic matter is a nourishing cycle (or ecological recycling).

The flow of energy is one-way, non-cyclical, while the motion of mineral nutrients is cyclical. The mass of the plant tissue generated in waters typically accounts for 80–95% and is an important part of plant growth. For several physiological processes (such as carbohydrate synthesis) and related physical functions (such as turgid plants), plants need water.

The polarity (which makes it an excellent solvent), the viscosity (which makes it possible for capillaries to transmit plant tissues), and the thermal properties (which enable plant tissue to cool down) fulfill many functions of the H_2O molecules. Water, soil nutrients, dioxide, oxygen and solar radiation are required for cultivating plants.

Water is the most restrictive: it affects species productivity and diversity in both natural and agricultural ecosystems. The water is moved from the ground to the roots of the plants and to the leaves through the tree's xylem tissue, where it is lost in the process of transpiration. This cycle is known as the soil-plant-atmosphere continuum (SPAC).

When water passes through the SPAC, water passes through a variegation of medial distances and uses different modes of transport, including the cell wall, the cell membrane and the airspace. There are three main ways of transporting water: diffusion, mass flow and osmosis. Water molecules spontaneously move from high-concentration regions to fewer concentration zones, i.e., a concentration gradient. This movement is quick over a short distance and leads, for example during water loss to the atmosphere, to a short-distance transport between the cells and stomachs from leaves. In groups of water molecules in mass flow like increasing the pressure forming a gradient, it is moved by an external force. Mass flow is the predominant mode in which water transport is conducted in stems over a long distance.

The flow of water through the soil and the cell walls of plants are also significant. The third mode of osmosis is the movement of water molecules, e.g., through the cytoplasmic membrane. Osmosis happens naturally in short-distance transport as a response to driving concentration forces (as in dissemination) and pressure gradients (as in mass flow).

The circulating forces of water movements are known as the water potentials as both osmotic movement (concentration) and mass movement (pressure). In pressure or suction units, water potentials are measured, i.e., the force per unit area necessary for the movement of a given amount of water. Kilopascals (kPa) are the most common soil water potential units used in field studies.

Thus, water movement in the SPAC depends on the water potential differences of soil, plant or atmosphere. The gradient of water latent is often directed from the root towards the spring, as transpiring blades that are atmospherically exposed have the smallest possible water potential. However, in conditions when the soil is excessively dry, this water latent gradient could be reverted causing water loss from the soil roots of the plant.

Any environmental factors that influence the transpiration of the leaf stomata, such as wind or increase in temperature may further reduce leaf water potential and accelerate water loss. Water cycle Water covers more than two thirds of the Earth's surface. It constitutes an important component of most life forms. It consists of water for up to 70% of plants and animals. Large amounts of water cycle in the atmosphere, oceans, land and the biosphere on Earth.

2.4.1. Water Cycle and Nutrients

This cycling of water is called the water or hydrological cycle. Water cycling is important for weather and climate determination, supports plant growth and facilitates life.

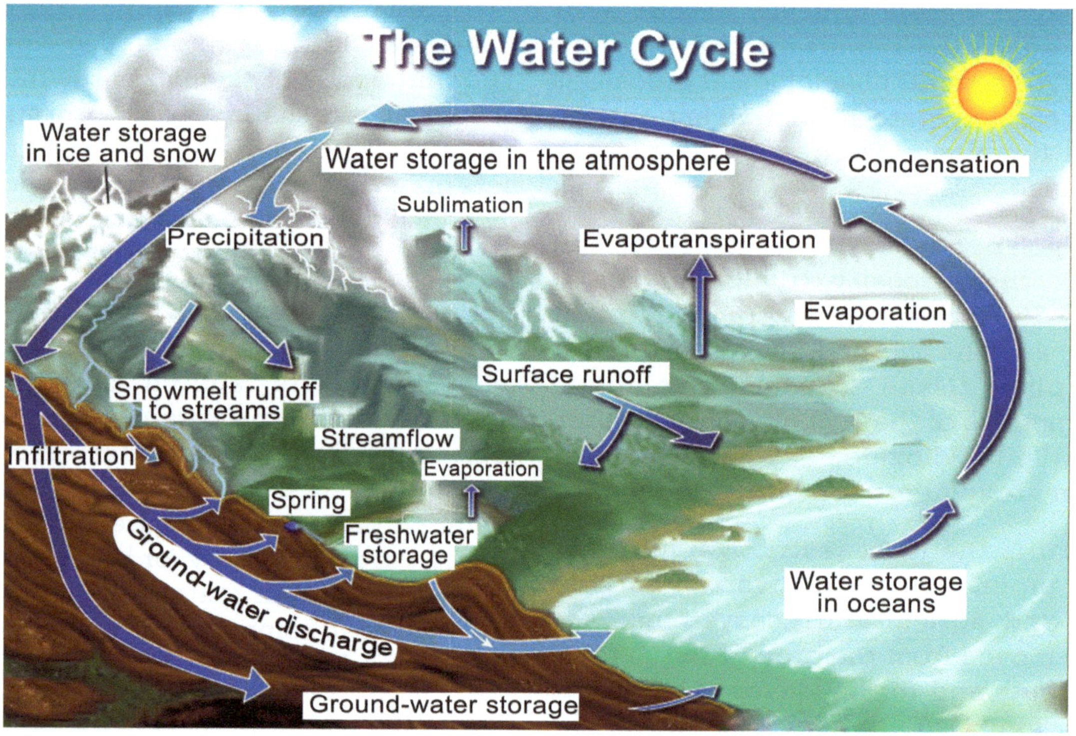

Figure 2.4. Water cycle.

Source: By John M. Even/USGS - USGS - http://ga.water.usgs.gov/edu/watercycleprint.html (English Wikipedia, original upload 27 April 2005 by Brian0918 en: Image: Water cycle.png), Public Domain, https://commons.wikimedia.org/w/index.php?curid=313135.

- *Evaporation:* The maximum quantity of the water evaporates out of the oceans with the highest amount of water. However, evaporation from lakes, rivers, streams and subsequent rain also occurs.
- *Transpiration:* This is a loss of water through plant surface (especially stomata). Transpiration is responsible for massive soil evaporation and complete evaporation. Evapotranspiration: evaporation and transpiration mechanisms are often referred to collectively as evapotranspiration.
- *Condensation:* The condensation is called the procedure of conversion of water vapor to liquid. Similar processes may have been observed when the dew falls

on a grass blade or cold glass. Water condenses into clouds in the atmosphere.

- *Precipitation:* Water comes back to earth by means of rain, sleet, snow, and ice (hail) precipitation. When precipitation rains, the majority of them go down into lakes and rivers, while a large part of them sinks into the ground.
- *Infiltration:* The process by which the water sinks into the soil is known as infiltration by soil or rock by which water passes. Water is filtered and purified during the sinking process into the surface of the Earth. The groundwater becomes increasingly purified depending on the type of soil and the penetration to which the water has sunk.
- *Freezing and Melting:* Some water freezes and is frozen in ice, in the form of glaciers and ice cubes. Water is also melted and returned to oceans and seas sometimes. Figure 2.4 shows the processes which are part of the water cycle.

2.4.2. Oxygen Cycle

Oxygen, together with nitrogen, is one of the main gasses in air. The following methods are used for oxygen re-cycling between the air and living organisms. Breathing and respiring: organisms like animals and plants take oxygen from the air during the breathing and gaseous exchanges. Oxygen is used to release energy from organic nutrients like glucose in cellular respiration. Photosynthesis: plants absorb air carbon dioxide to synthesize sugars, release oxygen, during photosynthesis. There are additional links between cellular respiration and photosynthesis in that the latter oxygen produces and oxygen consumes. Figure 2.5 shows the cycle of oxygen.

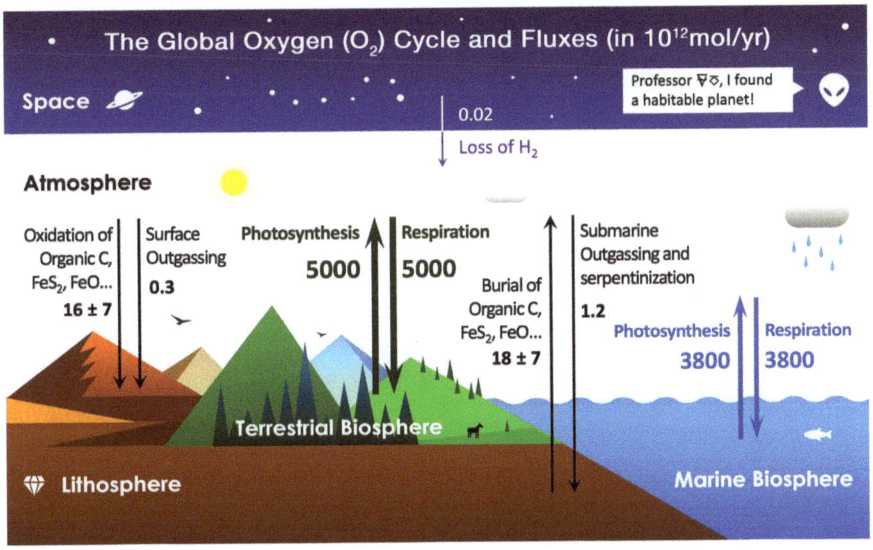

Figure 2.5. The oxygen cycle.

Source: By Pengxiao Xu - Own work, CC BY-SA 4.0, https://commons.wikimedia.org/w/index.php?curid=77951409.

2.4.3. Carbon Cycle

Carbon is a fundamental component of all organic materials and thus living organisms. The crust contains most of the carbon on Earth. The oceans and atmosphere are also part of other carbon reservoirs. Carbon transfers through such processes from one reservoir to another:

- *Combustion:* Wood and fossil fuel combustion transfers carbon as carbon dioxide through factory and automotive emissions to the atmosphere.
- *Photosynthesis:* In this process, plants take carbon dioxide and convert it into energy-rich organic molecules, such as carbon dioxide.
- *Metabolism:* Autotrophs translate carbon, such as fats, carbohydrates and protein, into organic molecules that animals can eat.
- *Cellular Breathing:* Animals eat organic carbon (carbohydrates) in food plants. Plants and animals decompose these organic elements and discharge energy, water and carbon dioxide in the course of cell respiration. During gaseous exchange, carbon dioxide is returned to the atmosphere.
- *Precipitation:* carbon dioxide can also precipitate in ocean sediments as carbonate.
- *Decay:* During the decay of all organisms, carbon dioxide gas is also released into the atmosphere.

The main processes for carbon cycling involving live organisms are photosynthesis and gaseous exchange. The carbon cycle is shown in the Figure 2.6.

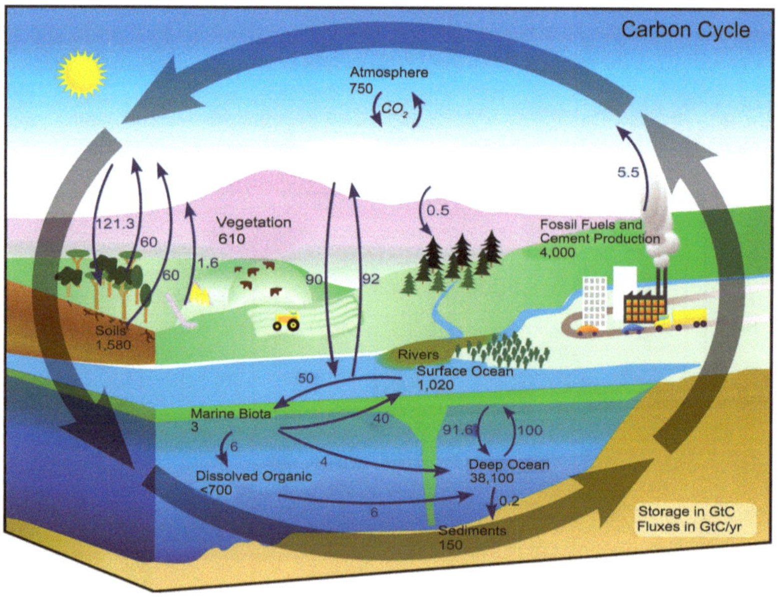

Figure 2.6. The carbon cycle.

Source: By Original: Kevin Saff Vector: FischX - Own work based on: Carbon cycle-cute diagram.jpeg (i.e., https://earthobservatory.nasa.gov/features/CarbonCycle), Public Domain, https://commons.wikimedia.org/w/index.php?curid=4134181.

2.4.4. Nitrogen Cycle

Most of the atmospheric gas is Nitrogen. Amino acids, proteins, and nuclear acids (DNA, RNA) are used as nitrogen that is essential for living organisms. Organisms do not have nitrogen gas in the air and therefore must be made available to plants and animals in a way that is absorbed. A few single-cell organisms can use it directly, such as bacteria. Nitrogen must be transformed into other forms for plants, for example, ammonia or nitrates. This procedure is called the fixation of nitrogen.

The nitrogen cycle includes the following stages:

- *Lightning:* Nitrogen can be directly changed to nitrate by lightning. This process increases the amount of nitrates which fall in rainwater onto the ground and act as fertilizers, causes the rapid growth of algae after thunderstorms.
- *Absorption:* Plants are absorbed in their roots by ammonia and nitrates.
- *Ingestion:* human beings and animals receive nitrogen from plants or animals.
- *Decomposition:* Bacteria and fungi break down plants and animals proteins and amino acids during decomposition.
- *Ammonification:* these decomposing bacteria convert nitrogenic amino acid products into ammonia.
- *Nitrification:* is the transformation by nitrifying bacteria of ammonia into nitrates.
- *Denitrification:* Bacteria convert ammonia and nitrate into nitrogen and nitric oxide in a process known as denitrification. To begin the cycle again, nitrogen is reverted to the environment. Figure 2.7 shows the nitrogen cycle.

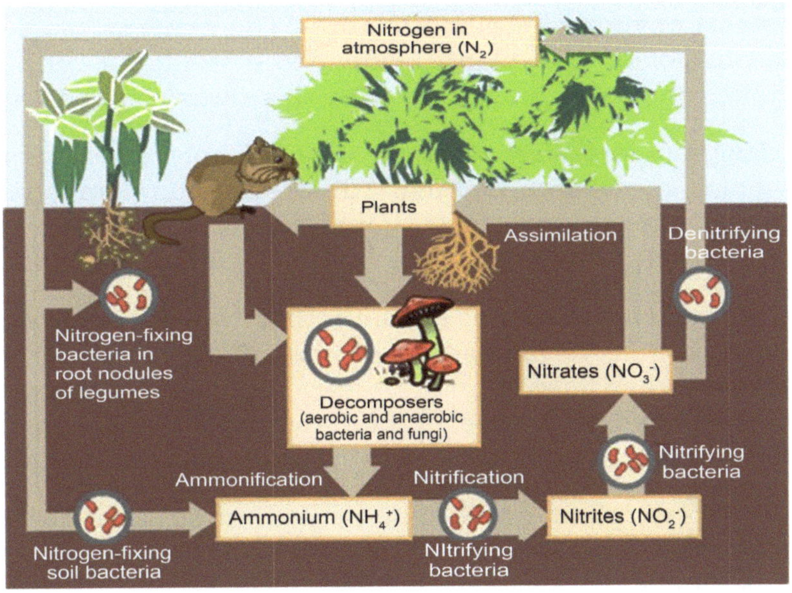

Figure 2.7. The nitrogen cycle.

Source: By U.S. Environmental Protection Agency - "Mid-Atlantic Integrated Assessment: Nitrogen," Public Domain, https://commons.wikimedia.org/w/index.php?curid=39327952.

2.5. PHYSIOLOGY IN STRESS ENVIRONMENTS

The current concept of stress in plants has evolved well in the last 60 years. Any unfavorable condition or substance that affects or perhaps prevents a plant's metabolism, growth or perhaps growth is considered stressful. Different natural and anthropogenic stressors are able to cause vegetation stress. You have to distinguish between long-term and short-term stress effects and minimal stress occurrences, partly offset by mechanisms of acclimatization, repair and adjustment, on the one hand, and chronic stresses or high stress which may ultimately lead to cell and plant death, on the other.

Energy is an absolute requirement for the sustainability of the structural organization throughout the organism. Energy is always needed to maintain such a complex order over time. The effect is a constant flow of energy through all biological organisms which provide the dynamic motor for key maintenance processes including transport and cell biosynthesis to protect their structure, organization and ability power to replicate and grow.

A stable state is maintained and a metastable state called homeostasis is achieved. Any environmental changes may interfere with homeostasis. Biological stress can be defined as homeostasis environmental modulation. The plant stress consequently has a bad impact on the physiology of a plant, leading to a sudden switch from an excellent to a suboptimal condition which interferes with this initial homeostatic condition.

Experimental design of stress assessment always involves measuring a physiological phenomenon in a plant species' suboptimal stress situation compared with measuring the same physiological phenomenon in optimum conditions of the same plant species. Plant stress is therefore a relative term. Plant stress can be split into 2 primary types.

Abiotic stress is actually a chemical insult that the environment (e.g., light, the plant or temperature) is able to impose on a plant. Biotic stress is a biological insult that plants, like diseases and insects, can be exposed in their lives (Figure 2.8). Some plants may be stressed and have one or perhaps more metabolic dysfunctions. The injury is able to be temporary and if stress is short and moderate, the plant can recover.

When stress is severe, plant death can be prevented from the flora, senescence and seeds. These plants are considered to be sensitive. Several plants, like short-lived or temporary desert plants, are completely relieved of stress. Drought Stress Drought is a major ecological stress which is responsible for different reasons, including low precipitation, salinity, low and high temperatures, high light intensity, etc.

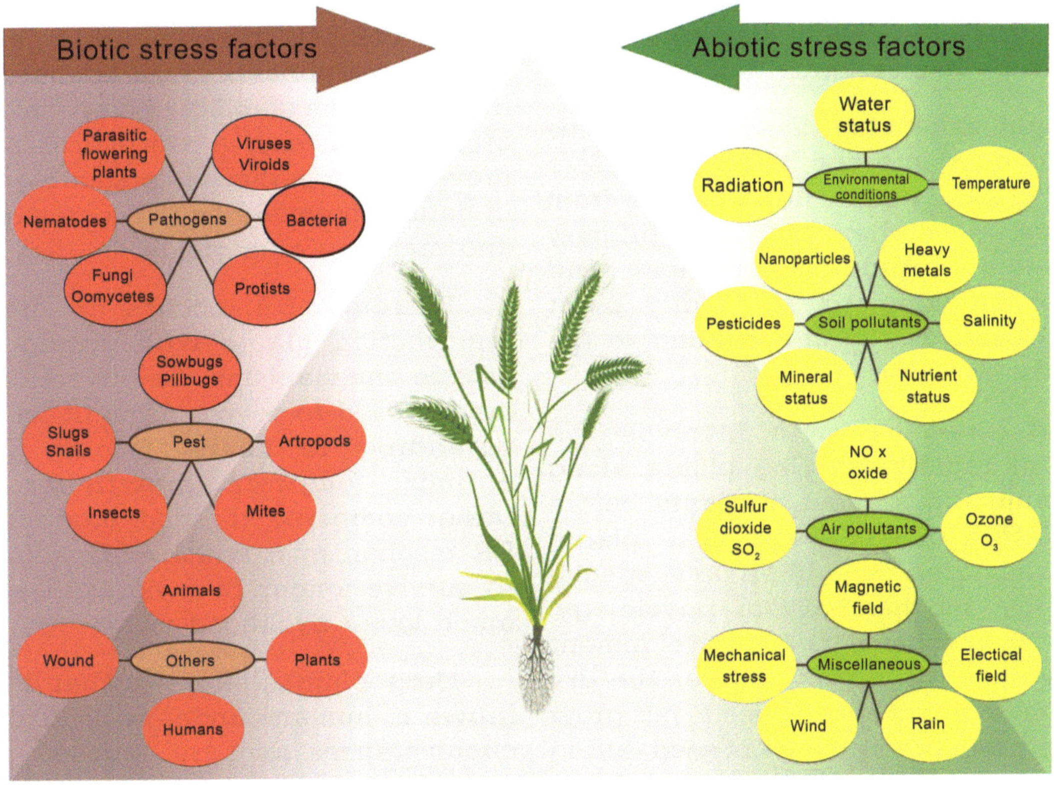

Figure 2.8. Impact of environmental stress on plant survival.

Source: Georgieva, M., & Vassileva, V. (2023). Stress Management in Plants: Examining Provisional and Unique Dose-Dependent Responses. International Journal of Molecular Sciences, 24(6), 5105. https://doi.org/10.3390/ijms24065105.

2.5.1. Drought Stress

Drought stress is a multi-dimensional stress that causes physiological, morphological, biochemical and molecular floral change. Many plants have enhanced their resistance mechanisms against drought stress, but they're unique and depend on the species. Mechanisms involving plant drought tolerance are usually part of the overall plan: the maintenance of cell homeostasis in water deficiency situations to be accomplished through increasing water intake from cells. Another common mechanism for annual plants to withstand drought is actually dryness prevention.

This mechanism is the primary strategy for plant growth under the conditions of drought and also the escape from stress. Probably the most common environmental factor in reducing crop productivity is the frequency and drought stress of severe drought is increasing as a result of global climate change.

The large diversity of plant species grown in extremely dry conditions across climate regions indicates in nature that plants have developed into a range of moral, biochemical and physiological adaptations to deal with drought stress. Drought resistance (DR) is a broader term used for plant species with adaptive characteristics to stay away from, prevent or perhaps tolerate drought stress.

Drought exhaustion is a plant species ability to complete its life cycle before drought starts. Plants don't suffer from dry stress, since they have two primary mechanisms, rapid phonological and plasticity development, to modify plant and reproductive growth according to the availability of water.

Rapid phenol growth involves the creation of rapidly growing plants along with a minimum number of plants before the depletion of soil water. The plants aren't considered to add in any special morphology, biochemistry or physiology. Plants with mechanisms for development plasticity show little growth over the dry season and few flowers and seeds but grow indefinitely and produce a lot of seed during the stormy season.

Drought avoidance' is the ability of plants to maintain (relatively) higher tissue water content despite their lower water in soil. The goal of this is to reduce water losses and optimize water absorption by way of a range of adaptive features. Water suppliers achieve greater water in tissue through increased rooting, hydraulic conductance, etc. Water absorption is maintained.

Water savers use water efficiently by reducing water consumption by sweating, absorption of radiation, etc. Drought tolerance (DT) is the ability of plants to withstand the low tissue water content with adaptive properties. The adaptive features include the maintenance of cell turgor through osmotic adjustment and cell elasticity, and increased protoplasm resistance.

Increasing and maintaining crop and crop yields yield stability are important for the food safety in normal and drought -conditions of a growing world population. Various components of DR are hard to solve in terms of the stability of crop yields. There are, nonetheless, a variety of escaping, preventing or perhaps tolerance mechanisms in natural populations that can improve DR and maintain crop yields.

Extreme DR occurs in nature in resurrection plants with strong drought escape mechanisms. Resurrection plants are exposed to severe drought for months, extending to years, forcing them to optimize long-term growth for survival rather compared to seed production. The DR mechanisms, which only allow plants to survive longer than normal, result in a much lower livelihood yield.

Crop plants, on the other hand, are grown in humans under ecological farming circumstances and are exposed to short-term random dryness stress, which means the harm caused by short-term drought stress needs to be reacted swiftly while still being cultivated in stressful environments. In plant lives that have been adapted to grow under extreme dry conditions, adaptive mechanisms for drought couldn't thus be put in place, leading to growth and/or yield penalties for crop plants both under drought and under normal conditions (Gunderson & Holling, 2002; Gann et al., 2019).

2.5.2. Water Logging Salinity and Alkalinity in Plants

Water logging hinders sustainable agriculture with substantial losses of yield in water logged plants. Plants are often exposed, under natural environmental conditions, to permanent or transient waterlogging. Several changes are caused by flooding to the soil, including oxygen levels, redox potential, and soil pH, in significant physiochemical characteristics. Hypoxia (O_2 deficiency) and anoxia (O_2 absence) create stressful conditions for plants growing in waterlogged soils. These poor oxygen conditions significantly impact plant growth, survival, and development. In response, plants may shift their metabolic processes from aerobic respiration to anaerobic fermentation. This adaptation allows them to survive in low-oxygen environments, but it can limit their growth and overall health. (Figure 2.9).

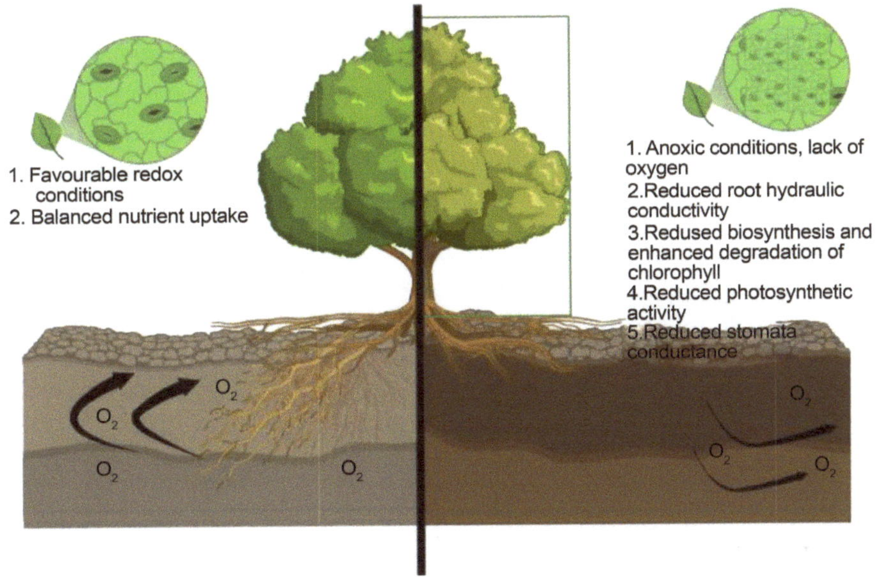

Figure 2.9. Waterlogging and its effects on tree growth.

Source: Topali, C., Antonopoulou, C., & Chatzissavvidis, C. (2024b). Effect of Waterlogging on Growth and Productivity of Fruit Crops. Horticulturae, 10(6), 623. https://doi.org/10.3390/horticulturae10060623.

The different research carried out in this field shows that most of the genes expressed under flood stress may be involved with the synthesis of enzymes that actively participate in the establishment of this fermentation process. This metabolite change occurs in plants, which continues to provide adenosine triphosphate (ATP). Plants show several responses, including impaired conductivity, net CO_2 assimilation, and hydraulic conductivity in water-borne conditions.

Additionally, in plants grown under the conditions of water supply, the oxidative damage caused by reactant oxygen species is common. These reactive oxygen species in return affect the integrity of the membrane and affect the effectiveness of Photo System II, thereby reducing net photosynthetic rates significantly. Such a disorder may also affect the reserves and translocations of carbohydrates in physiological mechanisms.

Effective use of carbohydrates could effectively discriminate sensitive and tolerant plant species in water logging. Water logging has also been known to have a negative effect on various processes for biochemical and physiological plants, causing deficits in important nutrients like nitrogen, magnesium, calcium and potassium. Additionally, plants that grow under flooded conditions show some morphological changes that come in the development of the adventure root, the initiating of hypertrophied lenticels and/or the development of aerenchyma. Additionally, these changes in physiology are a result of water logging. Their root tissue is breached by the absence of oxygen in the plant root region.

2.5.2.1. Symptoms and Causes

It typically takes place at the end of the roots, so the roots appear to be truncated. As a consequence, the plant has stalled development and growth. Lastly, the plant dies, so long as the anaerobic condition persists. Water normally does not last long enough for the plant to kill. Following an event, the plants start to breathe again. As long as soil conditions are moist, older roots near the plant is allowed by the surface to survive.

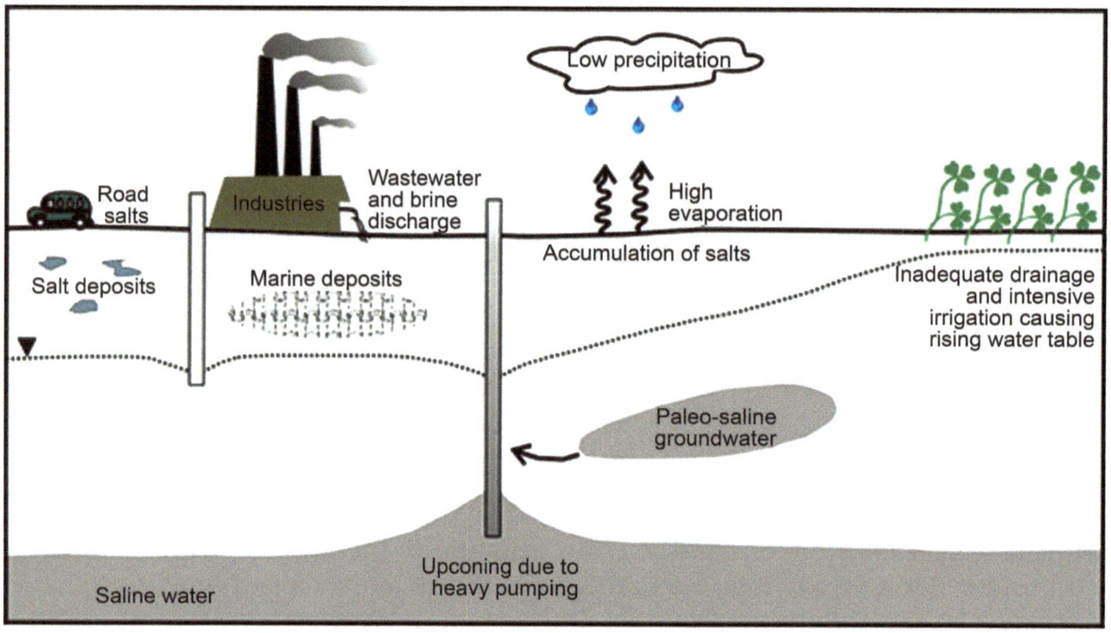

Figure 2.10. Illustrates the effects of salinity on trees and the area around it.

Source: Vengosh, A. (2003). Salinization and Saline Environments. In Elsevier eBooks (pp. 1–35). https://doi.org/10.1016/b0-08-043751-6/09051-4.

Nevertheless, dry conditions or further rot caused by water logging might weaken the plant to the degree that it's extremely poorly productive and can eventually die. Lots of farmers do not realize that a web site is blocked until the soil surface is actually filled with water. Nevertheless, the origins of plants may already be harmed and the yield potential severely affected.

If the soil profile of a plant or even the root area is saturated the waterlogging takes place. This happens if more rainfall causes soil to absorb or perhaps evaporate the atmosphere. Western Australia's Mediterranean climate makes cool, wet winters and hot dry summers much more rain than the atmosphere is able to evaporate every winter. The amount of excessive rain is particularly high in the higher precipitation regions of the South West.

- *Salinity and Alkalinity:* Saline and alkaline soils are widely distributed on all continents and under virtually any soil type in nearly all climate zones. Nevertheless, these issues are becoming most serious and pressing in irrigated areas in the subhumid, arid and semi-arid regions. Water harvesting and salt infestation, fertile and good once, has damaged quite large areas (Figure 2.10).

Heavy water percolation and irrigating canals, overflowing fields, substantial increases in the water table undergrowth, excess soil moisture, and overflow of rain, artesian water and inadequate drains causes the degradation of the planet. Indeed, water causes soil salinity or perhaps alkalinity mainly, and an insufficient attention to drainage has resulted in water and salt flora being blocked. Waterlogged soil becomes over time saline or perhaps alkaline, which doesn't just affect yield and crop growth but ultimately renders the soil unfit for just about any crop cultivation. Salt-impacted soils include various soil types, like salt, salt-alkaline and alkaline, acid sulfate and degraded soils. Such soils are broadly classified into two main groups as saline and alkaline soils.

The saline soils develop with neutral reactions under the influence of electrolytes of sodium salt and alkaline soil with alkaline hydrolytes. Both soils are commonly higher compared to normal soils in terms of electrolytes, which affect plant growth and soil fertility negatively. Nearly every acid is made up of salts found in numerous quantities of soils.

The primary types of soluble salts in soil and water leading to the formation of salt and alkaline soils include borates, nitrates, chlorides, sulfates, and carbonates (Gunderson & Holling, 2002; Gann et al., 2019).

2.6. CONSERVATION OF FOREST ECOSYSTEMS

Forests cover nearly one-third of the Earth's land area and account for more than 80% of the planet's biodiversity. However, forest habitats are continuing to decline, which threatens the functioning of these ecosystems and their ability to provide essential ecosystem services. Given the growing population pressure, conserving, and restoring forest ecosystems is crucial.

Understanding how biodiversity influences ecosystem functioning has led to increased efforts in ecological restoration. The primary goal is often to re-establish the connection between biodiversity and ecosystem processes. However, it is important to note that biodiversity does not always play a uniformly critical role in all types of ecosystem services.

Some ecosystem services aren't strongly related to or perhaps connected to biological diversity. It has to be clearer for the decision-making of forest management to find the connection between ecosystem services and biodiversity. Many experiments to clarify the links between biodiversity and ecosystem function for freshness or grassland microcosms have been carried out.

However, very little research has been undertaken in forested ecosystems. Moreover, although these experimental scientific studies found that a variety of features and ecosystem services relate to biodiversity, the high levels of biodiversity (i.e., species wealth or perhaps interactions) were generally not addressed. This is an overview from this BEF's perspective of the significant elements of forestry restoration.

Despite increasing efforts to maintain forestry and forest conservation, the range of forest habitat in the tropics in particular continues to decline, in particular through forest transformation into agriculture and land use related to urban population growth. Between 1980 and 2000, the intact forest clearing produced more than half the new agricultural land in the tropics.

Many damaged and secondary forests have finally been cleared for agrarian purposes, which is actually turning into a habitat of increasing importance for many forest species. The majority of the forest and segments of the forest are actually affected by decreasing habitat patches and reduced population size that in turn is able to lead to a decline in genetic variation and population viability.

Additional anthropogenic risks like fire should exacerbate the negative effects of forest fragmentation and isolation, especially in the context of global climate change. Cryptic deforestation changes forest selection and degradation of forests and plant communities, including biodiversity threats, vitality and capacities together with forest losses and fragmentation. At the same time, both the amount and quality of forests will be reduced and lots of forest environments are actually projected to be extinct massively. For a wide range of taxa, including liana and trees fats, dung beetles, lizards, epidemic arachnids, large mammals, amphibians, butterflies, birds, and liana, over a gradient of forest degradation from primary to secondary to plantations, biodiversity has been proven to significantly reduce.

The loss of forest biodiversity may present serious threats to the functioning of forest ecosystems (e.g., forest activities, processes or perhaps properties such as organic decomposition, soil nutrient cycling and water conservation), and the ability of a wood forest to provide ecosystem services.

2.6.1. Importance and Ways to Conserve the Forest

Forests are immensely valuable for several reasons. They are essential for our survival, providing critical resources such as wood, medicinal plants, and fibers, and offering ecosystem services like air and water purification, climate regulation, and habitat for biodiversity. Therefore, protecting forests is crucial. Their role as refuges for biodiversity and providers of ecosystem services cannot be overstated.

As the human population grows, sustaining the remaining forests and ensuring their continued functionality and biodiversity becomes increasingly important. Large-scale forestry restructuring, whether active or passive, is likely the only viable long-term solution to meet the growing demand for forested ecosystem services. This is especially important as many people depend on forests for their livelihoods.

Ways to conserve forests may include, but are not limited to:

- *Control of Deforestation*: While deforestation cannot be entirely avoided, efforts should focus on controlling it. This involves reducing the cutting of premature and young trees and preventing large-scale deforestation driven by business activities. Long-term strategies, such as selective logging and sustainable forestry practices, can help maintain forest cover and health.
- *Protect Against Forest Fires*: Forest fires are a significant threat, often caused by natural events or human activities. Controlling and preventing fires requires measures such as creating firebreaks, using fire-resistant chemicals, clearing dry leaves and debris, and implementing community fire prevention programs. Effective fire management is crucial for maintaining forest ecosystems.
- *Afforestation*: Afforestation involves planting trees in areas where there were none previously, or replanting trees in degraded areas. This helps increase forest cover, balance the ecosystem, and mitigate the effects of deforestation and pollution. Afforestation efforts aim to restore and enhance forest ecosystems through new plantations and manual tree transplants.
- *Better Farming Practices*: Practices such as slash-and-burn agriculture, overgrazing by livestock, and shifting cultivation can severely damage forests. Monitoring and improving these practices is essential. Techniques like Jhoom farming (also

known as swidden agriculture) can help. This involves rotating cultivation areas and allowing previously used land to naturally regenerate, which can reduce forest degradation and improve soil fertility.

As a result of the high concentrations of atmospheric CO_2 that are affecting and distressing the earth's climate, the demand for efficient forestry protection has become clear. Trees are crucial to combating global climate change, aimed at regulating carbon dioxide levels by absorbing gas, storing carbon and releasing oxygen. The stored forest carbon is released into the atmosphere and CO_2 levels are intensified once the trees are destroyed. Today, deforestation produces approximately 15–20% of global carbon emissions, making it necessary to bring down the global climate change rate by halting deforestation (Gunderson & Holling, 2002; Gann et al., 2019).

CASE STUDY

PLANT COMMUNITY DYNAMICS IN A COASTAL DUNE ECOSYSTEM

The coastal dune ecosystem is characterized by a unique assemblage of plant communities adapted to harsh environmental conditions, including high salinity, shifting sands, and strong winds. One prominent plant community in this ecosystem is dominated by marram grass (Ammophila arenaria), which plays a critical role in stabilizing sand dunes and preventing coastal erosion. The deep root system of marram grass allows it to access water and nutrients from deeper soil layers, while its extensive rhizome network helps bind the sand, creating a stable substrate for other plant species to establish. Over time, this initial colonization by marram grass paves the way for a more diverse plant community, including species like beach strawberry (Fragaria chiloensis) and various herbaceous perennials.

As the plant community matures, several ecological interactions influence its dynamics. Competition for resources such as sunlight, water, and nutrients occurs among the resident species, while herbivory by local fauna, such as rabbits and deer, can further impact plant abundance and distribution. Moreover, the presence of invasive species poses additional challenges, as they can outcompete native plants for resources and disrupt established ecological relationships. Restoration efforts aimed at preserving these coastal plant communities often focus on controlling invasive species and enhancing the resilience of native plants to environmental stressors, such as climate change and rising sea levels.

The study of plant community dynamics in coastal dune ecosystems is crucial for understanding ecological processes and informing conservation strategies. Monitoring shifts in plant composition and diversity can provide valuable insights into the health of the ecosystem and the impacts of human activities. Furthermore, understanding the relationships between different plant species and their environment can help guide the restoration of degraded areas, ensuring the long-term sustainability of these vital coastal habitats.

Discussion Questions

- How do environmental factors influence the composition and structure of plant communities in coastal dune ecosystems?
- What role do invasive species play in shaping the dynamics of native plant communities?

SUMMARY

Forest ecosystems are dynamic and complex communities of plants, animals, microorganisms, and their physical environment, interacting as an ecological unit. These ecosystems are primarily characterized by their dense tree canopies and extensive biodiversity, playing a crucial role in maintaining global ecological balance.

Key components of forest ecosystems include various types of vegetation (trees, shrubs, and groundcover plants), diverse wildlife (birds, mammals, insects, and microorganisms), and abiotic elements (soil, water, and climate). Forests provide numerous ecosystem services such as carbon sequestration, oxygen production, water filtration, and soil stabilization. They also offer habitat and food for countless species, contributing to biodiversity conservation.

Forest ecosystems are categorized into different types, including tropical, temperate, and boreal forests, each with unique climatic conditions and species compositions. These ecosystems face numerous threats, including deforestation, climate change, pollution, and habitat fragmentation, which can lead to biodiversity loss and disruption of ecological functions.

Sustainable forest management practices are essential to preserve these vital ecosystems. This includes conservation efforts, reforestation, and policies aimed at reducing deforestation. Protecting forest ecosystems is crucial not only for the environment but also for the economic and social well-being of human populations that depend on forest resources (Gunderson & Holling, 2002; Gann et al., 2019).

REVIEW QUESTIONS

1. What are the types and characteristic features of forest ecosystems?
2. Describe the structure of a forest ecosystem and its components.
3. Explain the concept of vertical stratification in plant communities. Why is it important?
4. Discuss the roles of water, oxygen, carbon, and nitrogen in nutrient cycling within forest ecosystems.
5. How do plants adapt physiologically to stressful environments like drought and waterlogging?

REFERENCES

1. Akinola, O. V., & Adegoke, J. (2019). Assessment of forest fire vulnerability zones in Missouri, United States of America. *International Journal of Sustainable Development and World Ecology, 26*(4), 251–257. https://doi.org/10.1080/13504509.2019.1632255.
2. Albrich, K., Rammer, W., Turner, M. G., Ratajczak, Z., Brazinuas, K. H., Hansen, W. D., & Seidl, R. (2020). Simulating forest resilience: A review. *Global Ecology and Biogeography, 29*(12), 2082–2096. https://doi.org/10.1111/geb.13138.

3. Ali, A. (1999). Climate change impacts and adaptation assessment in Bangladesh. *Climate Research, 12*(2), 109–116. https://doi.org/10.3354/cr012109.

4. Angeler, D. G., Baho, D. L., Allen, C. R., & Johnson, R. K. (2015). Linking degradation status with ecosystem vulnerability to environmental change. *Oecologia, 178*(3), 899–913. https://doi.org/10.1007/s00442-015-3314-7.

5. Behera, M. D., Murthy, M. S. R., Das, P., & Sharma, E. (2018). Modelling forest resilience in Hindu Kush Himalaya using geoinformation. *Journal of Earth System Science, 127*(1), 95. https://doi.org/10.1007/s12040-018-0990-5.

6. Berger, A. R., & Hodge, R. A. (1998). Natural change in the environment: A challenge to the pressure-state-response concept. *Social Indicators Research, 44*(2), 255–265. https://doi.org/10.1023/A:1006937228620.

7. Bernier, P., & Schoene, D. (2009). *Adapting Forests and Their Management to Climate Change: An Overview* (Information Service of FAO, Vol. 60, pp. 5–11). FAO.

8. Bone, C., Moseley, C., Vinyeta, K., & Bixler, R. P. (2016). Employing resilience in the United States Forest Service. *Land Use Policy, 52*, 430–438. https://doi.org/10.1016/j.landusepol.2015.12.024.

9. Boonstra, R., Andreassen, H. P., Boutin, S., Hušek, J., Ims, R. A., Krebs, C. J., Skarpe, C., & Wabakken, P. (2016). Why do the boreal forest ecosystems of northwestern Europe differ from those of western North America? *Bioscience, 66*(9), 722–734. https://doi.org/10.1093/biosci/biw093.

10. Bourgoin, C., Oszwald, J., Bourgoin, J., Gond, V., Blanc, L., Dessard, H., Van Phan, T., Sist, P., Laderach, P., & Reymondin, L. (2020). Assessing the ecological vulnerability of forest landscapes to agricultural frontier expansion in the Central Highlands of Vietnam. *International Journal of Applied Earth Observation and Geoinformation, 84*, 101958. https://doi.org/10.1016/j.jag.2019.101958.

11. Brown, E. D., & Williams, B. K. (2018). Resilience and resource management. *Environmental Management, 56*(6), 1416–1427. https://doi.org/10.1007/s00267-018-1071-2.

12. Cantarello, E., Newton, A. C., Martin, P. A., Evans, P. M., Gosal, A., & Lucash, M. S. (2017). Quantifying resilience of multiple ecosystem services and biodiversity in a temperate forest landscape. *Ecology and Evolution, 7*(24), 9661–9675. https://doi.org/10.1002/ece3.3560.

13. Carpenter, S. R., Westley, F., & Turner, M. G. (2005). Surrogates for resilience of social–Ecological systems. *Ecosystems, 8*(8), 941–944. https://doi.org/10.1007/s10021-005-0172-7.

14. Chuluundorj, O. (2006). A multi-level study of vulnerability of Mongolian pastoralists to natural hazards and its consequences on individual and household well-being. University of Colorado, Denver, CO, USA.

15. Cinco-Castro, S., & Herrera-Silveira, J. (2020). Vulnerability of mangrove ecosystems to climate change effects: The case of the Yucatan Peninsula. *Ocean & Coastal Management, 192*, 105196. https://doi.org/10.1016/j.ocecoaman.2020.105196.

16. Clements, F. E. (1905). *Research Methods in Ecology*. University Publishing Company.

17. Costanza, R., d'Arge, R., de Groot, R., Farber, S., Grasso, M., Hannon, B., Limburg, K., Naeem, S., O'Neill, R. V., Paruelo, J., et al. (1997). The value of the world's ecosystem services and natural capital. *Nature, 387*(6630), 253–260. https://doi.org/10.1038/387253a0.

18. Craven, D., Filotas, E., Angers, V. A., & Messier, C. (2016). Evaluating resilience of tree communities in fragmented landscapes: Linking functional response diversity with landscape connectivity. *Diversity and Distributions, 22*(5), 505–518. https://doi.org/10.1111/ddi.12395.

19. Cui, X., Gibbes, C., Southworth, J., & Waylen, P. (2013). Using remote sensing to quantify vegetation change and ecological resilience in a semi-arid system. *Land, 2*(1), 108–130. https://doi.org/10.3390/land2010108.

20. Danielson, T. M., Rivera-Monroy, V. H., Castañeda-Moya, E., Briceño, H., Travieso, R., Marx, B. D., & Farfán, L. M. (2017). Assessment of Everglades mangrove forest resilience: Implications for above-ground net primary productivity and carbon dynamics. *Forest Ecology and Management, 404*, 115–125. https://doi.org/10.1016/j.foreco.2017.07.014.

21. De La Serrana, R. G., Vilagrosa, A., & Alloza, J. A. (2015). Pine mortality in southeast Spain after an extreme dry and warm year: Interactions among drought stress, carbohydrates and bark beetle attack. *Trees, 29*(6), 1791–1804. https://doi.org/10.1007/s00468-015-1227-4.

22. De Lange, H. J., Sala, S., Vighi, M., & Faber, J. H. (2010). Ecological vulnerability in risk assessment—A review and perspectives. *Science of The Total Environment, 408*(16), 3871–3879. https://doi.org/10.1016/j.scitotenv.2010.04.032.

23. Delgado, A., Vidal, J., Castro, J., Felix, J., & Saenz, J. (2020). Assessment of surface water quality on the upper watershed of Huallaga River, in Peru, using grey systems and Shannon entropy. *International Journal of Advanced Computer Science and Applications, 11*(5), 437–444. https://doi.org/10.14569/IJACSA.2020.0110559.

24. Derroire, G., Balvanera, P., Castellanos-Castro, C., Decocq, G., Kennard, D. K., Lebrija-Trejos, E., Leiva, J. A., Oden, P. C., Powers, J. S., Rico-Gray, V., et al. (2016). Resilience of tropical dry forests—A meta-analysis of changes in species diversity and composition during secondary succession. *Oikos, 125*(10), 1386–1397. https://doi.org/10.1111/oik.02973.

25. Di Mauro, B., Fava, F., Busetto, L., Crosta, G. F., & Colombo, R. (2014). Post-fire resilience in the Alpine region estimated from MODIS satellite multispectral data. *International Journal of Applied Earth Observation and Geoinformation, 32*, 163–172. https://doi.org/10.1016/j.jag.2014.04.014.

26. Duveneck, M. J., & Scheller, R. M. (2016). Measuring and managing resistance and resilience under climate change in northern Great Lake forests (USA). *Landscape Ecology, 31*(4), 669–686. https://doi.org/10.1007/s10980-015-0284-2.

27. Dwomoh, F. K., & Wimberly, M. C. (2017). Fire regimes and forest resilience: Alternative vegetation states in the West African tropics. *Landscape Ecology, 32*(9), 1849–1865. https://doi.org/10.1007/s10980-017-0493-5.

28. Fang, C., Wang, Y., & Fang, J. (2016). A comprehensive assessment of urban vulnerability and its spatial differentiation in China. *Journal of Geographical Sciences, 26*(2), 153–170. https://doi.org/10.1007/s11442-016-1274-3.

29. Fang, O., Zhang, Q. B., Vitasse, Y., Zweifel, R., & Cherubini, P. (2021). The frequency and severity of past droughts shape the drought sensitivity of juniper trees on the Tibetan plateau. *Forest Ecology and Management, 486*, 118968. https://doi.org/10.1016/j.foreco.2021.118968.

30. Folke, C., Carpenter, S., Walker, B., & Scheffer, M. (2004). Regime shifts, resilience, and biodiversity in ecosystem management. *Annual Review of Ecology, Evolution, and Systematics, 35*, 557–581. https://doi.org/10.1146/annurev.ecolsys.35.021103.105711.

31. Gann, G. D., McDonald, T., Walder, B., Aronson, J., Nelson, C. R., Jonson, J., Hallett, J. G., Eisenberg, C., Guariguata, M. R., Liu, J., et al. (2019). International principles and standards for the practice of ecological restoration (Second edition). *Restoration Ecology, 27*(S1), S1–S46. https://doi.org/10.1111/rec.13035.

32. Gao, S., Liu, R., Zhou, T., Fang, W., Yi, C., Lu, R., & Luo, H. (2018). Dynamic responses of tree-ring growth to multiple dimensions of drought. *Global Change Biology, 24*(11), 5380–5390. https://doi.org/10.1111/gcb.14472.

33. Giglio, E. (2016). Landscape ecological method to study agriculture vegetation: Some examples from the Po Valley. *Annals of Botany, 6*(2), 95–110. https://doi.org/10.1093/aob/mcw012.

34. Gong, W., Hu, T. X., & Gong, Y. B. (2004). A brief discussion on the formation characteristics of fragile forest ecosystems and their management measures. *Sichuan Forestry Science and Technology, 3*, 5–10.

35. Guidi, C., Di Matteo, G., & Grego, S. (2018). An overview of proven climate change vulnerability assessment tools for forests and forest-dependent communities across the globe: A literature analysis. *Journal of Forestry Research, 29*(6), 1167–1175. https://doi.org/10.1007/s11676-018-0645-4.

36. Gunderson, L. H., & Holling, C. S. (2002). *Panarchy: Understanding Transformations in Human and Natural Systems*. Island Press.

37. Guo, B., Zang, W., & Luo, W. (2020). Spatial-temporal shifts of ecological vulnerability of Karst Mountain ecosystem—Impacts of global change and anthropogenic interference. *Science of The Total Environment, 741*, 140256. https://doi.org/10.1016/j.scitotenv.2020.140256.

38. Hart, S. J., Henkelman, J., McLoughlin, P. D., Nielsen, S. E., Truchon-Savard, A., & Johnstone, J. F. (2019). Examining forest resilience to changing fire frequency in a fire-prone region of boreal forest. *Global Change Biology, 25*(3), 869–884. https://doi.org/10.1111/gcb.14515.

39. Hodgson, D., McDonald, J. L., & Hosken, D. J. (2015). What do you mean, 'resilient'? *Trends in Ecology & Evolution, 30*(9), 503–506. https://doi.org/10.1016/j.tree.2015.06.012.

40. Holling, C. S. (1973). Resilience and stability of ecological systems. *Annual Review of Ecology and Systematics, 4*, 1–23. https://doi.org/10.1146/annurev.es.04.110173.000245.

41. Hou, K., Tao, W., Wang, L., & Li, X. X. (2020). Study on hierarchical transformation mechanisms of regional ecological vulnerability and its applicability. *Ecological Indicators, 114*, 106343. https://doi.org/10.1016/j.ecolind.2020.106343.
42. Hu, W. J., Chen, B., Arnupap, P., Zhang, D., Liu, X. M., Gu, H. F., Ajcharaporn, P., Chao, B. X., & Zheng, X. Q. (2020). Ecological vulnerability assessment of coral reef: A case study of Si Chang Island, Thailand. *Chinese Journal of Ecology, 39*(3), 979–989. https://doi.org/10.13292/j.1000-4890.202003.014.
43. Huang, M., Wang, X., Keenan, T. F., & Piao, S. (2018). Drought timing influences the legacy of tree growth recovery. *Global Change Biology, 24*(8), 3546–3559. https://doi.org/10.1111/gcb.14169.
44. Huang, Z., & Wan, J. (2020). Vulnerability assessment of safe operation of utility tunnel based on improved matter-element extension method. *Construction Science and Technology, 8*, 78–81.
45. Ibarra, J. T., Cockle, K., Altamirano, T., van der Hoek, Y., Simard, S. W., Bonacic, C., & Martin, K. (2020). Nurturing resilient forest biodiversity: Nest webs as complex adaptive systems. *Ecology and Society, 25*(2), 27. https://doi.org/10.5751/ES-11592-250227.
46. IPCC. (2001). *Climate Change 2001: Impacts, Adaptation, and Vulnerability*. Cambridge University Press.
47. IPCC. (2014). *Climate Change 2014: Impacts, Adaptation, and Vulnerability*. Cambridge University Press.
48. IUCN. (2008). *Ecosystem-Based Adaptation: An Approach for Building Resilience and Reducing Risk for Local Communities and Ecosystems*. International Union for the Conservation of Nature.
49. Johnstone, J. F., Allen, C. D., Franklin, J. F., Frelich, L. E., Harvey, B. J., Higuera, P. E., Mack, M. C., Meentemeyer, R. K., Metz, M. R., Perry, G. L. W., et al. (2016). Changing disturbance regimes, ecological memory, and forest resilience. *Frontiers in Ecology and the Environment, 14*(7), 369–378. https://doi.org/10.1002/fee.1302.
50. Keenan, R. J. (2015). Climate change impacts and adaptation in forest management: A review. *Annals of Forest Science, 72*(2), 145–167. https://doi.org/10.1007/s13595-015-0481-0.
51. Kneeshaw, D. D., Sturtevant, B. R., DeGrandpré, L., Doblas-Miranda, E., James, P. M. A., Tardif, D., & Burton, P. J. (2021). The vision of managing for pest-resistant landscapes: Realistic or utopic? *Current Forestry Reports, 7*(2), 97–113. https://doi.org/10.1007/s40745-021-00206-2.
52. Kumar, M., Kalra, N., Singh, H., Sharma, S., Singh Rawat, P., Singh, R. K., Gupta, A. K., Jumar, P., & Ravindranath, N. H. (2021). Indicator-based vulnerability assessment of forest ecosystem in the Indian Western Himalayas: An analytical hierarchy process integrated approach. *Ecological Indicators, 125*, 107568. https://doi.org/10.1016/j.ecolind.2021.107568.
53. Law, B. E., & Waring, R. H. (2015). Carbon implications of current and future effects of drought, fire and management on Pacific Northwest forests. *Forest Ecology and Management, 355*, 4–14. https://doi.org/10.1016/j.foreco.2015.04.021.

54. Lecina-Diaz, J., Martinez-Vilalta, J., Alvarez, A., Banque, M., Birkmann, J., Feldmeyer, D., Vayredm, J., & Retana, J. (2021). Characterizing forest vulnerability and risk to climate-change hazards. *Frontiers in Ecology and the Environment, 19*(2), 126–133. https://doi.org/10.1002/fee.2351.

55. Lennox, G. D., Gardner, T. A., Thomson, J. R., Ferreira, J., Berenguer, E., Lees, A. C., Nally, R. M., Aragao, L. E. O. C., Ferraz, S. F. B., Louzada, J., et al. (2018). Second rate or a second chance? Assessing biomass and biodiversity recovery in regenerating Amazonian forests. *Global Change Biology, 24*(11), 5680–5694. https://doi.org/10.1111/gcb.14453.

56. Li, B., Yang, Z., & Su, F. (2016). Vulnerability measurement of Chinese marine economic system based on set pair analysis. *Science in China Series D: Earth Sciences, 36*(1), 47–54. https://doi.org/10.1007/s11430-016-5287-4.

57. Li, J. D., & Xu, D. (2018). Evaluation on exploitation potential of forest health tourism based on AHP and fuzzy synthetic evaluation: A case study of Liaodong Mountain area. *Chinese Journal of Agricultural Resources and Regional Planning, 39*(1), 135–142. https://doi.org/10.7621/cjarrp.2018.01.017.

58. Liao, K. H. (2012). A theory on urban resilience to floods: A basis for alternative planning practices. *Ecology and Society, 17*(4), 48. https://doi.org/10.5751/ES-05379-170448.

59. Liu, M., Liu, X., Wu, L., Tang, Y., Li, Y., Zhang, Y., Ye, L., & Zhang, B. (2021). Establishing forest resilience indicators in the hilly red soil region of southern China from vegetation greenness and landscape metrics using dense Landsat time series. *Ecological Indicators, 121*, 106985. https://doi.org/10.1016/j.ecolind.2020.106985.

60. López-Jiménez, L. N., Duran-Garcia, R., & Dupuy-Rada, J. M. (2019). Recovery of structure, diversity and composition of a tropical semievergreen forest in Yucatan, Mexico. *Madera y Bosques, 25*, e2511587. https://doi.org/10.21829/myb.2019.2511587.

61. Lucier, A., Ayres, M., Karnosky, D., Thompson, I., Loehle, C., Percy, K., & Sohngen, B. (2009). Forest responses and vulnerabilities to recent climate change. *IUFRO World Series, 22*, 29–52.

62. Ma, X. Y. (2019). *Spatial Analysis, Evaluation and Optimization of the Yellow River Beach Area in Xinxiang City from the Perspective of Resilient Cities* (Unpublished doctoral dissertation). Beijing Forestry University, Beijing, China.

63. Mildrexler, D. J., Yang, Z., Cohen, W., & Bell, D. (2016). A forest vulnerability index based on drought and high temperatures. *Remote Sensing of Environment, 173*, 314–325. https://doi.org/10.1016/j.rse.2015.11.012.

64. Millennium Ecosystem Assessment Board. (2005). *Ecosystems and Human Well-Being*. Island Press.

65. Miyawaki, A. (2001). A vegetation-ecological study on regeneration and restoration of vegetation. *Vegetation Science News, 6*, 55–62.

66. Nemec, K. T., Chan, J., Hoffman, C., Spanbauer, T. L., Hamm, J. A., Allen, C. R., Hefley, T., Pan, D., & Shrestha, P. (2014). Assessing resilience in stressed watersheds. *Ecology and Society, 19*(4), 34. https://doi.org/10.5751/ES-06809-190434.

67. Nikinmaa, L., Lindner, M., Cantarello, E., Jump, A. S., Seidl, R., Winkel, G., & Muys, B. (2020). Reviewing the use of resilience concepts in forest sciences. *Current Forestry Reports, 6*(1), 61–80. https://doi.org/10.1007/s40745-020-00101-4.

68. Niu, W. Y. (1989). The discriminatory index with regard to the weakness, overlapness, and breadth of ecotone. *Acta Ecologica Sinica, 9*(2), 97–105.

69. Ouyang, H., & Ye, Q. (2016). A review on the evolution of resilient city theory: Concept, context and tendency. *City Planning Review, 40*(6), 34–42.

70. Ouyang, Z. Y. (1998). On the comprehensive management, development and breaking away from poorness of the ecologically fragile karst area in south-west China. *World Science and Technology Research and Development, 2*(1), 3–5.

71. Ozcan, O., Musaoglu, N., & Tyrke⊠, M. (2018). Assessing vulnerability of a forest ecosystem to climate change and variability in the western Mediterranean sub-region of Turkey. *Journal of Forestry Research, 29*(4), 709–725. https://doi.org/10.1007/s11676-018-0535-4.

72. Peng, Q., Lin, C. H., & He, T. F. (2006). The development of study on the characteristics of soil and water loss and its conservation in Guizhou karst mountain area. *Guizhou Science, 24*(2), 66–70.

73. Perie, C., & de Blois, S. (2016). Dominant forest tree species are potentially vulnerable to climate change over large portions of their range even at high latitudes. *PeerJ, 4*, e2218. https://doi.org/10.7717/peerj.2218.

74. Pimm, S. L. (1984). The complexity and stability of ecosystems. *Nature, 307*(5949), 321–326. https://doi.org/10.1038/307321a0.

75. Polsky, C., Neff, R., & Yarnal, B. (2007). Building comparable global change vulnerability assessments: The vulnerability scoping diagram. *Global Environmental Change, 17*(3–4), 472–485. https://doi.org/10.1016/j.gloenvcha.2006.11.009.

76. Prentice, I. C., Cramer, W., Harrison, S. P., Leemans, R., Monserud, R. A., & Solomon, A. M. (1992). A global biome model based on plant physiology and dominance, soil properties and climate. *Journal of Biogeography, 19*(2), 117–134. https://doi.org/10.2307/2845491.

77. Rana, I. A. (2020). Disaster and climate change resilience: A bibliometric analysis. *International Journal of Disaster Risk Reduction, 50*, 101839. https://doi.org/10.1016/j.ijdrr.2020.101839.

78. Reyer, C. P. O., Brouwers, N., Rammig, A., Brook, B. W., Epila, J., Grant, R. F., Holmgren, M., Langerwisch, F., Leuzinger, S., Lucht, W., et al. (2015). Forest resilience and tipping points at different spatio-temporal scales: Approaches and challenges. *Journal of Ecology, 103*(1), 5–15. https://doi.org/10.1111/1365-2745.12327.

79. Rivera-Monroy, V. H., Danielson, T. M., Castañeda-Moya, E., Marx, B. D., Travieso, R., Zhao, X., Gaiser, E. E., & Farfan, L. M. (2019). Long-term demography and stem productivity of Everglades mangrove forests (Florida, USA): Resistance to hurricane disturbance. *Forest Ecology and Management, 440*, 79–91. https://doi.org/10.1016/j.foreco.2019.03.015.

80. Sahana, M., & Ganaie, T. A. (2017). GIS-based landscape vulnerability assessment to forest fire susceptibility of Rudraprayag district, Uttarakhand, India. *Environmental Earth Sciences, 76*(16), 676. https://doi.org/10.1007/s12665-017-6895-4.

81. Schelhaas, M. J., Hengeveld, G., Moriondo, M., Reinds, G. J., Kundzewicz ter Maat, H. W., & Bindi, M. (2010). Assessing risk and adaptation options to fires and windstorms in European forestry. *Mitigation and Adaptation Strategies for Global Change, 15*(7), 681–701. https://doi.org/10.1007/s11027-010-9255-1.

82. Seidl, R., Spies, T. A., Peterson, D. L., Stephens, S. L., & Hicke, J. A. (2016). Searching for resilience: Addressing the impacts of changing disturbance regimes on forest ecosystem services. *Journal of Applied Ecology, 53*(1), 120–129. https://doi.org/10.1111/1365-2664.12511.

83. Shao, Y. W., & Xu, J. (2015). Understanding urban resilience: A conceptual analysis based on integrated international literature review. *Urban Planning International, 30*(3), 48–54. https://doi.org/10.12677/UPI.2015.30309.

84. Spasojevic, M. J., Bahlai, C. A., Bradley, B. A., Butterfield, B. J., Tuanmu, M. N., Sistla, S., & Suding, K. N. (2016). Scaling up the diversity–resilience relationship with trait databases and remote sensing data: The recovery of productivity after wildfire. *Global Change Biology, 22*(4), 1421–1432. https://doi.org/10.1111/gcb.13191.

85. Standish, R. J., Hobbs, R. J., Mayfield, M. M., Bestelmeyer, B., Suding, K. N., Battaglia, L., Eviner, V. T., Hawkes, C. V., Temperton, V. M., & Harris, V. A. C. J. (2014). Resilience in ecology: Abstraction, distraction, or where the action is? *Biological Conservation, 177*, 43–51. https://doi.org/10.1016/j.biocon.2014.05.018.

86. Thompson, I., Mackey, B., McNulty, S., & Mosseler, A. (2009). *Forest Resilience, Biodiversity and Climate Change: A Synthesis of The Biodiversity/Resilience/Stability Relationship in Forest Ecosystems* (Technical Series no. 43). Secretariat of the Convention on Biological Diversity.

87. Timmerman, P. (1981). *Vulnerability, Resilience and the Collapse of Society: A Review of Models and Possible Climatic Applications*. Institute for Environmental Studies, University of Toronto.

88. Truman, C. C., Strickland, T. C., Potter, T. L., Franklin, D. H., Bosch, D. D., & Bednarz, C. W. (2007). Variable rainfall intensity and tillage effects on runoff, sediment, and carbon losses from a loamy sand under simulated rainfall. *Journal of Environmental Quality, 36*(6), 1495–1502. https://doi.org/10.2134/jeq2006.0461.

89. Turkes, M., Musaoglu, N., & Ozcan, O. (2018). Assessing the vulnerability of a forest ecosystem to climate change and variability in the western Mediterranean sub-region of Turkey: Future evaluation. *Journal of Forestry Research, 29*(5), 1177–1186. https://doi.org/10.1007/s11676-018-0574-8.

90. Turner, B. L., Kasperson, R. E., Matson, P. A., McCarthy, J. J., Corell, R. W., Christensen, L., Eckley, N., Kasperson, J. X., Luers, A., Martello, M. L., et al. (2003). A framework for vulnerability analysis in sustainability science. *Proceedings of the National Academy of Sciences, 100*(14), 8074–8079. https://doi.org/10.1073/pnas.1231335100.

91. Wang, C. J., Zhang, Z. X., & Wan, J. Z. (2019). Vulnerability of global forest ecoregions to future climate change. *Global Ecology and Conservation, 20*, e00760. https://doi.org/10.1016/j.gecco.2019.e00760.

92. Wang, J. Y., Zhao, G. X., Wang, X. F., Wang, L. H., Liu, M. S., & Liu, T. (2004). On the vulnerability of China's ecological environment and its assessment. *Shandong Agricultural Sciences, 2*, 9–11.

93. Weißhuhn, P., Müller, F., & Wiggering, H. (2018). Ecosystem vulnerability review: Proposal of an interdisciplinary ecosystem assessment approach. *Environmental Management, 61*(6), 904–915. https://doi.org/10.1007/s00267-018-1040-0.

94. Wirehn, L., Danielsson, A., & Neset, T. S. S. (2015). Assessment of composite index methods for agricultural vulnerability to climate change. *Journal of Environmental Management, 156*, 70–80. https://doi.org/10.1016/j.jenvman.2015.03.014.

95. Wu, X., Zhang, J., Geng, X., Wang, T., Wang, K., & Liu, S. (2020). Increasing green infrastructure-based ecological resilience in urban systems: A perspective from locating ecological and disturbance sources in a resource-based city. *Sustainable Cities and Society, 61*, 102354. https://doi.org/10.1016/j.scs.2020.102354.

96. Xenarios, S., Nemes, A., Sarker, G. W., & Nagothu, U. S. (2016). Assessing vulnerability to climate change: Are communities in flood-prone areas in Bangladesh more vulnerable than those in drought-prone areas? *Water Resources and Rural Development, 7*, 1–19. https://doi.org/10.1016/j.wrr.2016.03.001.

97. Xiong, K. N., Li, P., & Zhou, Z. F. (2002). *A Typical Study on Remote Sensing-GIS of Karst Desertification in Guizhou Province*. Geological Publishing House.

98. Xiong, K. N., Mei, Z. M., Peng, X. W., & Lan, A. J. (2006). Integrated management of the karst rock desertification areas—A case study of Huangjiang karst gorge. *Guizhou Forestry Science and Technology, 1*, 5–8.

99. Xu, G. C., Kang, M. Y., He, L. N., Li, Y. F., & Chen, Y. R. (2009). Advance in research on ecological vulnerability. *Acta Ecologica Sinica, 29*(8), 2578–2588. https://doi.org/10.1016/j.chnaes.2009.04.015.

100. Yang, M. D. (1990). On the fragility of karst environment. *Yunnan Geographic Environment Research, 2*, 21–29.

101. Yang, M. D. (1998). On the geographical structure of karst landscapes and their environmental effects. *Guizhou Society of Environmental Science, 11*, e03159.

102. Yu, B. H., & Lu, C. H. (2011). Assessment of ecological vulnerability on the Tibetan plateau. *Geographical Research, 30*(11), 2289–2295. https://doi.org/10.1111/j.1745-5871.2011.00249.x.

103. Zemp, D. C., Schleussner, C. F., Barbosa, H. M. J., & Rammig, A. (2007). Deforestation effects on Amazon forest resilience. *Geophysical Research Letters, 44*(11), 6182–6190. https://doi.org/10.1002/2017GL074406.

104. Zhang, X. L., Yu, W. B., Cai, H. S., & Guo, X. M. (2018). Review of the evaluation methods of regional eco-environmental vulnerability. *Acta Ecologica Sinica, 38*(17), 5970–5981. https://doi.org/10.5846/stxb201709061191.

CHAPTER 3

Urban Forest Ecosystem

LEARNING OBJECTIVES

After studying this chapter, you will be able to:

- Understand the historical development of urban forests from the 19th to the 20th century.
- Analyze the sustainability challenges and strategies specific to urban forest ecosystems.
- Explore the management tasks involved in maintaining urban forest health and diversity.
- Evaluate the impact of urban sprawl, population growth, and water scarcity on urban forest ecosystems.
- Discuss community engagement and resistance in the context of urban forest management.

KEY TERMS FROM THIS CHAPTER

Community engagement
Sustainability
Urban sprawl
Pest and disease management
Urban forest ecosystem
Water scarcity

INTRODUCTION

Urban forests have been adopted in many urban areas and transformed their public and private core into something significant that would give better meaning and a sense of their existing environment. This chapter will provide definitions, the needs for applying this strategy, and enhancing an invaluable asset.

Urban or city forest includes native and exotic trees within cities or municipalities. These plants support the water and soil within the area. The locations, including balconies and roofs, campuses, creek embankments, gardens, green walls, parks, plazas, and streets are incorporated with different types of vegetation. City forests are trees positioned within the city boundaries, whether naturally occurring or planted.

City forests play a critical role in supplying ecosystem services, for example, carbon sequestration, filtering air, water and oxygen, habitat, nutrient cycling, and shade. The urban forest serves as the bridge to a landscape that, in most cases, seems to be lacking in most urban areas.

Urban forestry is the "management of trees, forests, and associated green spaces in urban areas to improve the quality of life for residents and to contribute to the overall health and sustainability of urban ecosystems" (Konijnendijk et al., 2006).

Urban forestry is distinct from horticulture and agriculture, encompassing the management of all trees within an entire city. It focuses on urban forests and their associated ecosystem services, addressing significant issues such as population growth, climate change, and urban heat islands.

The term "forestry" often evokes a particular image or meaning, which can differ from the concept of urban forestry. Urban forestry is not limited to simply planting trees in urban areas, nor is it related to the management of national parks or state forests. It does not involve converting parks and sports fields into forested areas, but rather integrates tree management into the urban environment to enhance ecological, social, and aesthetic benefits. (Table 3.1).

Table 3.1. *Traditional vs. Contemporary Urban Forest Approach*

Tree Management in Traditional Urban	**Contemporary Urban Forestry Model**
Trees seen as decorations.	Trees are considered as critical infrastructure.
Focusing on specific tree species.	Focus on the forest and overall canopy.
Trees are considered as low importance.	Trees give equivalent importance like other city infrastructure, for instance, roads and services.

Trees have no monetary worth.	The economical cost of trees is recognized.
Emphasis on smaller vegetations and ornamentals.	Focusing on the longevity of tree canopies.
Specific preservation of trees.	Complete forest management.
Designing purely to improve aesthetics.	Design based on Ecological importance.
Tree management is determined through legal boundaries.	Urban forest is an endless resource irrespective of boundary ownership.

Source: Pearlmutter et al. (2017).

The management of urban forests is often perceived as primarily the responsibility of local government, but in reality, it involves a broader range of stakeholders. Local communities, schools, organizations, businesses, industries, as well as state and federal governments, all play significant roles. Each citizen also contributes to or impacts the urban forest in various ways. The main focus of urban forestry efforts and recommended actions is on public spaces for which the city is directly responsible, but effective management requires the involvement and cooperation of all these entities.

The field of urban forestry at the University of Toronto traces its origins to research conducted in 1965 by Jorgensen in Canada. This pioneering work highlighted the environmental benefits of urban trees, including their recreational and amenity value. The research was supported by funding from the USDA Forest Service and the International Society of Arboriculture, which helped integrate urban forestry into U.S. urban policy.

In the early 1980s, the concept of urban forestry extended to the UK, where the "Forest of London" project was initiated to regenerate urban environments and address socioeconomic issues in various municipalities. This idea also emerged in the Netherlands during the mid-1980s. Subsequently, cities in Asia, Europe, and Scandinavia adopted and expanded upon the concept, conducting further investigations and studies to deepen their understanding of urban forestry.

The paper titled, "A theoretical account of urban forest sustainability in 1997" written by Clark et al., was known to be the primary record that referred to the principle of Urban Tree Sustainability. This resulted in a significant outcome in maintaining the sustainable forest at its upper limit degree of net economic, ecological, and societal values over time. This concept aligned with the following standards:

- The characteristics of urban forest sustainability are included in the report of the Brundtland Commission in 1987 in the description for sustainable development.
- Specify criteria for the vegetation resource to achieve sustainability of the urban forest.
- Set up standards and performance indicators for the vegetation resource that will provide a community framework for the management of resources.

Urban forestry has yet to be well-researched and implemented worldwide. There is still a reliance on the initial research conducted by the United States, Europe, Scandinavia, and Asia to supplement the effective implementation of the program.

There are extensive public environmental services that urban forests can offer together with a cultural and aesthetic resource that are significant and are now rapidly fading. In the case of the United States, approximately 7,900 ha. /year of forest cover in urban areas is waning (Nowak & Greenfield, 2012). Because of this, the sustainable management of the urban forest is essential and needed. For the last three decades, various experts from different fields such as environment, researchers, government officials and local communities have tackled this problem and put actions on the ground, which executed the expansion of urban forests concerning science and management (Seamans, 2013). Through its development, the concept of city forest has progressed and grown, which allows us to comprehend and learn about how to operate with the principles and functions of the urban forest in our environment (Weber, 2013).

The urban forest is an example of a complex system that we have to recognize that it has a dynamic function that lies within the city landscape and allows the city to function and experience in an integrative manner. At the most, we are starting to realize that attaining ambitious management goals must not limit the practice of urban forest within the field of arborists, horticulturists, and foresters. Urban forest works as a system, an area that needs full opportunities for expanded study, innovation, and collaboration of different fields. But due to a mixing of interdisciplinary fields and the ingress of foreign vocabulary, various theories and frameworks, and trade of another discipline resulted in difficulties in navigating and initiating the principles. In this case, there is a necessity to introduce and examine the fundamental principles and techniques as well as tasks that urban forests must focus on in providing hope and a point of exit for those fields of non-foresters that are trying to explore and research into this rapidly changing and complex area.

At a crucial level, a forest can be described as "a three-dimensional system that is ecologically dominated by trees and other woody vegetation that has dynamic interactions with other biotic and abiotic components within a specified landscape." On the other hand, the city forest is vaguely delineated. In practice, we can describe it into three main components: these are structural elements (from single tree species connected to green parks and forest canopy), a specific scale is quantified, services and values are significantly considered and prioritized for management purposes (Rundrup, 2005; Rowntree, 2008). Are trees isolated in a limited spot in urban areas? Are there emergent vegetation and saplings in open or vacant spaces? Or urban forest is just a vastness of land with a connected canopy of trees along its edge?

Urban forestry is a field of study and practice that is tested to outline the context, historical and cultural aspects of a metropolitan area (Konijnendijk, Ricard, Kenney, & Randrup, 2006). Based on the history of the United States, "Urban forestry" was recognized back in the late 1890s and applied in the field of arboriculture, which concentrates on the management of individual tree species or collection of highly manageable trees that are situated in gardens and parks. The application of the professional line of forestry started in the late 1800s in the United States and did not focus on city forests, which at that time were recognized as a fragmented system and managed using traditional methods and evaluated in broken manners. Moreover, the management of city forests lagged behind in studying the ecological importance compared to its rural counterparts. Equality in European countries, urban forests were focused on managing the urban woodlots and peri-urban forests which is closely related to the traditional practice of forestry, but located within the outlying areas of the city and therefore topographically and spatially controlled. At the start of the 1960s and 1970s, conversely, we have witnessed the increasing development of urban forest practices across the United States and expanding the urban forest's meaning and becoming more comprehensive as compared earlier.

At present, urban forest in the United States is termed by the U.S. Forest Service and other practitioners as "trees that are publicly and privately-owned situated in an urban area, which counts the individual trees along streetscapes and home backyards and vestige of forest stands" (Nowak, Noble, Sisinni, & Dwyer, 2001). If we consider this definition of urban forest, this will count roughly 25% of the United States landscape area is an urban forest. Until such time that the American Society for Foresters described urban forestry as the "art and science of trees and forest resources management in and around urban community ecosystem for the physiological, socioeconomic and cultural benefits that trees can provide in the society" (Helms, 1998). Given these two descriptions, we can see that the forest is described as the aggregate number of tees with multiple layers and complex functional parts.

The acknowledgment of forests in cities as a structure is conceivably best represented by the public portrayal of city forests as components of the green infrastructure (Benedict & McMahon, 2002). This meaningful challenge is to deepen our comprehension and understanding of city forests, including other factors of the ecosystem that function as a whole. These theories and visions necessitate different disciplines from the outside field of the science of ecology which involves architects, engineers, and landscapers who were exposed to buildings and developing the urban landscape. As we try to take on the wide-ranging definition of city forest in this book and execute this new integration at the community level, urban foresters are responsible for developing a new operational meaning that will direct future activities.

Foresters in general count on the civic which includes the knowledge of growing trees, reproduction and acting in response to the alteration in the environmental setting – as this helps foresters to define forest stands, comprehend growth and standard management for a long time (Smith, 1997). Forests are grouped according to their arrangement and composition: distribution of trees in a plot (vertical and horizontal), including diameter at breast height (DBH), age, and species characteristics and state. Features such as abiotic components like temperature, climate, availability of water, and the ability of soil nutrients are the determining factors of an area. – "These are the overall total of external factors immediate and accessible to the flora within a specific geographic plot" – and illustrate as the accepted capacity for assigning a type of forest in a given area. Managers of forest used their knowledge and silvics to change the forest composition and structure to achieve the desirable economic demand of ecological and socioeconomic targets (Nowak & Dwyer, 2007). Hence, the physical structure and composition of the forest have something to do with its function which reflects the biophysical factors and measured rules regarding management. Even city forests are described very similarly to the principles and complexity of the condition of built-up areas in a fragmented structure of the forest, including the augmented disturbances and alteration in the availability of resources, these factors can change the growth behavior and public status of individual trees. Many studies have proven that controlled spaces, excessive or low soil moisture, limited availability of nutrients, evaporative demands, the existence of pathogens, diseases and pests, and competition among other vegetation and nearby trees can inhibit the growth of trees (Rhoades & Stipes, 1999). The Urban environment is not entirely unreceptive to the growth of trees. The high concentration of carbon dioxide, stable, warm temperature, and various deposition of nutrients existing in cities can result in accelerated growth of trees (Gregg, Jones, & Dawson, 2003). Another contributing factor could be changing human aspects of forest demographics can result in the positive or negative output of the urban forest system (Jack-Scott, Piana, Troxel, Murphy-Dunning, & Ashton, 2013). Moreover, these identified physiological stresses mainly exist in forest arrangement and structure. The highest magnitude of intricacy and unpredictability is an essential tool in developing the proper policy, analyzing the scenarios needed for the management, area planning, and infrastructure restrictions would determine the best practices in managing the city forest based on selection, planting, seeding, and preservation. Ecology of urban is just the start map to show the variability of the site condition and describe the city space, but our understanding of the physiological characteristics of trees in response to urban stresses is still comparatively evolving as equated to the conventional forest ecology concept.

3.1. THE EVOLUTION OF CITY FORESTS

Is it significant for us to know the development of urban forests? Why? It is because this will give us information and connection to those individuals and people who gave meaning to this concept. This chapter will cover the historical development, roots in the nineteenth century and the changes during the twentieth century that made this all possible to discuss the existing matters for the management of the city forest.

3.1.1. Historical Development

It is complicated to trace back the structure of the forest from the beginning. To make it more comprehensible, hunters and gatherers are likely planting trees. One of the evidence that proves this is the useful tree composition in the Amazon basin. In Europe, trees are believed to be sacred groves, tree cults, holy trees and havens for the outlaws and oppressed (Figure 3.1).

Figure 3.1. *The Amazon forest.*

Source: BBC News, Brazil Builds Giant Amazon Observation Tower, http://www.bbc.com/news/world-latin-america-29199728.

Until the church of the early Christians ordered to destruct the trees and holy groves. The birch tree is one of the first trees that exist in the forest and gave birth to a number of other trees including Cipher (Figure 3.2).

Figure 3.2. *The birch tree.*

Source: Miller, Bob, Mesopotamia to Milwaukee: A Brief History of Trees in the City, 2014 http://www.isa-arbor.com/events/conference/presentationDescription.aspx?ID=1113.

The Agricultural Revolution came and put to an end the hunting and gathering period, which led to growing food, selecting plants that are useful, selecting animals that could be domesticated and the emergence of permanent settlements and villages. Over time these communities become cities. Trees during this period are valued due to their fruits. Egyptian hieroglyphs show the tree planting with balls of soil for over 4,000 B.P. Temple and gardens were planted for the elites and priests. In 1480 B.C., the royal garden of the Egyptians had grown palms, pomegranates, and sycamores until the Egyptian tapestry.

By 2600 B.C. during the Mesopotamia period, Babylon was walled and large temples, and a system of rectangular streets was established for the dwellings of different classes. Based on ancient texts there is also the existence of the hanging gardens. At the plaza area along the roadways going to the marketplace, you can witness that trees were cultivated. This was established to provide shade. The Roman law sets the penalties for anyone who injures a single tree. During the pre-Columbian American period, there are significant populations of Native Americans, and they have extensive agricultural communities. This is also the rise of the Maya, Inca, and Aztec civilizations that produced large cities and architectural monuments that implements that system of agroforestry and introduced tree planting and development of the gardens. The agroforestry would include palms and trees, nitrogen-fixing legumes, ornamental and fruit trees, house gardens and domestic crops and edible or regenerative root crops (Figure 3.3).

During the 13th century, Kublai Kahn in China required the planting of trees along all the public roads in and around Beijing. The trees serve as shading and road markers during the winter season. Until the 16th century, when Europe established its trees and parks which is before the trees during the Renaissance period intended for aesthetics, private gardens, and fruit production. Gardens are called residence villas and walled around. They also practice the Allées wherein trees are lined in the pathways of the gardens. These Allées were moved into the city area along its walls and moats. This is also the start of Baroque gardens and tree-lined canals in the Netherlands. In the 17th century, the Pall Mall game was played under the Allées. Tree-lined avenues are directed to Rome, and some Allées became urban streets where urban expansion took place. There is also a creation of amenities for the aristocrats. By the 18th century, the professionals and merchants demand amenities, housing, and recreation areas such as ways for avenues, squares in the London area and public plazas on the continent. Gardens became a fashion statement, and there is a creation of extensive public gardens for commercial use and imported plants from the colonies. The city walls were replaced with parks and promenades and gated parks to exclude the lower class. There were also landscaped resort towns. It is also during the same period that Napoleon's promenades and boulevards were known until there is the recognition of the English tradition for naturalistic parks and British-engineered streets. The French inclusion of trees along the streets of Paris.

Figure 3.3. *Hanging Garden of Babylon.*

Source: National Geographic News, Photo Gallery: New 7 Wonders VS. Ancient 7 Wonders, 2007 https://news.nationalgeographic.com/news/2007/07/photogalleries/seven-wonders/photo12.html.

3.2. ROOTS FROM THE 19TH CENTURY

During the 19th century, Europe recognized trees to sanitize the urban setting. It helps to eliminate the cause of miasma and disease in the form of poisonous air. Parks during this time calmed and civilized the lower classes and were convinced that open spaces are intended for amusement. This is also the time when the British House of Commons 1833 was built. The transformation of Paris was made by Napoleon III and Haussmann in 1852.

Suburban landscapes or the romantic landscapes were a reaction to the industrialization period. Trees give value to real estate, substantial park creation, tree-lined streets and garden squares. In cities of America, trees were planted, and five parks were designed in Philadelphia. Most of the of insurance companies insure those homes near tree zones. Until the 1800s New England faced the deforestation problem and trees were instead planted in woodlots, pastures, streets, town centers, and cemeteries. But this had led to the moral virtue of nature in American The natural landscapes, and park system, became the Olmstead City parks movement in the 19th century. Europe imported the concept of romantic aspect and blended their City with nature. Chicago was known for the Columbian Exposition in 1892 and called the Beautiful City, which developed boulevards, civic centers, street trees and parks and parkways until the emergence of urban forestry was known in the 1960s. Cities to develop Detroit Avenues, Squares of Jackson, Mississippi, recognize the Arbor Day in Nebraska and the first course that taught arboriculture at State College in Massachusetts. It tries to promote the environmental movement due to increasing environmental degradation considering that most of the American population lives in cities. As early as that period, urban parks and trees are known to save energy, lower the energy costs, raise the value of properties, make cities more livable, improve retail sales and affect the locations of businesses (Haase et al., 2004; Elmqvist et al., 2013).

3.3. CHANGES IN THE 20TH CENTURY

Urban forestry has evolved significantly over the 20th century, adapting to the growing complexities of urban environments. Looking to the future, urban forestry will focus on addressing several key problems:

- Climate Change Mitigation and Adaptation:
  - Enhancing urban green spaces to absorb carbon dioxide, reduce urban heat islands, and improve resilience to extreme weather events.

Figure 3.4. *Consequences of climate change.*

Source: NASA, The Consequences of Climate Change, 30 November 2017 https://climate.nasa.gov/effects/.

- Biodiversity Preservation:
  - Promoting diverse plant and animal species in urban areas to support ecological balance and provide habitat for wildlife amidst expanding urban development.
- Air and Water Quality Improvement:
  - Increasing tree cover and green infrastructure to filter pollutants, manage stormwater, and improve overall air and water quality in urban settings.
- Human Health and Well-being:
  - Designing urban forests to provide mental and physical health benefits through accessible green spaces, recreational areas, and stress reduction.

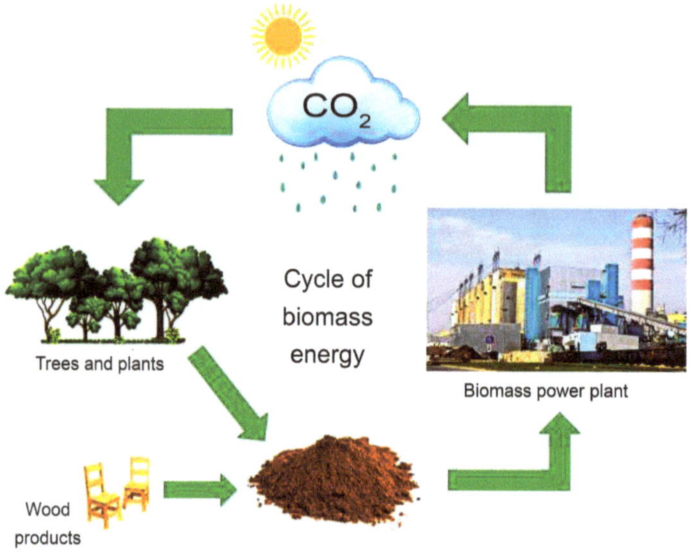

Figure 3.5. *Wastes from biomass fuels.*

Source: Humboldt State University, HSU wins $5.8 million federal grant for biomass research, 13 May 2014 http://biomassmagazine.com/articles/10393/hsu-wins-5-8-million-federal-grant-for-biomass-research.

- Sustainable Urban Development:
  - Integrating urban forestry practices with city planning and development to ensure long-term sustainability and the effective use of limited urban space.

Figure 3.6. *Philadelphia's urban by nature.*

Source: The Nature Conservancy, Pennsylvania Nature Cities, 2017 https://www.nature.org/ourinitiatives/regions/northamerica/unitedstates/pennsylvania/our-work/nature-in-cities-1.xml?redirect=https-301.

- Community Engagement and Education:
  - Involving local communities in the stewardship of urban forests and promoting environmental education to foster a sense of ownership and responsibility.
- Economic Benefits:
  - Leveraging urban forests for their economic contributions, including increased property values, job creation, and the potential for eco-tourism.

Figure 3.7. *Urban landscape.*

Source: Schifman, Laura, Urban Environmental Science – Why Is It Important and What is the big deal?, https://envirobites.org/2017/08/02/urban-environmental-science-why-is-it-important-and-what-is-the-big-deal-2/.

3.4. FOREST SUSTAINABILITY IN URBAN ECOSYSTEMS

The goal of the urban forester is to manage and administer the long-term viability of the urban forest ecosystem. The concept of sustainability in urban forests is not adequately explained, both in terms of scope and application. The difficult part is deciding how to describe sustainability, and whether it can even be characterized. A sustainable structure is one that survives or continues to exist. In the context of city forests, an arrangement would be stable throughout its duration, generating maximum profits as long as the forest is operational. Because there is no clear definition of sustainability, we can reasonably evaluate it by observing in reverse. In urban forests, we count the total number of trees cut as well as those that have naturally regenerated or been replant. Given this, we assess the development of a system that exists or continues. As a result, our understandings of sustainability are based on predictions about the future or the structures themselves. The following statements will describe a working sustainability model for city forests. We define detailed standards for assessing sustainability as well as measurable pointers for calculating those gauges. Furthermore, we acknowledge that sustainability is a process rather than a goal. As Goodland (1995) suggests, we consider environmental, social, and economic factors. Goodland (1995) believed that overall sustainability would be based on all three pillars: environmental, economic, and social. Maser (1994) defined sustainability as the intersection of what is socially and environmentally acceptable and desired by the current generation, recognizing that both will change over time. As a result, our course integrates floral resources (forest and its components) with the society that makes use of them. Furthermore, we recognize the complexities of both the supply chain and the program management that affect it. We also distinguish between societies' perceptions of ecological potentials and societal needs.

3.4.1. Defining Sustainability

In developing the sustainability model for city forests, researchers initially looked into how other systems become sustainable and how it was described and defined. Even though we have focused on the structures of the forest, other cases were reflected. However, some ideologies of sustainable structures were openly appropriate to city forests, while others need adjustment or were in a struggle with the environment of forests and forestry in urban areas. The Report of Brundtland Commission has commonly functioned as the initial idea for argument about sustainable structures. Sustainability in forestry was defined as Forest management that meets the needs of the current generation without conceding the capacity to meet the needs of future generations. This performs a principle of community stewardship, which incorporates the developing, cultivating and gathering of trees for valuable produce with the protection of environmental resources such as quality of air, water and soil, and habitat for flora and fauna (United Nations Economic Commission for Europe, 2005).

Both Webster (1993) and Wiersum (1995) studied this meaning from the standpoint of forest management. They acknowledged that problems of what needs to be continued and how to implement are unresolved sustainability matters. Wiersum (1995) recognized the past emphasis on upholding yield and its current expansion to sustainably managing. He recommended a necessity for attention to the concern of scale: the spatial size of the area. Salwasser (1993) defined sustainability as the capacity to generate and keep a preferred set of situations or things for some period in the future, not probably endlessly. Salwasser (1993) incorporated the three pillars of components (environmental, social, and economic), recognizing that sustainability is not merely a resource substance. He also emphasized that the aims and purposes of managing forests could not surpass the biological capability of the reserve, today and shortly. Sample (1993) concentrated more carefully on the management of forests, highlighting the essential for the common idea among varied landowners. In training on the management of ecosystems, Sample defined sustainable forestry as managing and practicing concurrently environmentally comprehensive, economic viability and social accountability. Some meanings of sustainable forestry are not straightforwardly related to urban locales. For instance, the explanation shown at the session on Sustainable Forestry counted in remarks about the capability for self-restoration. As the restoration of city forests must happen in a fixed, specific setting means, the use of such a meaning is unsuitable. Other definitions reflect the objective of sustainable woodlands in a way uneven, with our theory of city forests. Thompson, Pillsbury, & Hanna (1994) defined sustainability as agendas that produce preferred environmental and economic profits without waste, ineffective strategies and systems. However, these writers were concerned with city locales; their method was constrained to the agendas of forestry at a municipal level rather than processes or results in the citywide approach. Dehgi, Huffman, & Culver (1994) concentrated on California's indigenous Monterey pine woodland and delimited their explanation of sustainability to that structure. Furthermore, their concern was constrained to supporting the dynamic of natural genetic development. In the case of the American Forest and Paper Association (1995), it is principally directed at industrial woodland run-through and produces. This emphasis on industrial forestry appears to radically dissent with city environments. Given the patterns observed above, the part of individuals in sustainable structures (counting forests) is primarily recognized. Conversely, Botkin & Talbot (1992) as critiqued by Webster (1993) claimed for the tropical forests to meet sustainable development necessitates non-intrusion of anthropogenic. Overall, this notion is unsuited to city forests.

3.4.2. Moving Towards Sustainability in Urban Forests

To move the concept of sustainable development toward implementation and application in urban forest ecosystems, Webster (1993) mentioned numerous important points. These points are derived from the perspective of urban forest, such as, what are the objectives? What are the circumstances and ideals to be continued? In city landscapes, we concentrate on maintaining the net benefits of forests and flora at the highest level. We sustain the quality of the environment, conserve the resources, develop the economy, promote psychological health, protect the habitat of wildlife and enhance human wellbeing. What is the series of forest undertakings that influence on sustainable development? We recognize that a city forest needs a wide-ranging set of undertakings, initially managing the single-tree and tree stands for public awareness, educating the community, and developing a comprehensive management plan for the urban forest. On what geographical scale can sustainable development be applied? Political borders do not account for biological resources. The management of the ecosystem sets boundaries of scale based on watershed boundaries. Yet, cities maintain its social, political and economic units to be discreet. We must recognize the truth that political restrictions may be weightier to administer than environmental margins. City forestry agendas function within this geographical context. In this scheme and representation, research focuses on the city and its boundaries in a geographical manner. Though this method may disturb some of the biological actualities of timberland stands, it reasonably indicates the jurisdictional borders and distinctive managing units established in metropolises. The more collective alternative method is operational within ecosystems but is not without setbacks of meaning and extent (Fitzsimmons, 1996). What is the connection of sustainable development between new technology and forests in cities, effectually operated investigation and investment in the management of forests? City forests hold immense value, benefiting greatly from the integration of information, novel technologies, and resources. These urban green spaces provide critical environmental services, such as air purification, temperature regulation, and improved biodiversity, while also enhancing the well-being of city residents. Not only will the capacity to pick and cultivate trees in metropolises be enriched, but the capability to measure the profits accumulated by their occurrence will increase. Wiersum (1995) delivered a detailed examination at sustainability in forest structures, observing the lengthy history of the idea in the forest tradition. Several would debate that the idea of sustained produce is different from sustainable development. Gatto (1995) discourses this element in detail. Yet, Wiersum (1995) perceived the development of woodland sustainability in the direction of biological diversity, multiple uses, and dimensions of socioeconomic and climate change mitigation.

Wiersum (1995) précised four ideas contained with the management of sustainable forests (i) sustenance or maintenance of the ecological characteristics of the forest; (ii) production of valuable forest goods and services for the benefit of the society; (iii) humans are highly dependent on the forest; and (iv) social institutions safeguard forests are secured against undesirable outside societies. A parallel viewpoint on the management of sustainable forests was defined the quantifiable standards such as the desirable future situation, sustained yield, maintenance of the ecosystem, and stability of the city. Keene (1995) also mentioned that these values can be experienced in conventional forest management. Goods originating from woodlands in which sustainable forest management is performed may obtain a third-party authorization which is similar to the documentation of organically grown produce . Maser (1994), Wiersum (1995), and Charles (1994) all claimed that a sustainable forest would comprise issues from biological, social and economic perspectives. For instance, from the viewpoint of a resource of a fishery, sustainability is the real-time search for the goals of the community and institutions based on ecological and socioeconomic. In Maser (1994) opinion of sustainability in ecology, the objectives and requirements of people must mirror the perspective of the source to run into them. This indication may be worldwide for sustainable development and should positively be for city forests. This methodology can be straightforwardly utilized to cities, for we want city forests to provide environmental and socioeconomic benefits. We must not expense one objective in quest of an alternative. Trees lessen atmospheric pollutants all at once and they improve public health. But struggles would occur in explicit conditions (e.g., Planting vegetation under service lines or using hostile vegetation), overall, the comprehensive aims of city forest sustainability are well-suited combined with others. In this idea, when we direct on the suitable management of vegetation and city forests, somewhere management events take place with communal sustained objectives and intentions, we direct on supporting a far-reaching collection of values. We concur with Charles's (1994) conclusion that sustainability can only be achieved through comprehensive management at the local, community, and regional levels. Effective management must adapt to the dynamic nature and complexity of resources while respecting property rights. The concept of sustainable development, introduced by the Brundtland Commission, has led to a variety of interpretations. Although there is no universally accepted source for forestry-related sustainability, significant progress has been made in developing standards and indicators for assessing and achieving sustainability.

3.4.3. Features of Sustainable Urban Forests

Based on the specified features of sustainable structure and the detailed background of the city forest, here are the four ideologies to which the model of sustainability may be observed.

Sustainability is extensive in the overall objective. We may be able to describe the desirable functions for achieving a sustainable city forest, yet the design for the urban forest has not been optimized. Although we were aware that city forest has the potential to reduce atmospheric pollutants, we still do not know the appropriate design to maximize the benefits it can offer. Still, we admit the standing city forest to provide these functions at some point. Trees can improve human health, minimize the heat island effect in urban areas and remove contaminants present in the atmosphere. Still, there are costs incorporated in cultivating, maintenance, and tree removal within the city, in a sustainable point of view of the urban forest, these functions outweigh the costs connected to forest tending. A sustainable urban forest can provide stability of these net benefits through space and time. It is necessary to decide and recognize the general features of sustainable structures and develop courses from such structures in urban landscapes.

City forests mainly provide services rather than products. Sustainability mostly focuses on the products that the whole system can provide, e.g., sustained yield. Forests can provide different raw materials such as wood, fiber, food, and fisheries. In this example, outputs are goods. In disparity, products are limited output of city forests. The most significant outputs of the city forests are services such as removing atmospheric pollutants, reducing stormwater runoff, improving the state of water quality, reducing consumption of energy, protecting of habitats for wildlife and promoting social and psychological health and wellbeing. These services are accessible in two ways (i) direct (raising property value and providing shade) and (ii) indirect (enhancing the health and well-being of the community). During the planting period to achieve sustainability in an urban forest, the benefits of services must not be maximized by rather maintained to be balanced. Because maximizing the benefit from one component could create disputes with other services. For instance, maintaining the largest wildlife habitat could conflict with limiting the potential economic development in the same area and can trigger conflicts with humans.

Sustainable urban forest necessitates human involvement. One of the best features of natural systems is that they have the capability for self-repairs. Sustaining the different environments has taken full advantage of this point by gathering some limited fragments of these resources, frequently with periods of replacement or allowing for rest and renewal. In the Brundtland Commission Report, Maser (1994) and Charles (1994) stressed this crucial phase of the process. For instance, Goodland (1995) described environmental sustainability as the preservation of natural resources. Maser (1994) stated that a biologically sustainable forest is the basis for all phases of a sustainable structure. In forestry, a sustainable yield cannot possibly occur if a sustainable business, a sustainable society, or a sustainable public is deprived of a biologically sustainable supply. As Charles (1994) placed it in fisheries, if fish become extinct, nothing else will matter. Many of the city forests are a mixture of natural forest fragments and planted vegetation. The natural fragments may have some ability for self-restoration and repairs, predominantly in greenbelts and other unharmed stands. However, the established trees have fundamentally no capability to be restored in space. Consequently, we must agree, recognize and work with the truth that city forests, mainly in America may have an inadequate facility to recollect or substitute biological assets. This is principally the situation when we want the restoration to transpire in a way suitable for social profits. Undeniably, undesirable tree propagation may essentially have a net price for the control and suppression courses. Sustainable city forests cannot be detached from the undertakings of people. Such a movement can be together optimistic and harmful. In the concluding case, the formation and preservation of city infrastructure can be exceptionally damaging and disrupting. In core, we put on top cities over forests. The greater the burden, the less ordinary the forests look and perform. The contrary effects of individuals can be diminished by optimistic activities such as development, establishment, and management; All happening with shared obligation and collective idea. We cannot isolate sustainable city forests from the individuals who subsist in and near them. Actually, we need to combine the two as considerably as possible. The effects of this standard are distant realization. Initially, city forests necessitate dynamic, steady, ongoing management. The accretion of net profits can merely happen when sufficient and sensible attention is delivered. Next, administrators of trees (publicly owned and privately owned) must include the adjacent municipalities in resolutions and movements concerning city forests. We do not recommend renouncing accountability on the side of tree administrators; we promote involvement in it. Vegetation developing on private properties comprises the mainstream of city forests. While public property trees (mainly in the country park and along boulevards and further rights-of-way) have remained the enduring attention of city forestry, they involve only a percentage of the city forest. An appraised value of 60 to 90% of the vegetation in city forests is established on private property land in the United States. Hence, sustainable city forests are a huge step toward sustainable reserved woodlands. If we reflect beyond that trees perhaps are not uniformly scattered amongst all private property-owners, then we might also assume that a lesser number of property-owners and administrators might be accountable for an outsized portion of city trees. For instance, academies, corporate parks, business grounds, marketable real estate, independent semi-public organizations, services, etc., may be involved in managing enormous quantities of foliage. The achievement of any determination on sustainability must embrace their contribution and guarantee. However, small private property-owners, and specific housing assets, might also establish a substantial portion of communal foliage. Their influence on the city forest needs to be reflected in any work in the direction of sustainability.

3.4.4. Outlining Sustainable City Forests

Using these four ideologies leads to the understanding of a sustainable city forest as follows: The naturally transpiring and established plants in metropolitan areas provide residents with ongoing environmental, ecological, and socioeconomic benefits both currently and in the future. Implementing this concept in urban areas requires embracing three key notions:

- Recognition of Multifaceted Benefits:
  - People must acknowledge that city trees offer a broad spectrum of benefits beyond aesthetic value. These include environmental improvements like air purification, ecological contributions such as habitat creation, and socioeconomic advantages like increased property values and recreational spaces. Establishing, conserving, and sustaining trees should be viewed as integral to the current and future well-being of cities and their inhabitants, rather than as mere acts of goodwill or isolated projects.
- Community Involvement and Management:
  - To ensure that the benefits of city forests are preserved over time, their redevelopment must involve active participation and management by the community. As Nowak, Crane, & Stevens (2006) state, "individuals desire and necessity to direct the regeneration course, since the natural renewal does not come across most city requirements." Thus, the maintenance and growth of city forests depend on societal engagement and cannot be left to natural processes alone.
- Geographic and Political Boundaries:
  - Sustainable city forests exist within the specific geographic and political boundaries of a metropolis. This means that effective urban forestry management must consider local conditions and governance structures. Furthermore, sustainable city

forests are a collective asset of the community, regardless of ownership, and require coordinated efforts to manage and protect them effectively.

3.4.5. Resource Vegetation

The resource vegetation is the machine that initiates city forests. Its structure, scope, dispersal, and health delineate the boundary of profits delivered and outlays accumulated. As active creatures, city forests (and the foliage that develops them) alter over time as they cultivate, settle and expire. Thus, sustainable city forests must own a combination of species, ages and sizes that permit for stability of profits while foliage is established and detached (Table 3.2). The vegetation resource of an ecological city forest delivers a constant high concentration of net profits comprising of conserving energy, reducing pollutants in the atmosphere, increasing the value of property, minimizing runoff in storm water, and societal health. However, there are costs linked with the buildup of these uses. Trees that are defective, dead and failing may fall and harm or injure people and assets. Some vegetation may cause a health hazard from allergenic reactions. Others may compete with indigenous flora and restrict the role of naturally growing sections and structures.

Table 3.2. *Criteria of Vegetation Resource Sustainability in Urban Forest*

Canopy or Crown Cover	Attain crown cover well-suited to the community climate.	The epitome coverage of tree crown varies depending on the area and the climate of the local area (this will depend on the location of the community and take note of the optimal degree per crown cover for each city).
Age Distribution	Present an uneven age distribution.	A mixture of juvenile tree species is necessary in maintaining crown cover which is relatively constant. This will ensure the sustainability. There should be ongoing plans for tree planting during the extraction period of the matured foliage. Activities such as stock taking of vegetation will monitor this for easy reference or indicator. The private properties posed a bigger challenge on this stage and will require broader means of program monitoring.
Species Combination	Prefer species diversity	Diversity of plant species is important when dealing with the long-term health status of urban forests. A single tree species are much prone to pest infestation due to unpredictable weather pattern and this make take heavy toll on the trees in the city. It is recommended that 10% or less of the tree populace will composed of one tree species.

Native Vegetation	Achieve and preserve the regional biodiversity. It is important to uphold the ecological reliability of the wooded area and preserve corridors for wildlife to and from the city.	It is appropriate to preserve the native vegetations because this will add sustainability in the city forests. Indigenous tree species can easily adapt to the existing local climate and it can also give benefits to the innate flora and fauna at the same time. It is suggested to replant native species of plants in nursery as an alternative strategy. Introduced species of trees can possibly threaten the native species in the local area and can affect the regeneration capacity of *greenbe*lts and offcuts in the forest. Hostile species also necessitate programs for dynamic resistor.

Source: Raison et al. (2001).

Table 3.3. *Criteria of Urban Framework Sustainability in Urban Forests*

Cooperation of the Public Agencies	It is important that the city department operates in common targets and objectives.	Departments would include the public works, planning, fire, public utilities, school districts achieving the same goals and cooperating with the city councils and agencies.
Participation of the Significant Entities (e.g., Private Groups) and Influential Landowner	Landowners must support the citywide targets and objectives in terms of management plans of specific resources.	In urban areas, most of the private owners are the one managing and owning the urban forest. Their interests and adherence is very important when it comes to resource management plans and they must understand the values of urban forests. Take note that their involvement cannot be required.
Cooperation of Green Industry	Green industries must be managed by quality standards professionals and commits the citywide targets and objectives.	Commercial growers, landscape contractors and engineering professionals must implement green industries and considers the tremendous health impacts of the urban forests. Their full commitment is significant for to ensure the appropriate planning and implementation program for the city's urban forest.
Action Within the Neighborhood	Neighborhood must also actively participate in managing the city forest.	Neighborhood are the building blocks in cities and more often their individual action makes the biggest impacts in their quality of life. Given that most of the urban forest is privately owned (built-up or commercial) neighborhood plays a critical part in the city forest sustainability.

Interaction Among the Government, Businesses and Its Citizens	All of the constituents must participate and interact about the benefits they could get from the urban forest.	Involvement of all stakeholders is the crucial part in sustainability. It doesn't necessarily to result it as legislation, but rather achieve the common perception of the city forest and its value provided in the community and is part of the dialogue and cooperate among all stakeholders.
Wide-Ranging Cognizance from Trees Recognized as a Resource in the Community	The communal must be aware that trees have values.	It is fundamental for the community to understand that trees have values and people must elect officials who value tree resources. By doing this, those officials who value trees are more likely to maintain the quality standards of trees and provide adequate funds for project implementations.
Regional Cooperation	It is important to collaborate with neighboring regions and interact with other communities	Urban forests are borderless and city efforts must include the impacts to their neighboring regions and take into consideration their actions to easily address potential issues such as air quality and water quality.

Source: Binder et al. (2020).

3.4.6. Community Framework

A sustainable city forest shares the same vision across its landscapes for their woodland and perform to recognize that idea over definite aims and purposes. It is grounded in localities, public areas and private parcels. At one point, this necessitates that a community come to an agreement on the values of trees and work to make best use of them. In additional point, this collaboration needs that private property-owners recognize the important part of their foliage to communal well-being. Lastly, in a period of moderated administration provision, collaboration signifies allocating the monetary weight of tending for the city environment.

Table 3.4. *Criteria of Sustainable Urban Forest for Management of Resources*

Management Plan for the City	There is a need for the development of the plan for managing vegetation in both publicly and privately-owned properties.	A citywide plan for management will address significant issues of sustainability in the city forest. There must be a common idea for the society's urban forest community and fundamentals would comprise – vegetation planning, strategies, routine targets, and ethics for tending trees, innovative developments in vegetation planning and the conservancy and handling open spaces with other management specifications.

Capital Source	Generate capital sufficient to finance the citywide management plan.	Fundings must be accessible for both public and private owners since urban forest is a shared resource. Amount of fund must be available for educational purposes and community awareness for the value and benefits that urban forest can provide.
Staffing	Hiring and training suitable workforce to execute the management plan in the city.	Urban forest will be successful if the staffing members trained well and have ongoing schooling along with their least experiences and desirable documentations and accreditations.
Tools for Assessment	Development of tools to gather data regarding the forest in the city in a routinary manner.	These tools can be used, such as canopy cover assessment, inventories of trees, aerial mapping, GIS tools and other tools that have the capacity to monitor the urban forest trends in the city over succeeding periods.
Protection of Existing Trees	Conserving the existing flora to ensure the maximum function of the natural setting.	Protection to a prevailing stand of trees and auxiliary of cutting off trees are necessary to maintain the original setting of the natural environment. Monitoring and protection of existing trees often require policy vehicles supported by ordinances that will meet the standards and provide restrictions large removals of trees in both public and private properties especially during the development stage.
Site and Species Selection	Guidelines and specifications for species use is important and must be defined based on context.	Determine the planting sites and species selection is a critical phase in achieving sustainability. There must be an adequate space to allow the trees grow. Selecting the appropriate tree will reduce the maintenance costs and will prolong the urban forest's life span. Remember to avoid species that can cause allergic reactions to people.
Tree Care Standards	Make sure to adopt the professional standards for tree care.	Apply the tree care requirements and specification like tree pruning guidelines (ISA) and ANSI Z133 publications.
Protection of the Citizens	Take full advantage of the communal security measure and tree care.	Public areas must be safe enough for all its citizens. There must be regular inspected included in the management program to ensure safety and eliminate the potential tree hazards.
Recycling	There must be a system for tree waste.	Tree waste can be used as a compost such as chips, mulch, fuel, firewood and lumber.

Source: Nations (2018).

3.4.7. Managing Resource

In numerous ways, this factor is not merely managing resources, but the viewpoint of managing as well. On one point, a detailed guiding principle to safeguard remaining trees, control the selection of species, educate the workforce and employ criteria of tending directly on the tree source.

Table 3.5. *Resource Vegetation Criteria and Performance Indicators*

Standards	Routine Pointers				Main Objective
	Low	Moderate	Good	Optimal	
Crown coverage	On no account of valuation	Ocular inspection (i.e., photos)	Aerial photos used in tree sampling	Information available for the city's urban forest in GIS format.	Achieving the climate appropriate tree cover for the community.
Distribution of Tree Age within the community	On no account of valuation	Inventory of boulevard trees (comprehensive or test group)	Tree sampling in privately and publicly owned properties.	Citywide assessment using Geographic Information System (GIS).	Uneven distribution of trees age
Species mix	On no account of valuation	Inventory of street trees	Assessment of citywide species mix	Citywide assessment using Geographic Information System.	Promote species diversity
Native vegetation	No integration program	Volunteer practice of communal developments.	Necessities for the indigenous species of trees based on development provisions.	Preserving the regional biodiversity	Preserving and managing the regional biodiversity, maintaining ecological integrity and wildlife corridors from the city.

Source: Clark et al. (1997).

Equally, recognition of a wide-ranging plan for management and finance drivers of the city administration and its citizens permits common ideas to grow. Cities must identify that management methods will differ in its resource and its scope. An objective of preserving indigenous wildlife, the environment may best be succeeded somewhere there is a resilient indigenous forestry supply. For some municipalities, this is simply not possible. Likewise, managing of the city forest must happen in linking to the higher setting (for example neighboring woodlands). For instance, the preservation of integral riparian strips necessitates the assistance of the running organization of the tributary.

Table 3.6. *Community Framework Criteria and Performance Indicators*

Standards	Routine Pointers				Main Objective
	Low	Moderate	Good	Optimal	
Cooperation of public agency	Unawareness of issue	Educational resources and advices accessible to property-owners.	Well-defined basis for resources of trees and safeguarding of the privately-owned tree stand.	Property-owners work up a wide-ranging plan for the management of trees with budget requirement.	Significant private property-owners accept the city targets and purposes for the management of resource plans.
Involvement of private landholders and large institutions	No cooperation among segments and no observance to see the business ideals.	Overall collaboration among stakeholders	Well-defined basis for resources of trees and safeguarding of the privately-owned tree stand.	Property-owners work up a wide-ranging plan for the management of trees with budget requirement.	Significant private property-owners accept the city targets and purposes for the management of resource plans.
Cooperation of green industry	Absence of action	Isolated active groups	Specific cooperation arrangements, such as purchase certificates for right tree and right place.	Shared vision and goals, including the use of professional standards.	The green industry operates with high professional standards and commits the citywide goals and objectives.
Neighborhood action	Absence of action	Isolated or limiter active groups	Citywide coverage and interaction	All neighbors organized and cooperating	All the neighborhood level, citizen understands and participate in urban forest management.
Interaction among citizens– local administration and businesses	Disagreeing targets among stakeholders	Not any interactions among stakeholders	Informal for general cooperation	Formal interaction of the tree board with coordination	All constituents in the community interact for the benefit of the urban forest community.
Overall cognizance that community recognized trees as a resource.	Low – resources like trees are recognized as problems.	Moderate – resources like trees are significant to communities.	High – resources like trees recognized to offer services in the environment.	Very high - resources like trees as dynamic parts of ecosystem and setting.	The public comprehends the worth of trees as resources in the community.

Collaboration at the regional level	Stakeholders operate self-reliantly	Constituents impart parallel strategy actions.	Planning in regional level	Planning in regional level, organization and managing of plans.	Support collaboration and contact among regional members and groups.

Source: Clark et al. (1997).

Table 3.7. *Resource Management Criteria and Performance Indicators*

Standards	**Routine Pointers**				**Main Objective**
	Low	**Moderate**	**Good**	**Optimal**	
Citywide management plans	Absence of Plan	Existing proposal, but incomplete opening and execution.	Local administration plan executed and accepted	Citizen – local administrations – corporate resource plan for management, acknowledged and executed.	Build on and execute a plan for the management of forest trees on publicly and privately-owned properties.
Citywide capital	Funding by crisis management	Capital comes from the local inhabitants	Sufficient capital to make available net growth or inhabitants input.	Sufficient capital for privately and publicly owned domains to endure and sustain prospective gains.	Build on and preserve capital sufficient for the execution of the plan in managing the city forest.
Urban forest personnel	Absence of personnel	Absence of personnel	Licensed personnel and arborists	Specialized tree care experts.	Hire and educate suitable personnel to execute the citywide plan for managing the city forest.
Tools for assessment	No ongoing agenda of valuation	Incomplete listing	Comprehensive listing	Data on urban forests included in citywide GIS.	Progress approaches to gather data about the city forest routinely.

Source: Clark et al. (1997).

3.4.8. Accomplishing Sustainable Forests in Urban Settings

A sustainable city forest is created with public collaboration, quality maintenance, sustained finance and individual participation. It is formed and sustained over common ideas and collaboration with pervasive attention on making the best use of the profits and reducing costs. Taken collectively, there is recognition of the demand for communal ideas and accountability, for absolute interference with the resource and for agendas of maintenance that are continuing and receptive. The application of a model for forest sustainability in urban areas promotes passing on the customary direction of forest management in urban settings moving away from public trees to the combination of private and public foliage. Accomplishing sustainability for city forests includes convening each of these measures. To help in this mission, we ought to define gauges of achievement for each condition. A city that runs into the utmost point of each gauge for each standard would have the greatest instruments and capitals to attain sustainability. Our method of improving standards and pointers is modeled after what is established in the Santiago Contract which recommended standards and pointers for boreal and temperate forest preservation and sustainability. It acknowledged that both measurable and descriptive factors were necessary, for not all standards could be precisely calculated. Maser (1994) recommended that environmental sustainability includes four principles: (i) Availability for lasting stability between people and the supply, at present and in the future. (ii) Enhancing the intersection between social interest and environmental opportunities. (iii) Progressing evaluation tools for the supply and its productivities (services and benefits). (iv) Repairing ecosystems. Our model for city forest sustainability follows these four principles, employing them in a city setting. It distinguishes the environment of the public in municipalities and reassures involvement at an extensive level. The model also recognizes the essentialness of nurturing renewal, to assure the permanency of the supply. Managing sustainable forests in an urban setting is established upon a common idea for the supply, in which objectives and necessities are secured. As sustainability is an overall objective, we must be capable of evaluating our growth compared to well-defined principles (Haase et al., 2004; Elmqvist et al., 2013). Lastly, we understand that our activities, through such events as expansion, will impair woodlands and their operation and agree with the accountability of renewal. City trees and woodlands are measured as essential to the sustainability of towns as one. However, sustainable forests in cities are not natural, they are created. They do not ascend by chance, but result from a local communal obligation to their formation and administration. Attaining the obligation of a wide-ranging public, of several populations, cannot be uttered or constituted. It must stand up out of cooperation and admiration. While regulatory strategies play a role in managing city forests, fostering responsibility is perhaps more crucial for promoting learning, responsiveness, and positive encouragement. The primary challenge is to provide evidence that generates assurance and drives action. Financial resources are also essential for sustainable urban forestry, but the belief that funding will follow if education is effective remains significant. Despite current financial constraints, maintaining this perspective is important. Sustainable urban forests require a solid resource foundation. Although city arborists and foresters maintain confidence in their ability to manage forest resources, it is essential to acknowledge our limitations as well as our strengths. The optimal management of city forests, including the planning and placement of city trees, is a key area of focus. Our efforts must address issues such as the availability of nursery supplies, appropriate cultural practices, and the balance between site-specific concerns and the range of suitable tree species. (Haase et al., 2004; Elmqvist et al., 2013).

3.5. TASKS IN MANAGING THE URBAN FOREST ECOSYSTEM

There is a significant contribution that urban forests make to the solution of all of these pressing problems. It is necessary to implement a comprehensive adaptation of solutions that will produce a multitude of benefits through the implementation of actions. The effective adjustment of the building area must consider all of the green infrastructure that is already in place. This green infrastructure is extremely cost-effective and takes precedence over the results of the grey infrastructure. To make cities more robust, healthy, and livable, green infrastructure must take into consideration the fact that green corridors, open space, crown cover, and environmental services are the most valid instruments that can be utilized in urban areas (Figures 3.8 and 3.9). The shared critical challenges in urban areas are:

- Population of mature trees.
- Water scarcity and storage.
- Concerns of climate alteration.
- Life-threatening heat island effect in urban areas.
- Urban sprawl and growth.

Figure 3.8. *Urban green infrastructure.*

Source: CSIRO, Greening Our Cities, n.d. https://www.csiro.au/en/Research/LWF/Areas/Resilient-cities-21C/Green-infrastructure.

Figure 3.9. *Urban greening apartments.*

Source: CSIRO, Greening Our Cities, n.d. https://www.csiro.au/en/Research/LWF/Areas/Resilient-cities-21C/Green-infrastructure.

3.6. POPULATION OF MATURE TREES

When it comes to parks and boulevards, trees are recognizable symbols. Over the course of more than a century, the majority of the aging vegetation is on the verge of reaching the end of its life cycle. These trees are especially vulnerable to the effects of urbanization, droughts, and shifting cultural trends. The majority of these old trees are becoming less tolerant of the changes that are occurring right now (Figure 3.10).

Figure 3.10. *Urban trees in Melbourne.*

Source http://econews.com.au/40598/report-says-australia-needs-more-trees/.

For the purpose of determining which trees require removal and treatment, it is necessary to assess the available tree population. It is necessary to identify the trees that require replacement. It is necessary to conduct careful assessments when managing trees, and renewal does not simply involve replacing the old trees with new ones. The identification of the most suitable replacement tree that is also the most resilient tree, one that is able to withstand a variety of conditions and will provide multiple benefits to the community members and stakeholders, is another aspect of this process.

When compared to younger trees, older trees typically require a greater amount of resources to maintain. Some of the trees that are getting on in years lose their value over time and become a risk to the people who use these public areas during that time. As a result of the high probability that mature trees pose a threat to the general public, it is necessary to act and pay the associated costs.

In the vicinity of boulevards and main streets, trees that are symmetrical and uniform create a pleasant perspective. As an illustration, the majority of the trees in the city are deciduous, and the broad canopy of planes and elms. It is necessary to approach the stakeholders with caution to deal with the problem that has been exacerbated by this. It is necessary to maintain these avenues in a manner that is consistent to maintain the same aesthetics. The tree species that is suitable for replacement must be identical. The community may also experience difficulties in producing the same large trees at the same time as a result of this occurrence. One example is the Royal Parade, which necessitates careful planning and specialized attention to the elms and the planes that are getting on in years. The typical issue that arises in this region is the limited availability of water in this region, as well as the prolonged cycle of extreme heat. When it comes to replacing these trees and managing the loss at the same time, communities are currently facing a number of challenges.

An aging population has the potential to bring about a slow but steady transformation of the natural landscape. In addition to this, we need to view this as an opportunity to improve the state of the landscape and protect the vegetation that will be present in the future. The circumstances that may necessitate enhancement are those that are below the earth's surface, such as the water table, the structure of the soil, and inconsistency in the subversive services, and those that are above the ground, which are in conflict with infrastructure, access to stormwater, soil compaction, and mulching.

3.7. WATER SCARCITY AND SHORTAGE

When it comes to the growth of plants, water is an indispensable component. Dry conditions that last for an extended period of time can cause trees in urban areas to suffer severe damage to their health. These factors can also lead to significant increases in the mortality rate of trees. Because of the limited availability of water, the life expectancy of vegetation in city forests is approximately 23% within a period of approximately 10 years.

Trees that have reached maturity have the ability to restructure the urban warmth consequence, as well as provide cover and conserve energy. It is possible for vegetations to become visible in the soil if there is sufficient moisture available, which also contributes to the conservation of the environment. The process of sustaining the wetness in the ground through evapotranspiration serves a number of important purposes, one of which is to maximize the capacity of the trees to cool down the city.

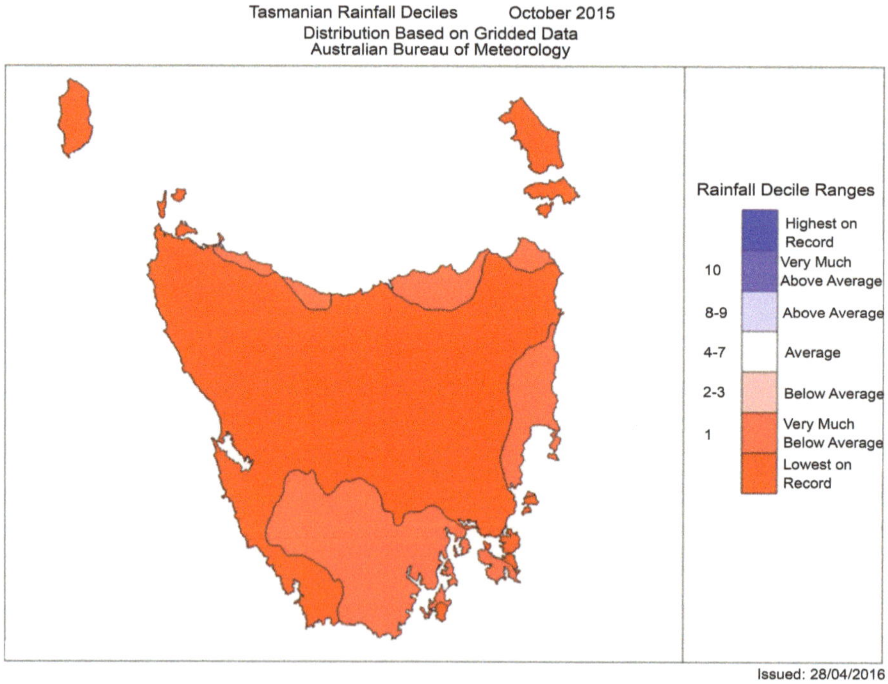

Figure 3.11. *A map showing the lowest rainfall decile in Tasmania recorded on October 2015.*

Source: Monash University, Climate change played a role in Australia's hottest October and Tasmania's big dry in 2015, 2016 https://theconversation.com/climate-change-played-a-role-in-australias-hottest-october-and-tasmanias-big-dry-in-2015-70389.

Adequate moisture in the soil is critical to maintaining the health of the existing vegetation. The drought for about 13 years in the case of the City has brought soil moisture at its lowest level and affected trees around the municipality. This scenario can be similar to trees subjected to irrigated landscapes. This can be a combination of watering restrictions due to low irrigation or low rainfall (Figure 3.11). The summer rain from 2010 to 2012 has been valuable to keep the soil moisture, yet not enough to address the depletion of water resources and pose a threat to tree health. There are several approaches done to replenish the soil moisture and maintain the levels of healthy growth through mulching, water barriers, water irrigation, watering from tanks and soil injection. It is essential to monitor the vegetation health and capacity of ground moisture because this will deliver a planned guideline for ensuring the health condition of our future forests. Another issue could be the lowest amount of water that can penetrate through the soil. Surface irrigation worsens the trees' weakness by urging shallow root systems. To access to better soil moisture, trees must develop deep root systems during low periods of rainfall. People must consider the long-term effect of low water futures. The reliance on potable water is very costly and not practical. There is a need for an alternative water source to maintain the healthy growth of vegetation. Another option to do this is to capture rainwater. Traditionally, many cities are using this method to catch the water, store it and reuse it. This is now a challenge for the densely built area to save stormwater for reuse during dry periods. There could be water-sensitive urban design of water tanks by placing them underground.

3.7.1.1. Concerns of Climate Alteration

The fundamental source of global warming is the enormous discharges of greenhouse fumes into the atmosphere. This has been the universal indicator to monitor climate change regarding global average temperature. For the last three decades, there had been an increase of 0.17 degree Celsius. Based on the worldwide average recorded from 2001 to 2010 it was notably 0.46 degree Celsius higher than the average recorded back from 1961 to 1990.

The effects of climate change are just very noticeable and increasingly conspicuous. The results include heat waves, increase in temperature, extreme weather events, and lower rainfall occurrences. There have been several frequencies of extreme floods worldwide due to higher storm intensity and intense heat that posed great damage, causing skin cancer and other weather-related diseases. The alarming part here is that it is becoming more prevalent than before.

Cities are not exempted from the risks posed by climate change. In fact, cities are very vulnerable to these severe conditions and affect so much the ecological and socioeconomic costs in the area. The forest in towns has enormous influences of climate alteration:

Vegetations are becoming more susceptible to the occurrence of pests and diseases which also affects the urban forest in its ability to survive and recuperate from these epidemics. Recently, the pine plantation in NSW suffered from an insect stem borer attack, fungi, and bark beetles. Climate change can affect the life cycle of the pests as well. Warmer temperatures can increase the development of some insects and their potential for reproduction. Warmer winters also benefit the survival of these pests due to an increase in over-wintering. Many of the pests and diseases favor the hot weather conditions and extend their presence in some geographical sites. Warm temperatures can influence vector spread and flight behavior. There are also introduced pests that discover a more beneficial condition to thrive in for the development of their population. Timberlands that are not under threat initially might also be susceptible to pests and diseases due to these changes.

Extreme weather conditions can significantly impact the health of vegetation. Such conditions can lead to a reduction in crown cover and, in severe cases, complete failure of the plants. High temperatures can scorch plants and their trunks, causing the canopy to wither. Storms may fragment foliage, split branches, and even uproot trees, further damaging the vegetation.

The lowering of rainfall frequency can kill many vegetations and species and will affect the total health of the forest in response to recurrent and harsh dearth.

The inundation can erode soil, increase salinity levels in the water, and instigate tree instability, mortality, and impairment to infrastructure. As a result, trees become weakened due to the disease and their capacity to endure drought is reduced.

3.7.1.2. Life-Threatening Heat Island Effect in Urban Areas

The city's heat island is a life-threatening heat and is very common worldwide. As cities become warmer than previously, most experience a 7 degree rise in temperatures compared to that of rural areas. This becomes a severe problem during the summer season.

The period of lengthy heat puts pressure on the cities and aggravates stress caused by weather affecting the vulnerable populations, the very young, the elderly, and worst, those with medical conditions. Heat waves kill many more communities compared to natural disaster events.

Extremely hot temperatures lead to the deterioration of vegetation combined with low soil moisture, resulting in the scorching of canopies and barks. The leading causes of urban heat island effects are the following:

- **Non-Impermeable Surfaces:** Many of the buildings and pavements are made of concrete and asphalt which can trap heat directly coming from the sun. The solar radioactivity is also mirrored in the building exteriors which causes greater absorption.
- Human products like motorized vehicles are major contributors of greenhouse gases that can trap heat in the atmosphere, especially during hot seasons. Another is the usage of cooling systems which generate more heat.
- Inadequate analysis of flora within the cities can result in lower cooling effects and fewer evapotranspiration processes.

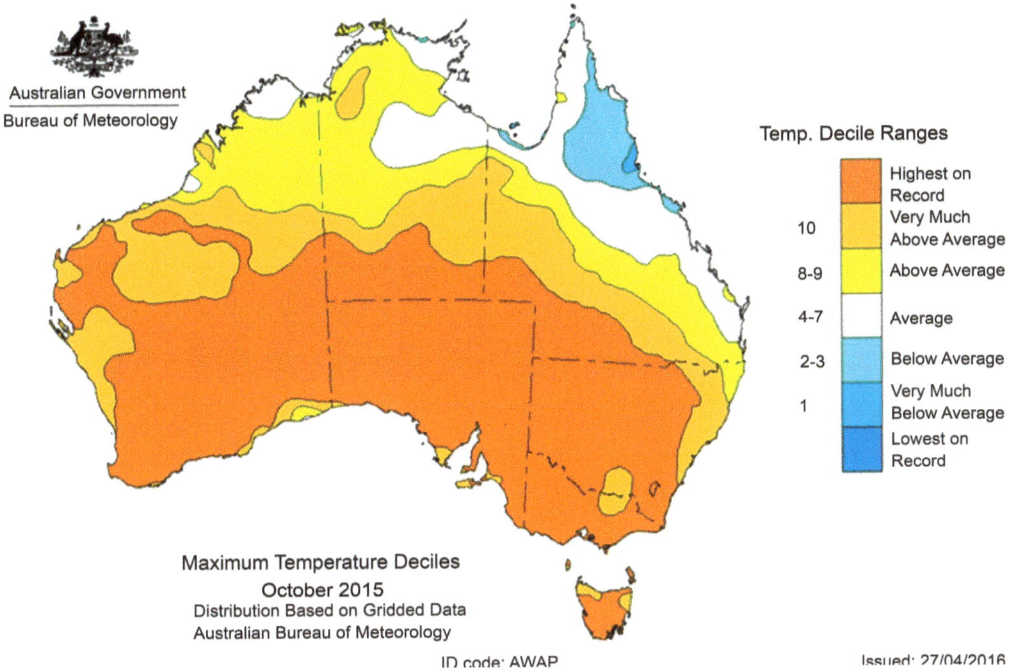

Figure 3.12. *Map showing the highest temperature recorded on October 2015 across Australia.*

Source: Monash University, Climate change played a role in Australia's hottest October and Tasmania's big dry in 2015, 2016 https://theconversation.com/climate-change-played-a-role-in-australias-hottest-october-and-tasmanias-big-dry-in-2015-70389.

Urban forests have proven highly effective in addressing various urban challenges and mitigating the effects of rising temperatures in cities (see Figure 3.12). Some of these challenges include:

- **Intensification of the Urban Heat Island Effect**: Climate change is expected to exacerbate the urban heat island effect, making cities hotter and increasing the demand for cooling solutions.
- **Lack of Canopy Coverage**: Many urban areas suffer from insufficient canopy coverage, which can worsen heat island effects and reduce overall urban resilience.
- **Time Required for Effective Tree Growth**: Newly planted trees typically require approximately 20 years to reach a size where they can significantly mitigate the impact of urban heat islands.
- **Need for Resilient Vegetation**: Urban vegetation must include a mix of robust and resilient species capable of surviving and thriving under increasingly warmer conditions.
- **Benefits of Enhanced Vegetation**: Expanding the city's green canopy helps improve water availability, enhance tree health, and maximize the benefits of evapotranspiration, which aids in cooling the urban environment.

3.8. URBAN SPRAWL AND GROWTH

Cities are major contributors to metropolitan footprints worldwide, leading to unsustainable resource use, excess production, and harmful emissions. Urban centers, however, benefit from extensive open spaces and robust public transportation networks, which can potentially reduce per capita energy consumption and mitigate climate change impacts such as water shortages and heatwaves.

Addressing the challenges of increasing population growth requires:

- Transformation of Infrastructure:
  - Adapting buildings and transportation networks to improve facility use and incorporate green infrastructure, enhancing the city's resilience to climate change.
- High-Quality Urban Design:
  - Implementing design strategies that promote active public and private spaces, providing protection from sunlight and weather, and ensuring efficient land use.
- Integration of Green Infrastructure:
  - Establishing trees and green infrastructure in developing areas, with urban forests serving as a central element for new growth and integration into the existing urban fabric.
- Carbon Management:
  - Focusing on carbon storage within urban areas and addressing the impacts of vegetation removal on carbon balance per hectare of municipal property.

Population growth in urban areas increases the significance of city forests, as it affects environmental well-being and leads to:

- Reduction of Vegetation on Private Lands:
  - The expansion of pavements and buildings typically reduces the amount of vegetation on private properties.
- Decrease in Permeable Surfaces:
  - Reduced porous surfaces limit water infiltration into the ground, contributing to increased runoff and potential flooding.
- Shade and Sunlight Issues:
  - Buildings may cast shadows over streets, hindering tree growth due to insufficient sunlight.
- Increased Pressure on Open Spaces:
  - Higher population densities put greater demand on public open spaces, leading to issues such as traffic congestion, lack of street parking, and challenges associated with compacted and hard surfaces (Haase et al., 2004; Elmqvist et al., 2013).

3.9. ROAD TO OUR FUTURE URBAN FOREST ECOSYSTEM

The next question you may have in mind is how to set out a vision to achieve a diverse, healthy and robust city forest that can improve cities livability and offer security to society.

The strategy will set out the priorities as a guide to future decisions, which will respond to the three themes to create a resilient landscape, healthy community, and a sustainable and livable city. The identified issues and challenges will undoubtedly directly affect the urban forest and necessities practical solutions (Figures 3.13, 3.14, and 3.15).

To ensure a forest's future, there must be a continuing idea and work assurance in the life cycle of trees. Development and management of urban forests require input from different fields of expertise, such as planning, urban design, engineering, landscape architecture, sustainability, economics, and the participation of the general public. A community's capacity to change and create a sense of place is necessary to capture and nurture the dynamics of managing urban forests in cities. Here are the research guides and tools for the development of urban forests:

- Collection of Ground Field Data:
  - Gather detailed data on trees and other environmental components, including species, health, and spatial distribution.
- Spatial and Temporal Mapping:
  - Use ArcGIS to create spatial and temporal maps that identify tree locations, predict potential tree loss, and estimate canopy cover changes over time.
- High-Resolution Aerial Imagery:
  - Utilize high-resolution aerial imagery to assess spatial heterogeneity, structural composition, and the current health of vegetation, as well as to evaluate the capacity for carbon storage.

Figure 3.13. *Cityscape in Brisbane, Queensland.*

Source: Ring, Chris, Brisbane in Queensland, June 2007 http://www.visitbrisbane.com.au/.

Figure 3.14. *Cityscape in Perth, Western Australia.*

Source: Greece, Christos, Perth in Western Australia, June 2007 http://www.skyscrapercity.com/showthread.php?t=777458&page=6.

Figure 3.15. *Cityscape in Sydney, New South Wales.*

Source: KruddPM, Sydney in New South Wales, June 2007 http://www.skyscrapercity.com/showthread.php?t=1911522.

3.10. CONCERNS ABOUT PESTS AND DISEASES

The introduction of non-native pests and diseases from overseas can pose significant risks to urban tree populations, potentially leading to widespread infestations and severe consequences for city forests. The impact of pests and diseases in urban areas is expected to increase due to several factors, including climate change, fluctuating temperatures, extreme weather events, and the movement of pests through internal travel. These factors contribute to the heightened risk of pests and diseases spreading to urban environments, making it essential to monitor and manage these threats proactively.

For instance, there are several pests and diseases recorded in Sydney that have negative impacts on their trees:

- Fig psyllid (*Mycopsylla fici*) (Figure 3.16).
- Australian honey fungus (*Amillaria luteobubalina)* (Figure 3.17).
- Fig leaf beetle (*Poneridia australis*) (Figure 3.18).
- Plane anthracnose (*Apiognomonia veneta*).
- Sycamore lace bug (*Corythucha ciliata*) (Figure 3.19).
- Pink wax scale (*Ceroplastes rubens*).
- Cuban laurel thrips (*Gynaikothrips ficorum*).
- Phytophthora dieback *(Phytophthora cinnamomi*).
- Fusarium wilt (*Fusarium oxysporum*).
- Myrtle rust (*Uredo rangelii*).
- Painted apple moth (*Teia anartoides*).
- Winter bronzing bug (*Thaumastocoris* sp.); and
- White rot (*Phellinus* sp.).

Figure 3.16. *Fig psyllid (Mycopsylla fici).*

Source: England, Dacre, Emerging Moreton Bay Fig Psyllid (Mycopsylla fici) on Host Plant (Ficus macrophylla), n.d.http://www.bowerbird.org.au/observations/41628.

Figure 3.17. *Australian honey fungus (Amillaria luteobubalina).*

Source: Canberra Nature Map, Australian honey fungus (Amillaria luteobubalina), n.d. Source: http://canberra.naturemapr.org/Community/Sighting/3377458.

Figure 3.18. *Figleaf Beetle (Poneridia australis).*

Source: Aussie Bugs, Figleaf Beetle (Poneridia australis), n.d.

Figure 3.19. *Sycamore lace bug (Corythucha ciliata).*

Source: Agriculture Victoria, Sycamore Lace Bug, 8 May 2017 http://agriculture.vic.gov.au/agriculture/pests-diseases-and-weeds/pest-insects-and-mites/sycamore-lace-bug.

The selection of tree species is crucial for managing risks associated with pests and diseases. A diverse composition of tree species can significantly reduce the likelihood of substantial losses from infestations. Pests that are difficult to eradicate can be costly to manage, often due to inadequate treatment options. While some tree species have natural defense mechanisms against pests and diseases, others may struggle in challenging environments due to factors such as limited water availability, compacted soil, shallow root systems, and lack of regular pruning.

Improving biodiversity in urban areas can help mitigate the risk of losing canopy cover due to pests and diseases. Cities should prioritize selecting tree species that are resilient and well-suited to their urban environment to ensure robust canopy cover. This approach not only enhances the city's green infrastructure but also contributes to greater ecological stability. (Haase et al., 2004; Elmqvist et al., 2013).

3.11. COMMUNITY RESISTANCE TO CHANGE AND CONVENING EXPECTATIONS

The city is composed of a different combination of population, living spaces, and properties. Each has a different perspective and varies in its interaction with urban forest. The business people may visit and engage with trees more often within the city. The city embraces different attitudes and interests towards trees. These are, in most cases, influenced by cultural background, income, education, age, and an individual's contact with the environment. At the most extreme, there is illogical anxiety towards trees. However, statistics show that there is a higher likelihood of dying by being struck by lightning compared to tree failure.

The owner's perception of their trees is often personal. The city council and education program should nurture an appreciation and create an understanding of the values of the trees in the environment. It is apparent that focusing on addressing such barriers is not about building awareness but behavioral change. In a practical sense, this may require considering different languages and venues such as community centers and religious institutions which are not a visible target.

The public needs to be well-informed and educated about effective techniques for managing urban trees, including the implications of tree removal on the long-term sustainability and viability of city forests. Addressing tree-related issues does not always require tree removal; often, proper pruning and maintenance, along with thoughtful redesign and repair of infrastructure, can address concerns with minimal impact on the trees.

It is also essential for individuals to stay updated about the changes in tree health. Trees, as living organisms, grow, age, and eventually die, requiring ongoing maintenance. In urban environments, aging trees can pose hazards if not managed properly, and their removal can be particularly emotional if they have been part of the community for decades or hold historical or sentimental value. Trees often contribute to community well-being and aesthetic value, and some individuals may invest significantly in maintaining aging trees to ensure they continue to benefit future generations.

Managing urban forests involves balancing these values and addressing the needs of the community. This effort often requires compensating for the needs of a few, ensuring that the urban forest remains a sustainable and valued part of the city. (Haase et al., 2004; Elmqvist et al., 2013).

CASE STUDY

FOREST ECOSYSTEM IN PORTLAND, OREGON

Portland, Oregon, is renowned for its commitment to sustainability and green urban planning, particularly through the development and maintenance of its urban forest ecosystem. Covering approximately 30% of the city's land area, Portland's urban forest includes over 200,000 street trees, 1,000 acres of parks, and numerous green spaces. The city has implemented an Urban Forest Management Plan to enhance the ecological health of its urban environment. This plan emphasizes the importance of urban forests in providing ecosystem services, such as air purification, stormwater management, and habitat for wildlife. Furthermore, the initiative encourages community involvement in tree planting and care, fostering a sense of stewardship among residents.

The city's diverse urban forest is not only vital for ecological balance but also contributes significantly to the social and economic well-being of its inhabitants. Research has shown that urban green spaces improve mental health, reduce stress, and promote physical activity among residents. Portland's urban forest also plays a crucial role in mitigating the urban heat island effect, which is becoming increasingly important as climate change leads to more extreme weather events. The trees provide shade, cool the air, and enhance the overall aesthetic appeal of the city, making it an attractive destination for both residents and tourists.

Despite these benefits, Portland faces challenges in maintaining its urban forest ecosystem. Issues such as invasive species, climate change, and urban development pose threats to tree health and biodiversity. The city has established programs to address these challenges, including public education campaigns about the value of native species and the importance of tree diversity. Ongoing research and community engagement are essential to adaptively manage the urban forest and ensure its resilience in the face of future environmental changes.

Discussion Questions

1. How can urban forests contribute to the overall resilience of a city in the face of climate change?
2. In what ways can community engagement in urban forestry initiatives lead to improved environmental, social, and economic outcomes for city residents?

SUMMARY

Urban forests, comprising trees, shrubs, and vegetation within cities, are crucial for enhancing urban life quality. These ecosystems include street trees, parks, private gardens, and other green spaces, offering significant environmental, social, and economic benefits. Environmentally, urban forests improve air quality, regulate climate, and manage water. Socially, they boost health and well-being, enhance city aesthetics, and foster community cohesion. Economically, they increase property values, reduce energy costs, and attract tourism.

Urban forest ecosystems face challenges such as urbanization pressure, environmental stressors, and resource limitations. Urbanization leads to the loss of green spaces, while pollution and climate change stress vegetation. Additionally, insufficient funding and expertise hinder effective management.

Sustainable urban forest management involves integrating green infrastructure into city planning, establishing protective policies, and promoting community involvement. Innovative practices, such as using technology for monitoring and adopting sustainable landscaping, are essential. Engaging communities in tree planting and maintenance, along with raising awareness about the benefits of urban forests, is crucial.

Therefore, urban forests play a vital role in creating sustainable, livable cities, providing essential ecosystem services and improving residents' quality of life. Effective management and investment in these green spaces are essential for a greener, healthier future.

REVIEW QUESTIONS

1. What are the historical roots of city forests, and how have they evolved over time?
2. Describe the changes in urban forest sustainability during the twentieth century.
3. What tasks are involved in managing an urban forest ecosystem?
4. Why is the population of mature trees in urban areas significant?
5. Discuss the concerns related to pests and diseases in urban forest ecosystems.

REFERENCES

1. Arowolo, A. O., Deng, X., Olatunji, O. A., & Obayelu, A. E. (2018). Assessing changes in the value of ecosystem services in response to land-use/land-cover dynamics in Nigeria. *Science of the Total Environment, 636*, 597–609. https://doi.org/10.1016/j.scitotenv.2018.04.290.
2. Baro, F., Chaparro, L., Gomez-Baggethun, E., Langemeyer, J., Nowak, D., & Terradas, J. (2014). Contribution of ecosystem services to air quality and climate change mitigation policies: The case of urban forests in Barcelona, Spain. *AMBIO, 43*(4), 466–479. https://doi.org/10.1007/s13280-014-0515-5.

3. Bolund, P., & Hunhammar, S. (1999). Ecosystem services in urban areas. *Ecological Economics, 29*(2), 293–301. https://doi.org/10.1016/S0921-8009(99)00013-0.

4. Briassoulis, H. (2019). Analysis of land use change: Theoretical and modeling approaches. In *WVU Research Repository*. Regional Research Institute, West Virginia University: Morgantown, WV, USA.

5. Bryan, B. A., Ye, Y., Zhang, J., & Connor, J. D. (2018). Land-use change impacts on ecosystem services value: Incorporating the scarcity effects of supply and demand dynamics. *Ecosystem Services, 32*, 144–157. https://doi.org/10.1016/j.ecoser.2018.07.006.

6. Camps-Calvet, M., Langemeyer, J., Calvet-Mir, L., & Gómez-Baggethun, E. (2016). Ecosystem services provided by urban gardens in Barcelona, Spain: Insights for policy and planning. *Environmental Science & Policy, 62*, 14–23. https://doi.org/10.1016/j.envsci.2016.01.005.

7. Costanza, R., De Groot, R., Braat, L., Kubiszewski, I., Fioramonti, L., Sutton, P., Farber, S., & Grasso, M. (2017). Twenty years of ecosystem services: How far have we come and how far do we still need to go? *Ecosystem Services, 28*, 1–16. https://doi.org/10.1016/j.ecoser.2017.09.008.

8. Costanza, R., de Groot, R., Sutton, P., van der Ploeg, S., Anderson, S. J., Kubiszewski, I., Farber, S., & Turner, R. K. (2014). Changes in the global value of ecosystem services. *Global Environmental Change, 26*, 152–158. https://doi.org/10.1016/j.gloenvcha.2014.04.002.

9. Council of Tree and Landscape Appraisers. (1992). *Guide for Plant Appraisal* (8th ed.). The International Society of Arboriculture: Atlanta, GA, USA.

10. Elmqvist, T., Fragkias, M., Goodness, J., Güneralp, B., Marcotullio, P. J., McDonald, R. I., Parnell, S., Schewenius, M., Sendstad, M., Seto, K. C., et al. (2013). *Urbanization, Biodiversity and Ecosystem Services: Challenges and Opportunities*. Springer: Dordrecht, The Netherlands. ISBN 978-94-007-7087-4.

11. Gómez-Baggethun, E., & Barton, D. N. (2013). Classifying and valuing ecosystem services for urban planning. *Ecological Economics, 86*, 235–245. https://doi.org/10.1016/j.ecolecon.2012.08.019.

12. Haase, D. (2013). Shrinking cities, biodiversity and ecosystem services. In *Urbanization, Biodiversity and Ecosystem Services: Challenges and Opportunities* (pp. 253–274). Springer: Dordrecht, The Netherlands.

13. Haase, D., Larondelle, N., Andersson, E., Artmann, M., Borgström, S., Breuste, J., Gomez-Baggethun, E., Gren, A., Hamstead, Z., & Hansen, R. (2014). A quantitative review of urban ecosystem service assessments: Concepts, models, and implementation. *AMBIO, 43*(4), 413–433. https://doi.org/10.1007/s13280-014-0502-x.

14. Kawaguchi, M., Hirabayashi, S., Hirase, K., Kaga, H., & Akazawa, H. (2021). Estimation of monetary value of street trees by i-Tree Eco using tree health surveys in Suita City, Osaka Prefecture. *Landscape Research Japan Online, 14*, 1–12. https://doi.org/10.5016/LRJ.2021.14.0_1.

15. Kim, G., Miller, P., & Nowak, D. (2015). Assessing urban vacant land ecosystem services: Urban vacant land as green infrastructure in the City of Roanoke,

Virginia. *Urban Forestry & Urban Greening, 14*(3), 519–526. https://doi.org/10.1016/j.ufug.2015.05.001.

16. Kiss, M., Takacs, A., Pogacsas, R., & Gulyas, A. (2015). The role of ecosystem services in climate and air quality in urban areas: Evaluating carbon sequestration and air pollution removal by street and park trees in Szeged (Hungary). *Moravian Geographical Reports, 23*(1), 36–46. https://doi.org/10.1515/mgr-2015-0004.
17. Kremer, P., Hamstead, Z., Haase, D., McPhearson, T., Frantzeskaki, N., Andersson, E., Kabisch, N., Larondelle, N., Rall, E. L., & Voigt, A. (2016). Key insights for the future of urban ecosystem services research. *Ecology and Society, 21*(2). https://doi.org/10.5751/ES-08394-210226.
18. Kremer, P., Hamstead, Z. A., & McPhearson, T. (2016). The value of urban ecosystem services in New York City: A spatially explicit multicriteria analysis of landscape scale valuation scenarios. *Environmental Science & Policy, 62*, 57–68. https://doi.org/10.1016/j.envsci.2016.01.008.
19. Larondelle, N., & Lauf, S. (2016). Balancing demand and supply of multiple urban ecosystem services on different spatial scales. *Ecosystem Services, 22*, 18–31. https://doi.org/10.1016/j.ecoser.2016.09.008.
20. Ning, Z., Chambers, R., & Abdollahi, K. (2016). Modeling air pollutant removal, carbon storage, and CO2 sequestration potential of urban forests in Scotlandville, Louisiana, USA. *iForest–Biogeosciences and Forestry, 9*, 860–867. https://doi.org/10.3832/ifor1870-008.
21. Nowak, D., Hoehn, R., Bodine, A., Greenfield, E., & O'Neil-Dunne, J. (2016). Urban forest structure, ecosystem services and change in Syracuse, NY. *Urban Ecosystems, 19*(4), 1455–1477. https://doi.org/10.1007/s11252-016-0568-4.
22. Nowak, D. J. (1994). Atmospheric carbon dioxide reduction by Chicago's urban forest; General Technical Report NE-186. Philadelphia, PA, USA: USDA Forest Service.
23. Nowak, D. J. (2002). Brooklyn's urban forest. Washington, DC, USA: USDA Forest Service.
24. Nowak, D. J. (2002). Compensatory value of urban trees in the United States. *Journal of Arboriculture, 28*(4), 194–199.
25. Nowak, D. J., Hirabayashi, S., Bodine, A., & Greenfield, E. (2014). Tree and forest effects on air quality and human health in the United States. *Environmental Pollution, 193*, 119–129. https://doi.org/10.1016/j.envpol.2014.05.028.
26. Nowak, D. J., Walton, J. T., Stevens, J. C., Crane, D. E., & Hoehn, R. E. (2008). Effect of plot and sample size on timing and precision of urban forest assessments. *Arboriculture & Urban Forestry, 34*(6), 386–390. https://doi.org/10.48044/jauf.2008.046.
27. Ordóñez-Barona, C., Devisscher, T., Dobbs, C., Aguilar, L. A. O., Baptista, M., Navarro, N. M., da Silva Filho, D. F., & Escobedo, F. J. (2019). Trends in urban forestry research in Latin America & the Caribbean: A systematic literature review and synthesis. *Urban Forestry & Urban Greening, 126544*. https://doi.org/10.1016/j.ufug.2019.126544.

28. Pickett, S. T. A., Cadenasso, M. L., Rosi-Marshall, E. J., Belt, K. T., Groffman, P. M., Grove, J. M., Irwin, E. G., Kaushal, S. S., LaDeau, S. L., & Nilon, C. H. (2017). Dynamic heterogeneity: A framework to promote ecological integration and hypothesis generation in urban systems. *Urban Ecosystems, 20*(1), 1–14. https://doi.org/10.1007/s11252-016-0634-x.

29. Selmi, W., Weber, C., Riviere, E., Blond, N., Mehdi, L., & Nowak, D. (2016). Air pollution removal by trees in public green spaces in Strasbourg City, France. *Urban Forestry & Urban Greening, 17*, 192–201. https://doi.org/10.1016/j.ufug.2016.04.009.

30. Song, P., Kim, G., Mayer, A., He, R., & Tian, G. (2020). Assessing the ecosystem services of various types of urban green spaces based on i-Tree Eco. *Sustainability, 12*(4), 1630. https://doi.org/10.3390/su12041630.

31. Szkop, Z. (2016). An evaluation of the ecosystem services provided by urban trees: The role of Krasinski Gardens in air quality and human health in Warsaw (Poland). *Environmental and Socio-Economic Studies, 4*(2), 41–50. https://doi.org/10.1515/ensec-2016-0007.

32. Tan, X., Hirabayashi, S., & Shibata, S. (2021). Estimation of ecosystem services provided by street trees in Kyoto, Japan. *Forests, 12*(3), 311. https://doi.org/10.3390/f12030311.

33. Tang, Y., Chen, A., & Zhao, S. (2016). Carbon storage and sequestration of urban street trees in Beijing, China. *Frontiers in Ecology and Evolution, 4*. https://doi.org/10.3389/fevo.2016.00076.

34. Uchida, K., Blakey, R. V., Burger, J. R., Cooper, D. S., Niesner, C. A., & Blumstein, D. T. (2021). Urban biodiversity and the importance of scale. *Trends in Ecology & Evolution, 36*(2), 123–131. https://doi.org/10.1016/j.tree.2020.11.004.

35. Ueno, H. (2010). City planning and formation of Kyoto in modern age. *Shisen, 111*, 30–44. (In Japanese).

36. Wang, J., Endreny, T. A., & Nowak, D. J. (2008). Mechanistic simulation of tree effects in an urban water balance model 1. *Journal of the American Water Resources Association, 44*(1), 75–85. https://doi.org/10.1111/j.1752-1688.2008.00120.x.

37. Wiens, J. A. (1989). Spatial scaling in ecology. *Functional Ecology, 3*(3), 385–397. https://doi.org/10.2307/2389612.

38. Wu, J., & Li, H. (2006). Concepts of scale and scaling. In *Scaling and Uncertainty Analysis in Ecology* (pp. 3–15). Springer: Dordrecht, The Netherlands. https://doi.org/10.1007/1-4020-3435-1_1.

39. Yi, H., Güneralp, B., Filippi, A. M., Kreuter, U. P., & Güneralp, İ. (2017). Impacts of land change on ecosystem services in the San Antonio River Basin, Texas, from 1984 to 2010. *Ecological Economics, 135*, 125–135. https://doi.org/10.1016/j.ecolecon.2017.01.009.

40. Zhao, C., & Sander, H. A. (2018). Assessing the sensitivity of urban ecosystem service maps to input spatial data resolution and method choice. *Landscape and Urban Planning, 175*, 11–22. https://doi.org/10.1016/j.landurbplan.2018.03.009.

CHAPTER 4

Forest Soil Dynamics

LEARNING OBJECTIVES

After studying this chapter, you will be able to:

- Identify the components of forest soils.
- Explain nutrient cycling dynamics in forest ecosystems.
- Understand factors affecting forest soil productivity.
- Explore methods to enhance forest soil productivity.

KEY TERMS FROM THIS CHAPTER

Decomposition
Forest soils productivity
Shortened rotation
Soil mineral
Soil-water
Timber stand improvement
Fertilization
Genetic improvement
Soil microorganisms
Soil organic matter
Species and site selection

INTRODUCTION

A forest ecosystem's development contributes to the regulation of the many ecosystem processes that are necessary for the maintenance of our environment. The soil is the largest pool that has the capacity to store carbon that is produced as a result of emissions, which can therefore contribute to the modification of the effects of climate change. In many different ways, soils provide support for both biotic and abiotic factors. It has been brought to our attention that a significant portion of our landscapes are not being transformed into alternative land uses (Ontl & Schulte, 2012). This chapter will provide an answer to the most important question that will be addressed, which is how landowners can protect and maintain the productivity of their land while also addressing the issue of climate change.

Forest soils serve a variety of purposes, all of which are influenced by the climatic conditions that are changing. These shifts in climate had a significant impact on the amount of timber that could be harvested from the forest. In forest areas, for instance, an increase in temperature has an effect on the population growth of beneficial insects and instigates the survival of both native and exotic pests. There is a growing threat to our native and indigenous species as invasive species develop resistance. According to Ramsfield, Bentz, Faccoli, Jactel, and Brockerhoff (2016), biological processes are impacted, which results in the death of certain trees and has an increased impact on biodiversity in general. Some factors, however, become restricted, which has an effect on the functions of the soil. By gaining an understanding of forest soils and the dynamics of these soils, we will be better able to comprehend how changes in climatic conditions can have a long-term impact on the ecosystems of our forests as a result of shifts in the environment.

4.1. COMPONENTS OF FOREST SOILS

Soils consist of various components such as organic matter, minerals, water, air, and soil organisms (Figure 4.1). Soil's properties are derived from biological, physical, chemical, and morphological processes that change with topography, gradually replacing parent materials over time. The five main components of forest soils form the spine of the terrestrial ecosystem.

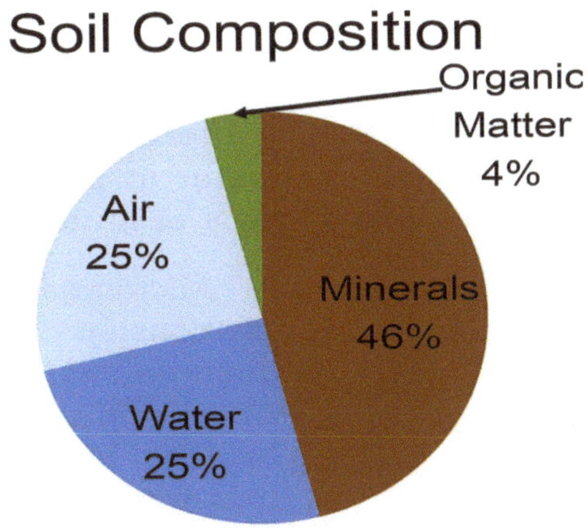

Figure 4.1. *Illustration of soil components.*

Source: https://earthsoils.com/wp-content/uploads/2010/06/soilcomp.jpg.

4.1.1. Soil Mineral

First, soil minerals account for 45% of total soil volume. Minerals have been around for a long time. Soils contain two types of minerals: primary minerals, which are soil particles found in silt and sand and formed similarly to the parent material, with a round or irregular shape. Secondary minerals are the excess of primary minerals formed during weathering. The latter emits ions and solidifies minerals in the form of silicate clay. Clay has significant chemical properties and can retain water in the soil for a long time. The presence of negative, neutral charges affects the soil's ability to hold essential nutrients such as cations, which contributes to the cation exchange capacity of the soil. The presence of minerals in the soil has an impact on the diversity of the soil's microbial community, which has yet to be fully recognized. Colin, Nicolith, Turpault, and Uroz (2017) discovered that minerals have a significant impact on the structure, diversity, and function of soil bacteria, which aids in the maintenance of soil and plant conditions in nutrient-poor environments. The weathering process makes minerals available for plant use. Minerals also serve as biogeochemical reserves, which are an essential component of nutrient cycling.

4.1.2. Soil-Water

Secondly, soil water is about 2 to 50% of the total soil volume. Water helps to transport soil minerals and nutrients into the soil to facilitate plant growth and support in the biological and chemical processes. The availability of soil water is important for the plant to survive. Clay and organic matter influence the water-holding capacity of the soil. But clays at some point have very fine micropores which makes it difficult to hold water. The ideal type of soils which are considered most productive are the silt loams and loams because their soil texture can hold ample amounts of water for plant growth. Forests play an important role in the global hydrological cycle. Forests with high water quality can be a good source of water supply that people can use.

Stabilizing the soil would also mean less erosion and reducing water quality problems. Forests absorb water from rainfall and the atmosphere and infiltrate water into the ground. Forests serve as a catchment to collect water for different uses and to support ecological needs and protect the lowlands from experiencing surface runoff and flood events (Calder, Hofer, Vermont, & Patrizio, 2007).

4.1.3. Soil Organic Matter

Organic matter comes from the dead remains of flora and fauna which can also support the soil to hold significant elements for plant growth. Soils with organic matter usually have a high cation exchange capacity, which makes the land more productive. The high percentage of organic matter in the soil also indicates high fertility. However, prolonged process of decomposition would also lead to unavailability of organic matter in plant use, which results in recalcitrant soil carbon storage.

4.1.4. Soil Microorganisms

Soil is the habitat of most microorganisms which makes up 1% of the soil volume. The soil can hold as much as 20,000 microbes which includes the communities of earthworms, bacteria, fungi, nematodes, etc. Microorganisms help to decompose vegetative materials and other raw organic matter. Decomposers feed on OM, air and water to produce new humus which is a type of nutrient for plant use.

Microorganisms also promote symbiotic relationship that allows plants to extract important nutrients needed by the plant. An example of this is the nitrogen-fixing bacteria. Mutualism also occurs in the plant root system between fungi and mycorrhiza. The growth of fungus helps to provide sugar in plant roots. Without the presence of these microorganisms, soils will just be limited to plant support.

Studies based on changes in soil biology in relation to landscape disturbance are very limited due to the high diversity and complexity of interaction occurring in forest soils. The USDA based in Puerto Rico studied thoroughly the changes of soil biota in accordance to gradient disturbance. We are all aware that invertebrates in tropical forests are very significant in the process of decomposition as compared to the substrates used in temperate forests. Small disturbances among arthropods can result in distinctive loss of litter decomposition and affect nutrient mineralization. Bacteria and fungi are wet forests are very sensitive to drying and can also affect the nutrient cycling. Tropical fungi in Puerto Rico show a very significant environmental sensitivity to the high turnover of terrain in the area (Gonzales & Lodge, 2017). Invertebrates help to accelerate decomposition in lowland dry forests, and it is also significant during the initial phase of log decomposition in mid-elevation wet forests.

4.1.5. Gases

This is a soil component that occupies the spaces between large pores. It makes up 2 to 50% of the soil volume. Oxygen is important to support the respiration processes happening in roots and microbes. Without gases, the soil will be waterlogged and restricted from gas exchange which will result in plant death. Soil gas is a good factor to determine if there is a presence of volatile compounds in the soil. This can also tell if the soil is contaminated or not. One indicator is the high vapor pressure and low water solubility (Wait, Ramsey, & Maney, 2015).

4.2. NUTRIENT CYCLING DYNAMICS

The plant requires sunlight, water, temperature and nutrients to grow and reproduce. Plant nutrients in the soil are available in chemical forms absorbed by the plants. Inorganic chemicals were dissolved for plant use. Understanding the biological and chemical courses is important for plant nutrients. Nutrients in plants are passed on to different organisms. Materials such as animal waste and litter fall are the main source of plant nutrients in forest environments. The plant requires 16 essential nutrients in the form of chemical elements to support plant growth. Elements like Hydrogen, Carbon and Oxygen from the atmosphere are needed in large quantities to support plant photosynthetic activities. Other elements needed by the plants are Nitrogen, Phosphorus, Potassium, Iron, Sulphur, Manganese, Copper, Zinc, Boron, Chlorine, Molybdenum, and Magnesium. Some elements like silicon or sodium are needed by a particular type of plant but are not mandatory because they may not or may have a beneficial effect on the plant growth. Legume plants usually require Cobalt in the process of nitrogen fixation. Essential nutrients vary depending on the amount needed by the plant. The primary macronutrients for crops are N, P, K about 22 to 68 kgs/acre, while secondary macronutrients like S, Ca and Mg are needed in 4.5 to 22 kgs/acre and the rest of the nutrients are < 0.5 kg/acre. Mineral nutrients are utilized by the plant in the form of solution. Soluble nutrients are produced from processes of chemical weathering, decomposition, application of fertilizers, atmospheric deposition of sediments and nitrogen-fixation in leguminous plant species. Minerals can be lost due to some events such as erosion, runoff, and losses of gas, leaching and removal of trees. Apart from other organic matter and microbes in the soil, plant nutrients are available in various forms or also known as nutrient pools. This is the process wherein nutrients are converted into soluble form for plant uptake. Nutrients in soluble forms are absorbed by the easily prior to soil loss. The presence of exchangeable cations helps the plant, restore its storage pools replenishing ion solution. Organic matter is a good source of N, S, P and B and other trace minerals needed by the plant. Some minerals are soluble types like sulfates and chlorides due to weathering reactions caused by organic acids becoming insoluble forms (mica, apatite, feldspar).

Iron oxides and phosphate are two examples of adsorbed ions that can bind organic materials on surfaces and slowly release them for a longer supply of nutrients or plant use. Chemical compounds with particular electric charge are called ions. A dissolved form of soil nutrients is available in the form of positively and negatively charged ions. Cation is positively charged, while Anion is negatively charged (Regents of the University of Minnesota, 2018).

There are nutrients that occur as cations. The soil has the capacity to store or hold these cations (Cation exchange capacity or CEC). The negatively charged ions in the soil are attracted by the positively charged which prohibits them from leaching down the soil horizons. The cations present in soil particles is known as exchangeable cations. The effect of negatively charged or anions in the soil makes sulfate, nitrates and chloride susceptible to leaching. The CEC is highly influenced by the clay content and organic matter in the soil. CEC can also determine the ability of the soil to hold nutrients, fertility status, and water holding capacity. Factors that can affect the CEC are soil pH level, types of soil (sand, silt, clay) and organic matter content. Sandy soils usually have a CEC range from 1 to 5 meq/100g, while clay soils (illite and smectite) have a CEC range from 25 to 100 meq/100g. Increasing clay content in the soil will help increase the CEC in sandy soil. Organic matter is also negatively charged which permits it to attract cations same with clay particles. CEC in organic matter can be as high level as 250meq/100g. The organic acid is dissociated in organic matter because of negatively charged ions. This process of dissociation still relies on the pH level occurring in the soil. CEC in organic matter is high and pH level dependent. One way to increase CEC in soil is by increasing organic matter content. Below is a photograph showing how CEC varies with soil depth (Figure 4.2). From the example below, at a depth between 0 to 10 cm, CEC is approximately 4.6 meq/100g while suddenly at depth of 20 to 30 cm, CEC decreases to 1.2 meg/100g. It is very remarkable that the presence of organic matter can highly influence CEC. At a depth between 40 to 50 cm, CEC increases to approximately 8.9 meq/100g due to the high content of clay (Efretuei, 2016).

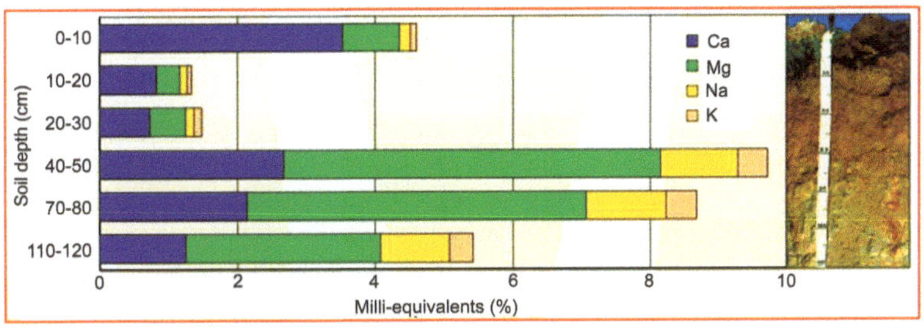

Figure 4.2. *Sandy soil CEC varies with soil depth.*

Sources: https://permaculturenews.org/2016/10/19/soils-cation-exchange-capacity-effect-soil-fertility and http://www.soilquality.org.au/factsheets/cation-exchange-capacity/).

Soil Organic matter is an important element to determine whether the soil is fertile or not. Plant nutrients and organic matter exist in different forms from living to dead residues and take time to decompose. Soil organisms complete the food web and carry out biological processes of nutrient cycling. Figure 4.3 below explains the process of carbon and nutrient cycling in the soil. Plant roots must also be given attention because these are the living biomass in the soil.

The nutrient cycle in plants has an important role in the plant growth. This process contributes not only to soil fertility but also to productivity. Soil fertility can be maintained if the nutrient cycle is efficient enough to occur on the soil food web system (Figure 4.4). Organic matter is the main high point of the soil nutrient cycling. Many of the plant nutrients follow the path of the carbon cycle. The organic residues are returned to the plant for plant use. Most organic carbon compounds are food for fungi, bacteria and other decomposers in the soil. Organic matter is broken down into a soluble solution for root and plant uptake.

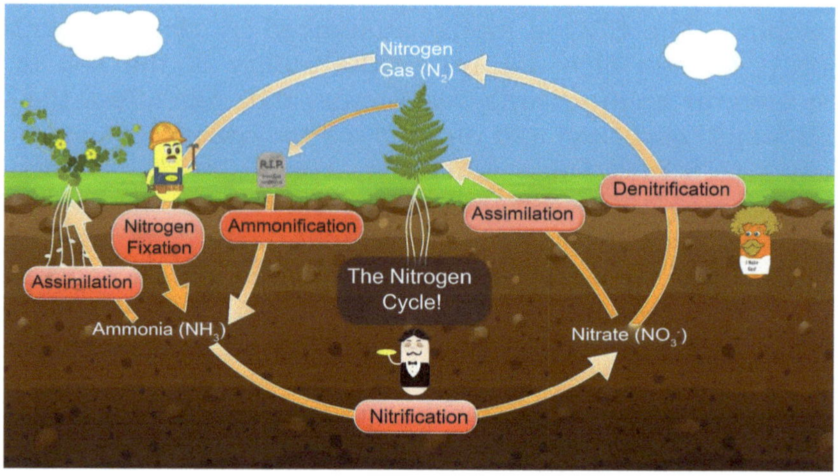

Figure 4.3. *Carbon and nutrient cycling.*

Source: https://i.ytimg.com/vi/uip4Q6t7yfQ/maxresdefault.jpg.

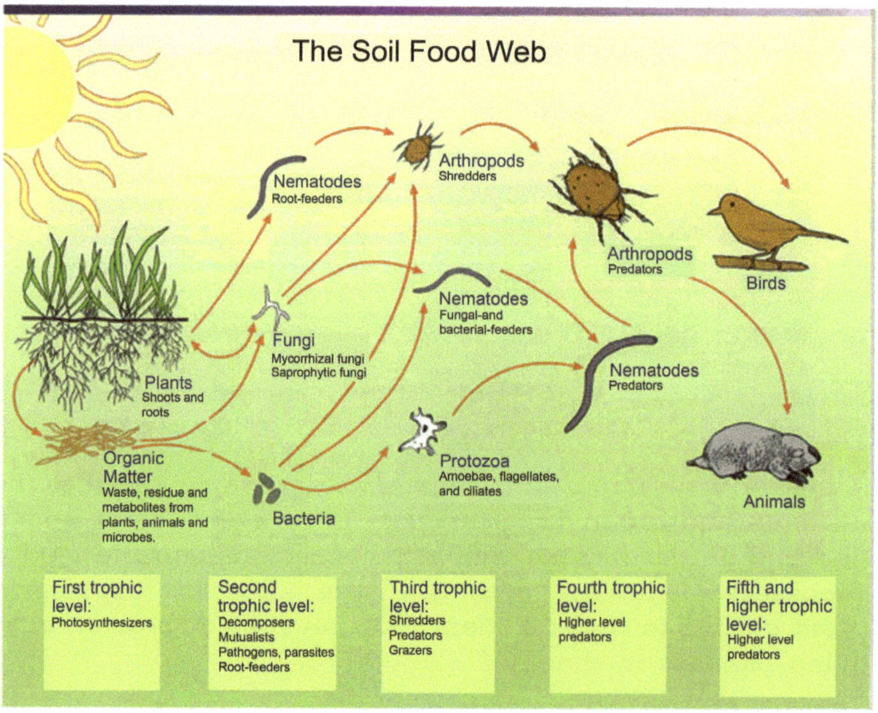

Relationships between soil food web, plants, organic mater, and birds and mammals
Image courtesy of USDA Natural Resources Conservation Service

Figure 4.4. *Schematic of the soil food web.*

Source: USDA.

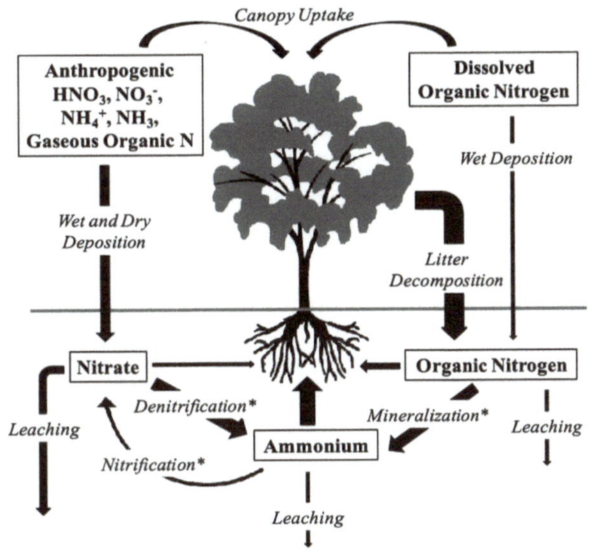

Figure 4.5. *Illustration of the nitrogen cycle.*

Source: https://www.researchgate.net/profile/Eric-Steig/publication/252552732/figure/fig2/AS:670016466870297@1536755782955/Simplified-schematic-of-nitrogen-cycling-in-temperate-nitrogen-limited-forests-The.png.

Another complicated cycle occurring in the soil-plant-animal community is the nitrogen cycle (Figure 4.5). Nitrogen is available in different forms for plant use. It can be in the form of organic or inorganic nitrogen compounds that require biochemical transformations performed by soil organisms to make the nitrogen solution available for plant use. Nitrogen can also be in the form of gases that freely move in the air. Nitrogen is said to be one of the limiting nutrients in terrestrial ecosystems and nitrate in soluble form is one of the most significant soil nutrients that plant needs, but at the same time can be a source of contaminants in nearby water systems. The nitrogen cycle is not 100% because some of the nitrogen is lost in the natural ecosystem.

Recently the deposition of atmospheric nitrogen has been accelerated due to anthropogenic activities (e.g., combustion of fossil fuels, nitrogen fertilization and nitrogen-fixing cultivation). Deposition of nitrogen is an important activity in the forest ecosystem, but if the deposition processes exceed beyond the capacity of the soil to hold nitrogen then it will result in nitrogen losses in the form of nitrate leaching or nitrogen oxide emission which can threaten the environment by cases of eutrophication and acidification. This is the biggest problem now of forest ecosystems located in North America and Europe. China is also suffering from atmospheric nitrogen deposition (Zhu, Zhang, Chen, & Mo, 2015).

4.2.1. Decomposition, Its Regulation, and Control

Decomposition occurs in forest soils, but the rate of decomposing matters also matters because smaller particles help to release carbon and nutrients for plant use and microbial activities. Decomposition can convert dead organic matter, into organic, inorganic nutrients and carbon dioxide. The process of decomposition requires a series of physical and chemical processes. The nutrient breakdown limits the percentage of carbon that an ecosystem can store. Such breakdown permits carbon for plant use and other processes for ecosystem function (e.g., the production of plants). An increase in decomposition affects the global carbon cycle (Norby, Cotrufo, Ineson, O'Neill, & Canadell, 2001). This not only increases the amount of carbon intake of plants, but also increases the release of carbon dioxide in the atmosphere. This is the reason why understanding decomposition is necessary. Decomposition can be regulated and controlled by different factors.

- **Temperature:** An increase in temperature also indicates an increase in rates of chemical reaction. Indirect manner, an increase in temperature can intensify the rate of evapotranspiration, which can prevent the substrates from working with water dependent flora and fauna, microsite reactions and extracellular enzymes. Even though that study claimed that an increase in temperature could only bring a small impact, if these small impacts are multiplied globally, then it could bring fluctuations to the global carbon cycle (Davidson & Janssens, 2006).
- **Quantity and Quality of Substrates:** The amount of substrates present in the soil can affect the rate of decomposition. Carbon substrate can be easily broken down as compared with other substrates. It only requires less energy. However, some carbon requires a longer period to decompose and requires more energy depending on complexity. For instance, humus, it takes time to decompose permitting the soil to acquire more carbon over a certain period of time. However, if the complex carbon substrate is high, decomposition rate also slows down. The substrates with high nitrogen content can hasten the rate of decomposition. Materials with higher lignin content usually slows down decomposition. Carbon, nitrogen and lignin are important components to regulate and control the rate of decomposition (Melillo, Aber, & Muratoe, 1982).
- **Composition of Soils:** The physical content of the soil can also affect the rate of decomposition. The high presence of clay particles can slow down the rate of decomposition by protecting carbon, which also allows more carbon build up in the soil. Some ions are also attracted to organic matter which can slow down the decomposition rate.
- **Disturbance:** Temporary exposure of the soil surface can increase the loss of soil moisture. This will also increase the rate of respiration in soils and high yield or organic matter. Disturbance can be natural (weathering) or anthropogenic events (soil cultivation or tillage).

4.3. FOREST SOILS PRODUCTIVITY

Soil Productivity refers to the soil's capability to support plant growth under normal conditions in the environment. The amount of organic matter developed in the soil reflects the vegetation growth. The volume of trees produced is measured to determine the productivity of the forest as practiced in forest management. Soil is an essential tool for sustainable forest management. The primary conditions in conserving soil productivity. Soil productivity determines the number of harvestable timber over the sustained duration of time. It also affects other life forms and attributes in the forest such as the provision of ecosystem services, wildlife habitat and biodiversity. Maintaining forest and forest soil productivity will help to meet the needs of the present and future generations without impeding the forest products and other functions of the forest. Soil productivity is highly inclined on the type of tree species that will grow in forest lands. Sustaining soil productivity will preserve the forest soil conditions to survive, regenerate and support long-term growth of vegetation in the forest. Maintaining soil productivity is more efficient than recovering the damaged forest areas using different mitigation measures. Most of the forest management activities and practices are designed in such a way as to minimize the negative impacts or keep at a minimum level the effects in the environment. Soil characteristics are usually divided into three. These soil characteristics are important to maintain forest sustainability.

- Physical properties as discussed in previous chapters (texture, structure, porosity, density, hydrology, and drainage).
- Chemical properties such as rate of nutrient cycling and soil pH level; and
- Biological properties that composed of different organisms that support plant growth.

Impacts on these soil properties can affect the vegetation growth in the forest. Physical properties are highly affected during periods of harvesting activities because soil nutrients are directly removed and modify the biological processes in the soil. However, little evidence has been proven that these impacts can decline vegetation growth (Crow, Dey, & Riemenschneider, 2006).

The productivity of forest soils was tested in British Columbia. The reason that led them to this research is due to the fact that anthropogenic impacts have not been acknowledged, much in relation to the productivity of forest soils. We are aware that timber harvesting and land preparation can deteriorate forest soils due to adverse impacts of disturbance. Soil compaction will be discussed further in this section on how nutrient removal can diminish soil fertility. Unfortunately, there are only a few experiments that support this argument. Most of the examples show that the survival rate is low, resulting in losses of soil productivity over time (Powers, Alban, Miller, & Tiarks, 1990). Real data were used to validate the effects of forest activities in relation to forest soil productivity. International collaboration was conducted to address the large-scale issue of the U.S. Forest Service back in the 1980s involving potential treatments that can be used to ameliorate the disturbance of forest soil productivity. Studies concluded that porous soil and the removal of organic matter can have a long-term negative effect on site productivity. British Columbia was selected for this procedure because the area has a very wide climatic variation experiencing a long duration of winter and a short duration of the summer. The occurrence of wild fire is also another factor that shapes the landscapes. This area is exposed to several intensive forestry and mechanical procedures. The common climax tree species in the area are subalpine fir and white spruce. Lodgepole pines are popular in the southern region as a maturing climax species and others, such as aspen, Douglas fir and paper birch. All selected sites have medium-textured glacial till soil surface and represent the zonal climatic condition. Soil types range from loam to silt loam with a mixture of 40% or less gravel. Treatments done involve testing the organic matter OM1 – only stems of trees were eradicated; OM2 – whole tree harvesting; and OM3 – logging slash were utilized along with entire removal of forest floor. The result shows that compaction treatment on OM1 shows no signs of compaction; compaction treatment on OM2 has about 2 to 3 centimeters of mineral depression, while compaction treatment on OM3 shows heavy soil compaction and coarsely causes heavy traffic due to heavy forestry equipment about 5 centimeters depth compaction. The result of OM3 greatly reduced the sprouting of aspen species as observed near the Dawson Creek. Sedges rapidly occupy the compacted site with exposed mineral soils, even faunal species shows decline in numbers. Serious removal of surface soil and leaving bare mineral soils will prohibit conifers from growth in the area for many years (Sanborn, 1996).

4.3.1. Potential Impacts and Protection Measures on Physical Properties of Forest Soils

Soil physical properties are significant factors that determine the composition of tree species and growth rate. These properties are affected by the depth of the rooting system, water available for plant use, the amount of oxygen and soil gasses and the extent of water movement within the soil profile. Different impacts can happen during the harvesting of timber and forest management activities involve soil compaction, displacement, puddling, and rutting. Most of the disturbances are caused by heavy equipment during the operations. Vehicles like big tractors compress the soil resulting in soil compaction.

4.3.1.1. Soil Compaction, Displacement, and Soil Erosion

Heavy equipment can push the topsoil downward into deeper soil layers on rolling to steep slopes. During compaction, soil density increases, while soil volume decreases, which eliminates macropores in forest soils. Pore volume and spaces are important factors in soil water and air. Macropores help to regulate the rate of water and gas movement. The initial trips of tractors increase soil density and the vibrations of the machine cause compaction (Sakai, Nordfjell, Suadicani, Talbot, & Bollehuus, 2008).

Compacted soil can decrease the rate of tree growth because soil aeration is poor and less oxygen is available for respiration. The buildup of toxic gases and carbon dioxide will increase in the soil, which can result in root injury. Microorganisms will also not survive in soil lacking oxygen. Compaction can greatly affect the root system by intensifying the soil struggle to root infringement. Decreasing pore spaces also means low infiltration capacity and low availability of water for plant use. Rainfall will more or less result in surface runoff, overland flow, and massive soil erosion and worsened sedimentation. Soil recovery from compaction takes long-term or even decades, depending on the severity of the damage. In cold regions, freezing-thawing processes can loosen the soil properties rapidly, which can decrease the density of compacted soil gradually (Wortman & Jasa, 2009).

Figure 4.6. *Forest roads compacted.*

Source: http://labmodules.soilweb.ca/compaction/.

Grazing animals can also cause soil compaction when allowed to trample soils in forest lands. Grazing pressure can result in heavy soils during the wet season, especially if livestock continuously graze the forest soils. The damage begins when the layer of the forest floor is being mixed and trampled due to heavy livestock. The compacted soil slows down the infiltration rate, increase surface runoff and expect the incidence of soil erosion (Sharrow, n.d.). The decline in forest health is caused by reducing forest litter and organic matter affecting the fertility of the soil. Tree roots can be directly damaged creating wounds for the insects to attack the trees. Even the seedlings and saplings are removed, which can be a good start for the forest to regenerate. Cattle can also distress vegetation by compacting the herbaceous layer which helps the soil to maintain its moisture. Some non-native plants can be carried by cattle and invade the native vegetation in the forest areas. Livestock must be allowed to graze the woodland that provides quality trees and needed plant species. Mechanical equipment is frequently involved in soil displacement. Such displacement can also remove soil organic matter and nutrients essential for plant growth. Site techniques must move the extracted soil surface away from areas with seedlings or sources of regeneration. Prescribed proper fire management to reduce slash prior to planting. Fire if controlled is a good tool for conditioning soil flooring.

The surface layer is an important part of forest productivity because this layer is composed of organic matter that holds water and nutrients and supports the biological community. Surface horizon works as a cushion to filter the organic and inorganic substrates and also resist extreme temperature. Organic matter influences soil aeration, provides an area for seedling germination and starts of rooting system. Conserving soil organic matter also means maintaining the productivity of the site. Removal of soil organic matter can be damaging to the forest in general. Sandy soils in most cases, are easily compacted by heavy equipment, removing the surface and damaging vegetation by abandoning it and making it vulnerable. In hilly areas, sandy soil can collapse because of the pull of gravity and is easily eroded by water or wind. Displacement of soil surface can prohibit revegetation and decrease soil productivity.

Soil erosion is a major result of unsustainable forest management practices. Erosion is very minimal in areas that are fully covered with vegetation. Vegetative cover protects the soil from eroding. Unfortunately, the clear-cutting method can remove the entire forest cover which is often done in steep terrains resulting in accelerated soil erosion. Some type of soils is highly erodible when exposed. These open areas can also pave opportunities for invasive species to occupy the area, making the soil more erodible and does not hold soil similar to what native vegetation can do (Poltorak, Labelle, & Jaeger, 2018).

4.3.1.2. Rutting and Mud Puddle

Rutting occurs when heavy load equipment is used on soil with low strength, creating a deep slump under wet conditions. Rutting can directly damage the tree root system, causing wounds and cuts, compacting the soil, shifts soil layer resulting in poor aeration and a low rate of infiltration. Rutting distresses the hydrological processes, creating water flows and diverting water movement away from the desired site causing soil erosion and sediment loading. Soil rutting occurs along with soil compaction and mud puddles (Figure 4.8). This activity can damage soil structure due to churning and squeezing of wet soils using heavy equipment. Puddling is often observed in furrows or channels holding water (Figure 4.7). Soil particles gradually dissolve in water and settle once dried. As a result, small particles are crusting of the soil surface. Puddled soils could also prohibit revegetation like soil compaction (Michigan Government, 2007).

Soils that are vulnerable to soil compaction and rutting are fine-textured and medium-textured soils. These soils usually have poor drainage. Soils with shallow parent material are also prone to compaction because the soil and rock surface are easily trampled by heavy equipment. Use of heavy equipment must be limited during the dry season to minimize damage to physical soil properties. Soils in cold regions can solidify when frozen which makes it relatively unaffected by compaction, so operation during the winter is an alternative for mistier areas.

Figure 4.7. *Puddling.*

Sources: https://encrypted-tbn0.gstatic.com/images?q=tbn:ANd9GcT3DRujnQICOPDQiAzeN8F6fmLfp96nC93bTg&s.

Figure 4.8. *Mud puddle.*

Sources: http://thesethingshappentootherpeople.blogspot.com.au/2014/05/sunshine-mud-and-euphoria-on-trails.html.

However, wetlands in Southern Wisconsin are difficult to freeze for an adequate period to complete harvesting. Organic soils usually have a low bulk density, which makes it prone to rutting and compaction. Unfrozen wetlands run by vehicles can result in serious damage. Water usually flows horizontally in organic soils of wetland which transpires approximately 12 inches from the top. Rutting can also divert water flow and prevent wetlands to receive enough water supply to support regeneration. The proper timing of forest management activities is important to preserve the forest soils. Forest managers must limit the use of heavy equipment and control harvest activities in small patches if possible. Operating season must also be studied because climatic conditions may vary from one place to another. Use low-pressure ground vehicles and operate during desirable times of the year. Proper construction of road systems would also minimize soil compaction on forest soils and lessen the impact on soil productivity (NSW Government, n.d.; Queensland Government, 2013).

4.3.2. Potential Impacts and Protection Measures on Chemical Properties of Forest Soils

Soil chemical properties often include soil pH level and nutrients in soil. Management of forest soil can affect nutrient state by nutrient removal due to vegetation extraction and disturbance of soil surface due to harvesting activities and preparation of the site.

- **Nutrient Cycling:** This is the process wherein nutrients travel into, beyond and inside the environment. Forest areas have its own source of natural inputs that produce nutrients available for plant use. During the course of the forest life cycle, contributions are very substantial. Soils can generate and build up nutrients through the process of weathering and disposition (Johnson & Lindberg, 1993). Nutrients are lost during timber harvesting, soil displacement, leaching and volatilization. If the outputs of nutrients are greater than the inputs, there is a greater chance that nutrition depletion will take place. Nutrient depletion happens with short cycles, clear-cutting, nutrient-demanding species and sites that have naturally low nutrient stocks. If nutrient losses are minimal and the land is rich in nutrient stocks, then harvest removal will not cause nutrient limitations for at least hundreds of years. However, areas with sandy soils which can hold very little nutrient capital can suffer several rotations once an area is disturbed. Shallow soil cannot hold much water because of the small root system. The root system can affect nutrient uptake and moisture supplies. Soil scientists presented that deeply rooted trees with 95% fine roots and transpired through 40 inches down the soil indicates availability of almost all soil nutrients (Dobson, 1995). Forest growth is correlated with root depth and depth of parent materials. Commonly, a root system that emerges 10 inches and below in depth have the highest chance of failure to grow trees. Wetlands have unique mixtures of soil nutrients. Most of the soils are saturated near the surface and mineral weathering can slow anaerobic conditions in underground water. Many types of these soils only get nutrients coming from runoff. The low level of nitrogen and phosphorus replacement in peatlands is due to intensive harvesting activities. People are still uncertain about the exact amount of nutrient losses during harvesting activities. But it is very well known that trees accumulate more nutrients as compared to other elements and removing these trees also indicates nutrient removal. Some of the nutrient demanding species are Oaks, Aspens and northern hardwood species. Harvesting wood for biofuels can remove soil nutrients, leading to nutrient depletion (Potter, Ragsdale, & Berish, 1987).

Similar with the experiment conducted in British Columbia whole tree harvest which involves removal of biofuels will still result in depletion of soil nutrients. Nutrients are available in different quantities. Most of these nutrients are found in trees. Species of trees like Aspen as nutrient demanding plant species that store Calcium and Magnesium on their bark in greater amounts as compared with conifers. Seasonal alteration of nutrients are important factor used to determine the perfect time for timber to be harvested. Seasonal harvesting can be restricted to protect wildlife species during their nesting period. The factors affected by the removal of trees would include material removed, tree attributes and season of harvest. If forestry equipment will be used prior to the season of nutrient regeneration then stand would be able to withstand and recover back. Wisconsin is one of the areas suffering from low nutrient content in the soil. Their only option to manage the disturbed areas is to condition the soil using fire management. To retain soil nutrient, slash is redistributed on the area, avoiding whole tree harvesting, particularly on nutrient-demanding species, add nutrients in the form of fertilization and avoid short cycle duration for timber harvesting (Wisconsin Department of Natural Resources, 2018).

4.3.3. Potential Impacts and Protection Measures on Biological Properties of Forest Soils

Biological soil components perform several activities in the soil to keep and conserve its biogeochemical processes. Forest soils are the habitat of thousands of soil organisms. These organisms take part in processes such as decomposition, recycling, and nutrient availability which support tree growth and enhance metabolism. The survival of soil organisms is dependent on several soil attributes which include temperature, aeration, pH level, organic matter, and supply of nutrients. Mycorrhiza is an example of soil fungi that supports nutrient-poor sites by growing on the hairs of rooted trees to help the tree absorb nutrients. Some of the trees that have symbiotic relationships with mycorrhiza are black cherry, black walnut, pines, oaks, firs, spruces, maples, birches, ashes willows and beeches. The removal of forest floor can increase the soil temperature, which is not desirable for the survival of mycorrhizal fungi.

Water infiltration is significant to support the density of vegetation cover. The forest cover intercepts and collects moisture from rainfall and atmosphere to prevent potential runoff. Damaged forest areas are prone to massive soil erosion. Displaced surface soil, overly grazed forest floors, and compacted soils can decrease the soil's capacity to infiltrate water. Some of the non-native or invasive species introduced or grown in bare mineral soils can also reduce infiltration of soil moisture. They can block the sunlight needed by the native plant's understory.

Some non-native plant species can also alter the chemical composition of the soils affecting the biological activities including mycorrhizal fungi. The species of tree called Buckthorn found in the southern Wisconsin can alter the soil C: N ratio to favors the growth of other invasive species. This species can also reduce water infiltration and water holding capacity, and lessen the cycling of nutrients and organic matter in the soil. Garlic mustard is another species that can release chemicals into the soil and interfere with tree regeneration (Crow, Dey, & Riemenschneider, 2006).

Figure 4.9. *Fire used to control invasive species.*

Source: https://images.nationalgeographic.org/image/upload/t_edhub_resource_key_image/v1638892040/EducationHub/photos/controlled-burning.jpg.

Forest management activities can highly affect the physical, chemical, and biological properties of the soil. Changes in soil chemistry must be given attention because it can alter soil biological processes. Many organisms living in the soil are beneficial to plant growth. The physical and chemical characteristics of the soil also help and protect these biological organisms. Similar to British Columbia, the forest lands in Wisconsin have different variety of soil types. Topography varies across landscapes and unsustainable past management affects the site condition at present. Alteration in soil composition can also change the tree species growing. Site conditions are not the same in all areas, forest managers must evaluate carefully the soil and terrain landscapes for every site. For invasive species, their presence can deteriorate forest soil productivity. Before introducing new species in the area, be sure to check its limitations and consider assessing the response on how to handle the invasive species once it proliferates to avoid soil disturbance (Figure 4.9).

4.4. ENHANCEMENT OF FOREST SOILS PRODUCTIVITY

Forest management activities are proven to cause negative impacts on the productivity of forest soils. Opportunities to address the problems are possible to enhance the soil condition. This simple guideline will give ideas on how to enhance the forest soils. Sustainability meets the needs of the present population without impeding the needs of the future population. Sustainable management practices are necessary to sustain our forest resources. To enhance soil productivity here are the examples (Crow, Dey, & Riemenschneider, 2006).

4.4.1. Fertilization

Application of fertilizers is very common in agricultural areas to improve crop yields. In forestry, this is also applicable to increase the opportunities of the vegetation to increase its fiber outputs. Minnesota is one of the areas that practices fertilization in their forest lands. They find fertilization as a biological and economic means to enhance forest soil productivity in the area. Forest areas near the road markets are more economical and practical due to reduced transportation costs. Fertilization not only increases the yield but also improves the timber volume, adding to merchantable size.

Analyzing nutrient content must be done prior to fertilization. Application of unwanted fertilizer can result in the emergence of pests if not applied properly.

4.4.2. Genetic Improvement

For the past 20 years, Minnesota also uses genetic materials to improve their timber yield and keep soil productivity. This method helps them control the occurrence of field diseases and improves seedling growth for commercial distribution. This also helps to preserve some of the species with genetic variability like Norway red pine and Jack Pine.

4.4.3. Shortened Rotation

Nutrient depletion happens when forest landscapes are constantly exposed to mechanical forest activities. Forest activities must not be done on a regular basis to allow soil to regenerate and recover from the impacts of timber harvesting. Rotation age or cycle must be longer to stabilize the flow of energy and nutrients and reduce nutrient depletion.

4.4.4. Prevent Flooding

Maintaining proper soil drainage would also minimize the incidence of flooding. Forest lands have its own natural ability to hold water. Site preparation that involves bedding and molding in tree plantations can be done to support tree growth and maintain soil-water in the landscape. This process can also improve aeration and protect the root zone.

4.4.5. Control Competition

Controlling vegetation competition can also enhance soil productivity. Many of the desirable species would be able to survive and increase growth yield. Herbicides can be used to control species composition but must be done on a rotation basis to achieve the most harvest yield.

4.4.6. Timber Stand Improvement

Improving the stand by thinning can be useful to maintain the productive capacity of the forest area. Thinning can be done to generate a few stems and provide more sources for fibers. Thinning can also help produce large trees at early tree cycles and also enhance volume growth to increase the tree value which can provide better income for landowners.

4.4.7. Species and Site Selection

The condition must match the species planted in the area. For instance, aspen is an example of nutrient nutrient-demanding plant species. That only means that it needs more nutrients to survive over time. If this type of species is planted in low-nutrient capital sites, it will cause nutrient depletion and low quality produced of timber. To increase site productivity, species selection is important (Millennium Ecosystem Assessment, 2005)

CASE STUDY

SOIL MICROORGANISMS IN URBAN FORESTS OF PORTLAND, OREGON

Portland, Oregon, known for its vibrant urban forest ecosystems, provides a unique context to study soil microorganisms and their roles in sustaining forest health. Urban forests in this city, which encompass parks, green spaces, and tree-lined streets, are vital for enhancing biodiversity, improving air quality, and mitigating urban heat. Recent studies have focused on the microbial communities in the soil beneath these trees, revealing a rich diversity of bacteria, fungi, and archaea that contribute significantly to nutrient cycling, organic matter decomposition, and overall ecosystem functioning. This case study examines the microbial dynamics within Portland's urban forests, emphasizing their contribution to soil health and plant growth.

Findings:

Research conducted in various urban forest sites across Portland has shown that soil microorganisms exhibit remarkable adaptability to urban environments. The presence of diverse microbial communities is crucial for maintaining soil fertility, as these microorganisms break down organic matter, releasing essential nutrients for plant uptake. Notably, mycorrhizal fungi form symbiotic relationships with tree roots, enhancing nutrient absorption while improving the resilience of trees against drought and disease. Furthermore, studies have indicated that urban soil microorganisms can vary significantly based on factors such as soil compaction, pollution levels, and land use, highlighting the importance of managing these ecosystems to foster healthy microbial populations.

Implications

The health of soil microorganisms is intrinsically linked to the resilience of urban forests. Understanding the complex interactions between these microorganisms and their environment is essential for urban forest management and conservation strategies. By preserving the diversity and functionality of soil microbial communities, cities like Portland can enhance the ecological services provided by their urban forests, including carbon sequestration, stormwater management, and habitat for wildlife. This case study underscores the need for integrated urban forestry practices that prioritize soil health and microbial diversity to ensure the sustainability of these vital green spaces.

Discussion Questions

- How can urban forest management practices be adjusted to enhance the diversity of soil microorganisms?
- What role do soil microorganisms play in mitigating the impacts of urbanization on forest health?

SUMMARY

Forest soil dynamics encompass the intricate processes that govern the interactions between soil components, nutrients, microorganisms, and vegetation within forest ecosystems. Chapter 4 explores the multifaceted nature of forest soils, highlighting their critical role in sustaining ecosystem health and productivity. Key components of forest soils include mineral particles, organic matter, water, gases, and a diverse array of microorganisms. These components interact to influence soil structure, nutrient availability, water retention, and biological activity. Nutrient cycling, facilitated by decomposition processes, plays a pivotal role in maintaining soil fertility and supporting plant growth.

The productivity of forest soils is influenced by various factors, including physical properties (e.g., soil structure), chemical properties (e.g., nutrient content), and biological properties (e.g., microbial activity). Human activities such as deforestation, pollution, and improper land management can adversely impact these properties, leading to soil degradation and reduced productivity.

Strategies for enhancing forest soil productivity include fertilization, genetic improvement of tree species, and sustainable management practices like shortened rotation periods and species-specific site selection. Protecting forest soils from degradation is crucial for ensuring the long-term sustainability of forest ecosystems and their capacity to provide essential ecosystem services.

REVIEW QUESTIONS

1. What are the main components of forest soils?
2. How does decomposition contribute to nutrient cycling in forest soils?
3. What are some potential impacts on the physical properties of forest soils, and how can they be protected?
4. Discuss the methods for enhancing forest soil productivity.
5. Why are species and site selection important for improving forest soils?

REFERENCES

1. Antoniadis, V., Robinson, J., & Alloway, B. (2008). Effects of short-term pH fluctuations on cadmium, nickel, lead, and zinc availability to ryegrass in a sewage sludge-amended field. *Chemosphere, 71*, 759–764. https://doi.org/10.1016/j.chemosphere.2008.01.025.
2. Athira, M., Jagadeeswaran, R., & Kumaraperumal, R. (2019). Influence of soil organic matter on bulk density in Coimbatore soils. *International Journal of Chemical Studies, 7*, 3520–3523.
3. Bae, J., & Ryu, Y. (2015). Land use and land cover changes explain spatial and temporal variations of the soil organic carbon stocks in a constructed urban park. *Landscape and Urban Planning, 136*, 57–67. https://doi.org/10.1016/j.landurbplan.2014.12.004.

4. Baldocchi, D., Tang, J., & Xu, L. (2006). How switches and lags in biophysical regulators affect spatial-temporal variation of soil respiration in an oak-grass savanna. *Journal of Geophysical Research: Biogeosciences, 111*(G2). https://doi.org/10.1029/2005JG000063.

5. Barroso-Tagua, R., Alvarez, D., Huera, T., Changoluisa, D., & Bravo, C. (2017). La fertilidad del suelo como un servicio ecosistémico en cultivo de cacao (*Theobroma cacao* L.), en la provincia de Napo. In R. Alemán, H. Reyes, & C. Bravo (Eds.), *Libro de Memorias: Simposio Internacional Sobre Manejo Sostenible de Tierras y Seguridad Alimentaria* (pp. 99–106). Universidad Estatal Amazónica.

6. Bautista-Cruz, A., del Castillo, R. F., Etchevers-Barra, J. D., Gutiérrez-Castorena, M. d. C., & Baez, A. (2012). Selection and interpretation of soil quality indicators for forest recovery after clearing of a tropical montane cloud forest in Mexico. *Forest Ecology and Management, 277*, 74–80. https://doi.org/10.1016/j.foreco.2012.03.003.

7. Bautista, R. J. D., Baeza, A. T., Acosta, S. D. C. R., Morales, P. S., Alcántara, A. G., Rivera, A. A., & Hernández, R. S. (2019). Almacenamiento de carbono y agua en un área periurbana de Tabasco. *Revista Terra Latinoamericana, 37*(2), 197–209.

8. Bech, J., Roca, N., Tume, P., Ramos-Miras, J., Gil, C., & Boluda, R. (2016). Screening for new accumulator plants in potential hazard elements polluted soil surrounding Peruvian mine tailings. *CATENA, 136*, 66–73. https://doi.org/10.1016/j.catena.2015.08.012.

9. Benipal, G., Harris, A., Srirajayatsayai, C., Tate, A., Topalidis, V., Eswani, Z., Qureshi, M., Hardaway, C. J., Galiotos, J., & Douvris, C. (2017). Examination of Al, As, Cd, Cr, Cu, Fe, Mg, Mn, Ni, Pb, Sb, Se, V, and Zn in sediments collected around the downtown Houston, Texas area, using inductively coupled plasma-optical emission spectroscopy. *Microchemical Journal, 130*, 255–262. https://doi.org/10.1016/j.microc.2016.08.020.

10. Bikindou, F. D. A., Gomat, H. Y., Deleporte, P., Bouillet, J. P., Moukini, R., Mbedi, Y., Ngouaka, E., Brunet, D., Sita, S., Diazenza, J. B., et al. (2012). Are NIR spectra useful for predicting site indices in sandy soils under Eucalyptus stands in Republic of Congo? *Forest Ecology and Management, 266*, 126–137. https://doi.org/10.1016/j.foreco.2011.10.006.

11. Blum, W. E. (2005). Functions of soil for society and the environment. *Reviews of Environmental Science and Bio/Technology, 4*(1), 75–79. https://doi.org/10.1007/s11157-005-2256-0.

12. Burbano Orjuela, H. (2018). El carbono orgánico del suelo y su papel frente al cambio climático. *Revista de Ciencias Agrícolas, 35*, 82–96.

13. Cable, J. M., Ogle, K., Lucas, R. W., Huxman, T. E., Loik, M. E., Smith, S. D., Tissue, D. T., Ewers, B. E., Pendall, E., Welker, J. M., et al. (2010). The temperature responses of soil respiration in deserts: A seven-desert synthesis. *Biogeochemistry, 103*(1–3), 71–90. https://doi.org/10.1007/s10533-010-9448-z.

14. Calzolari, C., Tarocco, P., Lombardo, N., Marchi, N., & Ungaro, F. (2020). Assessing soil ecosystem services in urban and peri-urban areas: From urban soils survey to providing support tool for urban planning. *Land Use Policy, 99*, 105037. https://doi.org/10.1016/j.landusepol.2020.105037.

15. Canadian Environmental Protection Agency (CEPA). (2007). *Canadian Soil Quality Guidelines for the Protection of Environmental and Human Health*. National Guidelines and Standards Office.

16. Canedoli, C., Ferrè, C., Abu El Khair, D., Padoa-Schioppa, E., & Comolli, R. (2019). Soil organic carbon stock in different urban land uses: High stock evidence in urban parks. *Urban Ecosystems, 23*, 159–171. https://doi.org/10.1007/s11252-019-0855-3.

17. Chávez-Aguilar, G., Burrola-Aguilar, C., González-Morales, A., & Pérez-Suárez, M. (2020). Almacén de carbono orgánico del suelo y abundancia ectomicorrízica bajo dos especies de coníferas en el Nevado de Toluca, México. *Agro Productividad, 13*. https://doi.org/10.32854/agroprod.v13n1a3.

18. Chen, M., Boyle, E. A., Switzer, A. D., & Gouramanis, C. (2016). A century long sedimentary record of anthropogenic lead (Pb), Pb isotopes and other trace metals in Singapore. *Environmental Pollution, 213*, 446–459. https://doi.org/10.1016/j.envpol.2016.02.025.

19. Cram, S., Cotler, H., Morales, L. M., Sommer, I., & Carmona, E. (2008). Identification de los servicios ambientales potenciales de los suelos en el paisaje urbano del Distrito Federal. *Investigaciones Geográficas, 66*, 81–104.

20. Davidson, E., Savage, K., Verchot, L., & Navarro, R. (2002). Minimizing artifacts and biases in chamber-based measurements of soil respiration. *Agricultural and Forest Meteorology, 113*, 21–37. https://doi.org/10.1016/S0168-1923(02)00100-0.

21. Denegri, G., Rodríguez Vagaria, A., Mijailoff, J., Mársico, J., & Acciaresi, G. (2018). Bosques urbanos: Su aporte al turismo en la costa atlántica norte de Argentina. *Estudios y Perspectivas en Turismo, 27*(2), 316–335.

22. Díaz-Limón, M. P., Cano-Santana, Z., & Queijeiro-Bolaños, M. E. (2016). Mistletoe infection in an urban forest in Mexico City. *Urban Forestry & Urban Greening, 17*, 126–134. https://doi.org/10.1016/j.ufug.2016.03.002.

23. Dovletyarova, E. A., Mosina, L. V., Vasenev, V. I., Ananyeva, N. D., Patlseva, A., & Ivashchenko, K. V. (2017). Monitoring and assessing anthropogenic influence on soil's health in urban forests: The case from Moscow City. In *Adaptive Soil Management: From Theory to Practices* (pp. 531–557). Springer. https://doi.org/10.1007/978-3-319-46324-0_31.

24. Edmondson, J. L., Davies, Z. G., McCormack, S. A., Gaston, K. J., & Leake, J. R. (2011). Are soils in urban ecosystems compacted? A citywide analysis. *Biology Letters, 7*, 771–774. https://doi.org/10.1098/rsbl.2011.0174.

25. Estrada-Herrera, I. R., Hidalgo-Moreno, C., Guzmán-Plazola, R., Almaraz Suárez, J. J., Navarro-Garza, H., & Etchevers-Barra, J. D. (2017). Soil quality indicators to evaluate soil fertility. *Agrociencia, 51*, 813–831.

26. Etchevers, B., & Jorge, D. (2005). *Manual Para La Determinación De Carbono En La Parte Aérea Y Subterránea De Sistemas De Producción En Laderas* (No. CP FE, FOLLETO 678). Colegio de Postgraduados.

27. Fang, C., & Moncrieff, J. (2001). The dependence of soil CO_2 efflux on temperature. *Soil Biology and Biochemistry, 33*(2), 155–165. https://doi.org/10.1016/S0038-0717(00)00125-5.

28. Fenn, M. E., Castro-Servín, J. M., Hernández-Tejeda, T., Krage, N., Goodson, C., & Meixner, T. (2002). Heavy metals in forest soils, vegetation, and drainage waters in the Basin of Mexico. In *Urban Air Pollution and Forests: Resources at Risk in the Mexico City Air Basin* (pp. 194–221). Springer.

29. Galicia, L., Gamboa Cáceres, A. M., Cram, S., Chávez Vergara, B., Peña Ramírez, V., Saynes, V., & Siebe, C. (2016). Almacén y dinámica del carbono orgánico del suelo en bosques templados de México. *Terra Latinoamericana, 34*(1), 1–29.

30. Gee, G. W., & Bauder, J. W. (1986). Particle size analysis. In *Methods of Soil Analysis, Part 1* (pp. 383–411). ASA and SSSA.

31. Grand, S., Rubin, A., Verrecchia, E. P., & Vittoz, P. (2016). Variation in soil respiration across soil and vegetation types in an alpine valley. *PLoS ONE, 11*(10), e0163968. https://doi.org/10.1371/journal.pone.0163968.

32. Hansson, K., Heijden, G., Laclau, J. P., Augusto, L., & Ranger, J. (2014). Chemical fertility of forest soils: Basic concepts. *Revue Forestière Française, 66*, 413–424. https://doi.org/10.4267/2042/54820.

33. Hansson, K., Laclau, J. P., Saint-André, L., Mareschal, L., van der Heijden, G., Nys, C., Nicolas, M., Ranger, J., & Legout, A. (2020). Chemical fertility of forest ecosystems. Part 1: Common soil chemical analyses were poor predictors of stand productivity across a wide range of acidic forest soils. *Forest Ecology and Management, 461*, 117843. https://doi.org/10.1016/j.foreco.2020.117843.

34. He, S., He, Z., Yang, X., Stoffella, P. J., & Baligar, V. C. (2015). Soil biogeochemistry, plant physiology, and phytoremediation of cadmium-contaminated soils. *Advances in Agronomy, 134*, 135–225. https://doi.org/10.1016/bs.agron.2014.11.005.

35. Hernández-Guillén, A., Rojas-García, F., & Benavides-Meza, H. (2013). Estimación del contenido y captura de carbono en la segunda sección del Bosque de Chapultepec, Distrito Federal. In F. P. Pellat, J. W. González, M. Bazán, & V. Saynes (Eds.), *Estado Actual Del Conocimiento Del Ciclo Del Carbono y Sus Interacciones En México: Síntesis a 2013*. Instituto Tecnológico y de Estudios Superiores de Monterrey.

36. Huang, N., Song, X. P., Black, T. A., Jassal, R. S., Myneni, R. B., Wu, C., Wang, L., Song, W., Ji, D., Yu, S., et al. (2020). Spatial and temporal variations in global soil respiration and their relationships with climate and land cover. *Science Advances, 6*, eabb8508. https://doi.org/10.1126/sciadv.abb8508.

37. Huaraca-Fernandez, J. N., Pérez-Sosa, L., Bustinza-Cabala, L. S., & Pampa-Quispe, N. B. (2020). Organic amendments in the immobilization of cadmium in contaminated agricultural soils: A review. *Información Tecnológica, 31*, 139–152. https://doi.org/10.4067/S0718-07642020000100014.

38. Hyun, J., Kim, Y. J., & Yoo, G. (2023). A method for soil quality assessment in the metropolitan greenery: A comprehensive view of ecosystem services and soil functions. *MethodsX, 10*, 102102. https://doi.org/10.1016/j.mex.2023.102102.

39. Hyun, J., Kim, Y. J., Kim, A., Plante, A. F., & Yoo, G. (2022). Ecosystem services-based soil quality index tailored to the metropolitan environment for soil assessment and management. *Science of The Total Environment, 820*, 153301. https://doi.org/10.1016/j.scitotenv.2022.153301.

40. Juhos, K., Madarász, B., Kotroczó, Z., Béni, Á., Makádi, M., & Fekete, I. (2021). Carbon sequestration of forest soils is reflected by changes in physicochemical soil indicators—A comprehensive discussion of a long-term experiment on a detritus manipulation. *Geoderma, 385*, 114918. https://doi.org/10.1016/j.geoderma.2020.114918.

41. Kandel, T. P., Lærke, P. E., & Elsgaard, L. (2016). Effect of chamber enclosure time on soil respiration flux: A comparison of linear and non-linear flux calculation methods. *Atmospheric Environment, 141*, 245–254. https://doi.org/10.1016/j.atmosenv.2016.06.030.

42. Kaninga, B. K., Chishala, B. H., Maseka, K. K., Sakala, G. M., Lark, M. R., Tye, A., & Watts, M. J. (2019). Review: Mine tailings in an African tropical environment—Mechanisms for the bioavailability of heavy metals in soils. *Environmental Geochemistry and Health, 42*, 1069–1094. https://doi.org/10.1007/s10653-019-00368-6.

43. Kerven, G. L., Menzies, N. W., & Geyer, M. D. (2000). Soil carbon determination by high temperature combustion—A comparison with dichromate oxidation procedures and the influence of charcoal and carbonate carbon on the measured value. *Communications in Soil Science and Plant Analysis, 31*, 1935–1939. https://doi.org/10.1080/00103620009370379.

44. Khaledian, Y., Brevik, E. C., Pereira, P., Cerdà, A., Fattah, M. A., & Tazikeh, H. (2017). Modeling soil cation exchange capacity in multiple countries. *CATENA, 158*, 194–200. https://doi.org/10.1016/j.catena.2017.07.015.

45. Legout, A., Hansson, K., van der Heijden, G., Laclau, J. P., Mareschal, L., Nys, C., Nicolas, M., Saint-André, L., & Ranger, J. (2020). Chemical fertility of forest ecosystems. Part 2: Towards redefining the concept by untangling the role of the different components of biogeochemical cycling. *Forest Ecology and Management, 461*, 117844. https://doi.org/10.1016/j.foreco.2020.117844.

46. Leon, E., Vargas, R., Bullock, S., Lopez, E., Panosso, A. R., & La Scala, N. (2014). Hot spots, hot moments, and spatio-temporal controls on soil CO_2 efflux in a water-limited ecosystem. *Soil Biology and Biochemistry, 77*, 12–21. https://doi.org/10.1016/j.soilbio.2014.05.029.

47. Liao, K., Xu, S., & Zhu, Q. (2015). Development of ensemble pedotransfer functions for cation exchange capacity of soils of Qingdao in China. *Soil Use and Management, 31*, 483–490. https://doi.org/10.1111/sum.12191.

48. Liao, K., Xu, S., Wu, J., Zhu, Q., & An, L. (2014). Using support vector machines to predict cation exchange capacity of different soil horizons in Qingdao City, China. *Journal of Plant Nutrition and Soil Science, 177*, 775–782. https://doi.org/10.1002/jpln.201300294.

49. López-López, S. F., Martínez-Trinidad, T., Benavides-Meza, H. M., García-Nieto, M., & Ángeles-Pérez, G. (2018). Reservorios de biomasa y carbono en el arbolado de la primera sección del Bosque de Chapultepec, Ciudad de México. *Madera y Bosques, 24*, 1–14.

50. Mañon de la Cruz, R., Orozco Hernández, M. E., & Mireles Lezama, P. (2018). Evaluación de los servicios ambientales del Parque Metropolitano Bicentenario, Toluca, México. *Revista Iberoamericana de Ciencias, 4*, 6–21.

51. Millennium Ecosystem Assessment. (2005). *Ecosystems and Human Well-Being* (Vol. 5). Island Press.

52. Mishra, G., Sulieman, M. M., Kaya, F., Francaviglia, R., Keshavarzi, A., Bakhshandeh, E., Loum, M., Jangir, A., Ahmed, I., & Elmobarak, A. (2022). Machine learning for cation exchange capacity prediction in different land uses. *CATENA, 216*, 106404. https://doi.org/10.1016/j.catena.2022.106404.

53. Morel, J. L., Chenu, C., & Lorenz, K. (2014). Ecosystem services provided by soils of urban, industrial, traffic, mining, and military areas (SUITMAs). *Journal of Soils and Sediments, 15*(8), 1659–1666. https://doi.org/10.1007/s11368-015-1092-5.

54. O'Riordan, R., Davies, J., Stevens, C., Quinton, J. N., & Boyko, C. (2021). The ecosystem services of urban soils: A review. *Geoderma, 395*, 115076. https://doi.org/10.1016/j.geoderma.2021.115076.

55. Pouyat, R. V., Page-Dumroese, D. S., Patel-Weynand, T., & Geiser, L. H. (2020). *Forest and Rangeland Soils of the United States Under Changing Conditions: A Comprehensive Science Synthesis*. Springer Nature. https://doi.org/10.1007/978-3-030-39359-7.

56. Pouyat, R. V., Yesilonis, I., & Golubiewski, N. E. (2009). A comparison of soil organic carbon stocks between residential turf grass and native soil. *Urban Ecosystems, 12*(1), 45–62. https://doi.org/10.1007/s11252-008-0059-x.

57. Raciti, S. M., Hutyra, L. R., & Finzi, A. C. (2012). Depleted soil carbon and nitrogen pools beneath impervious surfaces. *Environmental Pollution, 164*, 248–251. https://doi.org/10.1016/j.envpol.2012.01.047.

58. Romero, A., & García, F. (2013). Estimación del contenido de carbono en la zona ecológica y cultura Bosque de Tlalpan, Distrito Federal. In F. Paz Pellat, J. W. Wong González, M. Bazan, & V. Saynes (Eds.), *Estado Actual Del Conocimiento Del Ciclo Del Carbono y Sus Interacciones En México: Síntesis a 2013* (pp. 149–154). Programa Mexicano del Carbono Colegio de Postgraduados Universidad Autónoma de Chapingo, Instituto Tecnológico y de Estudios Superiores de Monterrey.

59. Saavedra-Romero, L. D., Alvarado-Rosales, D., Martínez-Trinidad, T., & Hernández-de la Rosa, P. (2020). Propiedades físicas y químicas del suelo urbano del Bosque San Juan de Aragón, Ciudad de México. *Terra Latinoamericana, 38*, 529–540. https://doi.org/10.28940/terra.v38n3a4.

60. Sarstedt, M., & Mooi, E. (2019). Regression analysis. In *A Concise Guide to Market Research: The Process, Data, and Methods Using IBM SPSS Statistics* (pp. 209–256). Springer.

61. Scharenbroch, B. C., Lloyd, J. E., & Johnson-Maynard, J. L. (2005). Distinguishing urban soils with physical, chemical, and biological properties. *Pedobiologia, 49*, 283–296. https://doi.org/10.1016/j.pedobi.2005.01.001.

62. Sefati, Z., Khalilimoghadam, B., & Nadian, H. (2018). Assessing urban soil quality by improving the method for soil environmental quality evaluation in a saline groundwater area of Iran. *CATENA, 173*, 471–480. https://doi.org/10.1016/j.catena.2018.10.006.

63. Setälä, H., Francini, G., Allen, J., Jumpponen, A., Hui, N., & Kotze, D. (2017). Urban parks provide ecosystem services by retaining metals and nutrients in soils. *Environmental Pollution, 231*, 451–461. https://doi.org/10.1016/j.envpol.2017.08.043.

64. Sulieman, M., Saeed, I., Hassaballa, A., & Rodrigo-Comino, J. (2018). Modeling cation exchange capacity in multi geochronological-derived alluvium soils: An approach based on soil depth intervals. *CATENA, 167*, 327–339. https://doi.org/10.1016/j.catena.2018.05.005.

65. Ulusoy, Y., Tekin, Y., Tümsava⊠, Z., & Mouazen, A. M. (2016). Prediction of soil cation exchange capacity using visible and near infrared spectroscopy. *Biosystems Engineering, 152*, 79–93. https://doi.org/10.1016/j.biosystemseng.2016.09.003

66. Vargas-Terminel, M. L., Flores-Rentería, D., Sánchez-Mejía, Z. M., Rojas-Robles, N. E., Sandoval-Aguilar, M., Chávez-Vergara, B., Robles-Morua, A., Garatuza-Payan, J., & Yépez, E. A. (2022). Soil respiration is influenced by seasonality, forest succession and contrasting biophysical controls in a tropical dry forest in Northwestern Mexico. *Soil Systems, 6*(4), 75. https://doi.org/10.3390/soilsystems6040075.

67. Vargas, R., & Allen, M. F. (2008). Diel patterns of soil respiration in a tropical forest after Hurricane Wilma. *Journal of Geophysical Research: Biogeosciences, 113*(G3). https://doi.org/10.1029/2007JG000620.

68. Vogel, H. J., Bartke, S., Daedlow, K., Helming, K., Kögel-Knabner, I., Lang, B., Rabot, E., Russell, D., Stößel, B., Weller, U., et al. (2018). A systemic approach for modeling soil functions. *SOIL, 4*(1), 83–92. https://doi.org/10.5194/soil-4-83-2018.

69. Wang, E. H., Zhao, Y. S., Chen, X. W., Zhou, Y. Y., Chai, Y. F., & Wang, Q. B. (2009). Effect of heavy machinery operation on soil aggregates character in Phaeozem region. *Chinese Journal of Soil Science, 40*, 756–760.

70. Wang, J., Feng, X., Anderson, C. W., Xing, Y., & Shang, L. (2012). Remediation of mercury contaminated sites—A review. *Journal of Hazardous Materials, 221*, 1–18. https://doi.org/10.1016/j.jhazmat.2011.10.016.

71. Weissert, L., Salmond, J., & Schwendenmann, L. (2016). Variability of soil organic carbon stocks and soil CO_2 efflux across urban land use and soil cover types. *Geoderma, 271*, 80–90. https://doi.org/10.1016/j.geoderma.2016.02.015.

72. Welles, J., Demetriades-Shah, T., & McDermitt, D. (2001). Considerations for measuring ground CO2 effluxes with chambers. *Chemical Geology, 177*, 3–13. https://doi.org/10.1016/S0009-2541(00)00315-8.

73. Wiesmeier, M., Urbanski, L., Hobley, E., Lang, B., von Lützow, M., Marin-Spiotta, E., van Wesemael, B., Rabot, E., Ließ, M., Garcia-Franco, N., et al. (2018). Soil organic carbon storage as a key function of soils—A review of drivers and indicators at various scales. *Geoderma, 333*, 149–162. https://doi.org/10.1016/j.geoderma.2018.07.026.

74. Xu, X., Wang, C., Sun, Z., Hao, Z., & Day, S. (2023). How do urban forests with different land use histories influence soil organic carbon? *Urban Forestry & Urban Greening, 83*, 127918. https://doi.org/10.1016/j.ufug.2023.127918.

75. Yin, X., Wang, X., Wu, H., Takahashi, H., Inaba, Y., Ohnuki, T., & Takeshita, K. (2017). Effects of NH_4^+, K^+, Mg_2^+, and Ca_2^+ on the cesium adsorption/desorption

in binding sites of vermiculated biotite. *Environmental Science & Technology, 51*, 13886–13894. https://doi.org/10.1021/acs.est.7b03784.

76. Zacháry, D., Filep, T., Jakab, G., Varga, G., Ringer, M., & Szalai, Z. (2018). Kinetic parameters of soil organic matter decomposition in soils under forest in Hungary. *Geoderma Regional, 14*, e00187. https://doi.org/10.1016/j.geodrs.2018.e00187.

77. Zhang, P., Cui, Y., Zhang, Y., Jia, J., Wang, X., & Zhang, X. (2016). Changes in soil physical and chemical properties following surface mining and reclamation. *Soil Science Society of America Journal, 80*(6), 1476–1485. https://doi.org/10.2136/sssaj2016.06.0172.

78. Ziter, C., & Turner, M. G. (2018). Current and historical land use influence soil-based ecosystem services in an urban landscape. *Ecological Applications, 28*(3), 643–654. https://doi.org/10.1002/eap.1746.

CHAPTER 5

Hydrological Processes in Forest Ecosystems

LEARNING OBJECTIVES

After studying this chapter, you will be able to:

- Understand the role of forests in the hydrological cycle.
- Identify how roots contribute to hydrological processes.
- Explain the effects of forests on hydrological processes.
- Evaluate land treatment methods for hydrological improvement in forested areas.

KEY TERMS FROM THIS CHAPTER

Annual runoff
Floods
Infiltration procedure
Roots and hydrological significance
Erosion and sedimentation
Hydrological cycle
Rapid infiltration land treatment
Water quality

INTRODUCTION

Water is essential for human survival and that of other living organisms. Water is essential for life on Earth. Forests have long been recognized as important sources of water for drinking and other purposes. Forested catchments play an important role in supplying a large amount of water for all domestic, agricultural, and industrial needs. Trees and forests supply water by recycling moisture in the atmosphere via transpiration, resulting in increased rainfall. In a watershed, forests play an important role in water storage and movement throughout the ecosystem. As the primary water consumer, forest vegetation has an impact on the balance of available river water resources. Furthermore, by evaporating the water stored in the forest canopy and floor, the soil surface, and bodies of water, such as wetlands, from the atmosphere, as well as through plant transpiration, water can reach the atmosphere. Changes in forest vegetation can have a significant impact on the hydrological cycle and environmental function of the entire ecosystem of forestry. Forest canopies intercept precipitation when it falls as rain or snow. Approximately 60% of rainfall on a forest canopy is intercepted and lost to the atmosphere, with the remainder falling on the ground. When water flows through the landscape, some of it rushes into rivers and bodies of water. Water that does not run off returns to the atmosphere via evaporation and sweating. Non-evaporated water aids in the recovery of vegetation, soil moisture, and groundwater. Water absorbs into the ground through a process known as infiltration. Water penetrates the soil to add moisture and replenish its water content. Plant roots absorb some of the water that penetrates the ground. Water then moves from the roots to the floors, where it is converted to vapor and released into the atmosphere. Transpiration is the process by which water moves from the soil to the atmosphere through the plants. Forests serve an important function in creating a dynamic buffer between land and atmosphere, which aids in the interplay of both as well as the storage of water throughout the river basin from these hydrological processes in forested environments. Forests have an important hydrological function that can have an impact on water resources if disrupted or removed. For example, human activities such as replacing forests with waterproof surfaces or removing forest flooring cause rivers to flow directly into wetlands or streams, rather than into the ground or the atmosphere. The loss of forest areas not only disrupts the water cycle, but it also causes severe erosion, reduces nutrient intake, and disrupts the food and habitat of aquatic ecosystems. In essence, woodland alteration has a direct influence on the hydrological cycle; thus, understanding the impact of human activities on forests is critical for improving land and watercourse management. Finally, an attempt is made to solve real-world problems involving hydrology systems by applying fundamental knowledge in hydrology, ecology, atmospheric science, and other related fields. This chapter is structured to facilitate information sharing among disciplines.

Our goal in this chapter is to present current ecological concepts regarding how forest vegetation structure and function influence hydrological processes. We prioritize certain aspects of the interaction between forest trees and hydrology, particularly transpiration, over others, such as forest canopy humidity interception, to cover this topic in a single chapter.

5.1. THE ROLE OF FORESTS IN THE HYDROLOGICAL CYCLE

For a variety of reasons, future demand and pressure on Earth's water and wood resources will rise. Water demand in developing countries will increase, as will per capita consumption in developed areas, as populations grow. The increase in wood requirements is frequently met through land-foresting schemes and natural forest clearing, both of which provide consistent high-quality water supplies in a variety of locations. Global climate change may lead to less consistent precipitation. There are strong links between deforestation and climate change. When there is a critical competition for land and water resources, as well as an urgent need for a sustainable environment, the performance of forests within the hydrological cycle must be objectively and critically evaluated. The impact of forests on hydrology is clearly of interest, and it manifests itself in two ways. First, forests and their manipulation affect flow and groundwater levels and dynamics, and second, forests and their management have a significant impact on water quality. Before considering the role of forests in various aspects of water quantity and quality, it is critical to first understand the hydrological cycle of vegetation. You can focus on parts of the cycle where there are significant differences between different types and forests of vegetation. Focusing on hydrological cycle processes, it is possible to recognize where physiological and physical features of forests play an important role in directly affecting and changing hydrology as forest cover increases and decreases. Managers, developers, and planners recognize the drainage basin as the most convenient unit for managing water resources. Nonetheless, detailed studies on a smaller scale frequently provide a clear understanding of the hydrological functioning of a catchment.

5.1.1. Forest and the Hydrological Cycle

We cannot protect our water resources unless we first understand how water moves through the environment. Today, the same water on which we rely has dispersed throughout the hydrological cycle's numerous procedures. The cycle shown in Figure 5.1 describes how water moves on and above the Earth's surface as ice, vapors, and fluids.

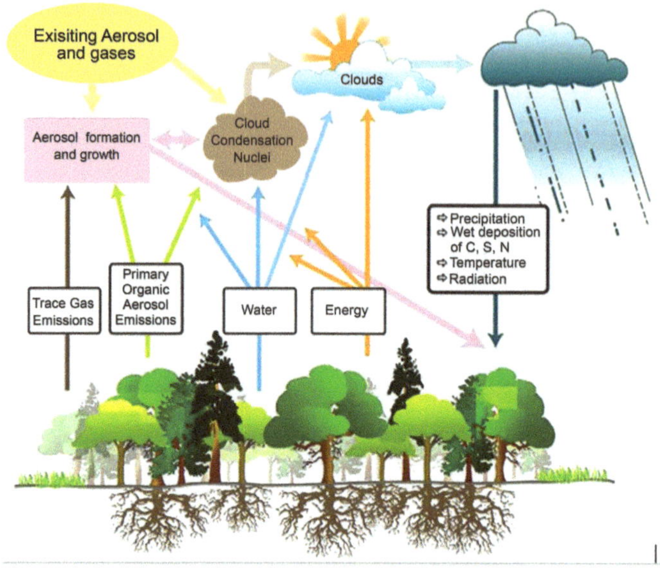

Figure 5.1. *The importance of forests in the hydrological cycle.*

Source: www.flickr.com.

Natural processes like human activities and plant transpiration have an additional impact on this particular cycle. The hydrological cycle is the primary building block of freshwater resources, which make up only 2.5% of the world's water. Of these, about 70% are in the form of ice and a continuous cover of snow, 30% are stored underground and 0.3% is made up of freshwater lakes and waterways in the planet. Consequently, the total amount of human and ecosystem fresh water supplies amounts to less than 1% of all freshwater resources, and under 0.015% of all water on the planet Earth. It is this little portion of the Earth's water that is so heavily reliant on our food production, ecosystem maintenance, drinking water and the industry. During the hydrologic cycle, the water is moved through many processes shown in the figure between the surface, subsurface and the atmosphere. Briefly, water travels from the surface area of the Earth to the atmosphere as water vapor through evaporation from surface water and transpiration and rinse into plants (the process of changing water from liquid to gas). Transpirations represent 62% of worldwide, sustainable fresh water and also the transportation of water through vegetation and soil. The transpiration results are referenced as "green water" during the hydrological cycle, whereas the substance water above and below the ground is described as "blue water." The steam accumulated during the evapotranspiration processes represents 10% of water that is fresh in the cycles and the world in the atmosphere in a "global dynamic envelope." This vapor condenses then in clouds and because of this return to the planet's surface by precipitation. The precipitation is the primary source of fresh water after which, by transpiration or evaporation, the water is either returned to the environment, recharging the groundwater or even providing the surface and subsurface rinse which eventually flows into larger water bodies.

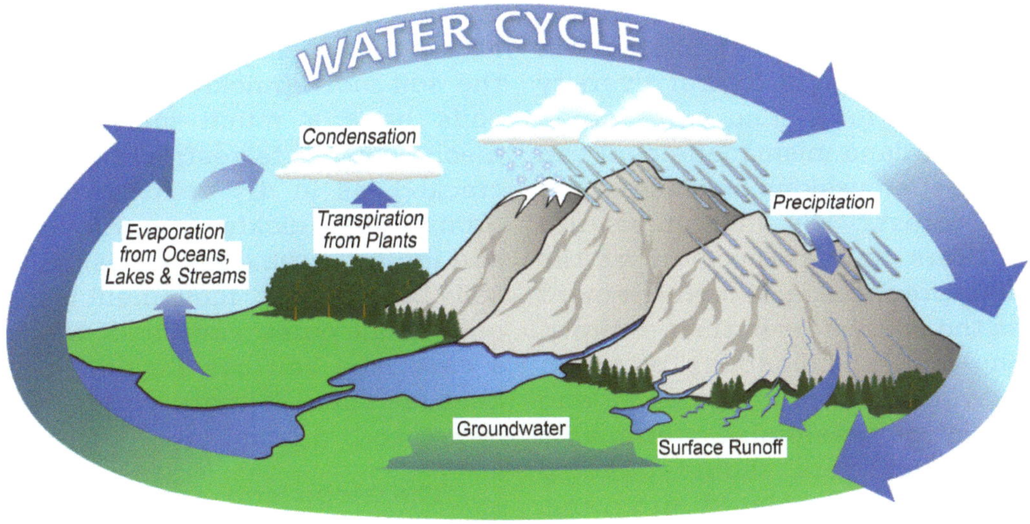

Figure 5.2. *Illustrates the interrelationships among environmental elements of the global water cycle.*

Source: contrainformacion.es.

Given the essential importance of water to life on Earth (Figure 5.2), it's essential to think about just how the hydrological cycle is linked to different ecosystems. One example that is important is the forest-wetland relationship. These ecosystems are far too frequently seen as completely different, instead of a linked unit, which plays an important part in the hydrological cycle as well as the preservation of the water resources. An even better understanding of their role in the hydrological cycle allows us to take these ecosystems better into account when formulating water resource policies and management practices. Forests play a significant part in the hydrological cycle through an immediate effect on the evapotranspiration rates as well as the way water is routed and stored. Forest soils absorb water easily and, as a result, surface runoff rarely occurs outside stream channels in forested areas, causing important water catchments to form under forests. Not only do these catchments help to store precious fresh water, but they also improve water quality since its forest cover reduces erosion and prevent rainwater from becoming pollutant-free within enriched soils on forest beds. Given the key role of forests in the hydrological cycle and in fresh water natural supplies, it's not surprising that a recent study showed that about one-third of the biggest cities in the world enjoy a sizable share of their drinking water produced from forested areas. The proportion increases to roughly 44% when water sources from distant, other and forested forests are managed as a top priority of water supply. The proportion increases to roughly 44% as one of the world's most successful coastal ecosystems, its strong ties with both towns and habitats that extend beyond their hydrological importance. For example, forested riparian wetlands play a significant role as flood buffers. While Mississippi wetlands could maintain approximately 60 days of fluvial drainage, they reduced the flood storage capacity to just 12 days of landing - a decrease of 80% by removing them to create levees and canals. Forest wetlands have huge biodiversity value, as a wide range of natural communities are produced by the varied habitat types. The connection between fish and trees in the Amazon's overloaded forests is a prime example of the unique biodiversity resulting from wooded wetlands. Mismanagement of forests can have negative effects on biological diversity and water in both forest and surrounding wetlands and wetland mismanagement can have adverse repercussions because of the surrounding forests. It's thus important for policymakers to take account of the impacts that practices and management have across the ecosystem in their entirety. For the long-term sustainability of water resources, watershed management and effective forest are essential. Governments and other stakeholders should develop and implement policy-related programs that encourage holistic, multi-stakeholder and cross-disciplinary approaches linking woodlands, water, human beings and the environment.

5.1.2. Soil Organic Matter

Water reaches the soil mainly as rainfall, though the primary form of precipitation in some cooler regions is snow. In the store and distribution of the water that reaches it, the ground plays a crucial role. The earth has a crucial part in the hydrological cycle to keep the biosphere, as it supplies humans, animals, and plants with water.

Three main effects are done by water reaching the ground surface:

i. Flow across its surface and reaching the streamlet, rivers, and lakes.
ii. Enter the ground and fill pores in part from which the water supplies the roots and the above-ground parts of the plant from which it fills its position.
iii. Pass right into the aquifers and rock bodies below the soil. Both rain and snow falls into the soil.

The nature of the soil has a significant impact on the flow of water, such as snow and rain. Sandy soils are porous, which allows snowmelt and rainwater to penetrate the soil more quickly and easily. Clay-rich soils have a much more complex flow mechanism. If terrestrial soils are dry and cracked, as in dry weather, water on the ground will quickly pass through the earth and infiltrate the aquifers below. When clay soils become wet, movement is limited, and only small pores can cross water. In such a case, the soil can become waterlogged because there is more water than the soil can absorb quickly.

The slope and ground of the ground play an important role in determining whether precipitation moves into the ground or across the surface area. If the rate of precipitation exceeds the rate of infiltration, water flows across the surface and into the water systems it serves. The soil is frequently referred to as the "unsaturated area," because it is only intermittently saturated with water, and the soil pores are rarely full. A saturated zone will form at some depth in the soil, with the pores completely saturated with water. In this saturated area, the term "water table" refers only to the top surface. Water will slowly enter the aquifers, streams, and rivers of this saturated region. In some soils, the water table is close enough to the dirt level.

5.1.2.1. Soil Organic Matter: Formation and Turnover

The organic soil is formed and transformed in an assortment of phases and soil humus consisting of living plants, animal and dead roots, microbial biomass and plants. As an outcome of microbial activity, the soil humus is thought to form, but the exact mechanisms of humus formation are little known. Humus has numerous aromatic rings with attached organic acid groups and phenolic, but there's minimal understanding of the majority of the chemical structure. Some tempting models for the whole molecular structure have been proposed, but a big part of soil humus, not consistently molecular in weight or perhaps repeated units of structure, is believed to be amorphous by numerous scientists. Humus accounts for 60–80% of the entire organic matter of soil and has relatively high nitrogen content, and is a major nitrogen reservoir in many ecosystems. Humus closely attaches to the inorganic soil matrix, thus decomposing very slowly, since most soil microbes and their enzymes are unattainable. The carbon cycling in soils is a synthesis of processes that result in the production and rapid turnover and accumulation of humus deeper in the soil profile of the vast majority of plant life in the soils.

The bulk of decomposition takes place close to the surface area of the soil where litter input from plants is concentrated. Gravity is mainly conveyed to the surface area of the ground, where initial release and decomposition of nutrients occurs. Because roots tend to develop into the surface soil, most root litters are primarily produced in the surface soil. In the process of transformation to stabilized soil organic matter, a proportionally greater amount of over ground plant litter carbon is lost as root CO_2. Soil mixing by dissolved organic oxidation and matter initiates the transfer of deeper carbon surface to the profile of the soil. Always a bit less compared to 20 cm is usually around half of the soil's organic carbon, but deep carbon is generally older and more recalcitrant than surface carbon. Soil disruption increases the decomposition of soil's natural carbon by exposing microbial attack to new soil surfaces and increasing soil oxygen content. At all scales, the mechanism used to stimulate disruption it's essentially the same. Disruption interrupts the soil's aggregates, which leads to more oxygen and microbial colonization of organic matter within it. This disturbance effect is accountable for increasing the decomposition rate of forest soils because of the introduction of European earthworms into the Northeast USA. This explains also the reason why organic matter is lost rapidly from grasslands and forest soils after conversion to farming, depletion of the reserves of soil nutrients and destruction of soil structural integrity. Nevertheless, this doesn't explain the reason why disturbances in the very first couple of years lead to an extremely rapid decreasing organic soil matter and, later, a much slower decline. To explain this observation, we should recognize the natural soil is a diverse mix of microbial residues, animals, and plants, which vary from one microbial to another. It may be conceptualized as a number of fractions that comprise a continuum based on decomposition rate by defining the soil organic matter. This may be extended by developing simulation models that describe the processes of turnover and formation of organic matter mathematically. 2 to 3 organic materials pools that are kinetically defined with different turnover rates are models probably which best represent these processes. These pools tend to be defined as a pool that is small with a quick turnover rate (labile pool) as well as a bigger pool and a reduced turnover rate of 1 to a number of pools.

The turnover rates are measured in months to years by the labile soil organic matter pool. While the addition of new plant residues may increase the size of the labile carbon pool quickly, once the dirt is disturbed also carbon is quickly lost in the labile pool. First affected by other changes and land distortions in land management are the labile pool of organic matter. In a couple of years of the tillage of native forest and grasslands grounds soils organic carbon is lost from this particular pool. Lipid soil organic pools have typically been measured by turnover times over the years. They're a crucial source of plant-based nutrients and mineralizable nitrogen as well as provide the soil microbial populations with a healthy way to obtain food. Native grassland soils are full of labile carbon intermediately, which accounts for their prolonged fertility when cultivated. Every 100 to 1,000 years, stabilized organic soil material turns and carbon losses from this pool are gradual. This pool comprises the majority of soil humus and makes up 60–90% of the natural soil in nearly all soils. This pool is connected to the colloidal characteristics of humus and is involved in soil structure stabilization. In the last several decades significant progress has been made in understanding the dynamics of soil organic matter in temperate ecosystems. The integration of the functional and structural qualities of soil organic matter and its association with soil biological processes were very important advances. Research indicates that the turnover of organic soil and consequently the potential for nutrient cycling and storage of carbon is associated with soil aggregates' stabilization and formation. Probably the most labile form of organic carbon in biological activity cycles quickly in the soil, accounting for essentially a tiny proportion of the whole organic carbon soil (3.5%). It's been thoroughly documented the value of the labile pool to ensure soil productivity as well as to manage the provision of nutrition for plants. The labile pool's time of turn is very fast, nonetheless, it is able to have a limited direct impact on soil carbon stock. A sizable proportion of organic biologically active carbon within soil is physically protected against decomposition by occlusion within soil aggregates. In a time scale that is ideal for assessing the effect of climate and land use change on the ecosystem's function, this intermediately labile pool of organic carbon is turning thus and around an important pool of biological matter which is to monitor long term changes in the potential of soil storage as a relative short-run indicator.

5.1.3. The Role of Roots in Hydrological Significance

Plants soak up nutrients and water from the earth, and spread it across the stalks and leaves through the roots. The surfaces of the leaves are evaporated. The evaporation of water from plant leaves is essentially transpiration. Transpiration also consists of a process called guttation, which is water loss, largely through water stomata, in liquid form, from the uninjured leaf or perhaps stem of the plant. Scientific studies indicate that about 10% of the atmospheric moisture is released via transpiration by plants. Most of the remaining 90% is provided by evaporation from oceans, other water bodies and water (lakes, rivers, streams). Plants put roots in the earth to draw water and nutrients and into the stems and leaves. Much of this water is returned into the air through transpiration (where the entire process is referred to as evapotranspiration in conjunction with evaporation). Transpiration rates vary widely depending on conditions in the weather conditions, like temperature, humidity, intensity and availability of sunlight, precipitation, type of saturation and soil, wind, land slope, and human use and water divertissement. Sweat may contribute to the loss of humidity in top of the soil area during dry seasons, which can influence crop and food crops. The links between water, wetlands and forests show just how important ecosystems are to protect their ecological character, in addition to ecological services and the freshwater so vital to human activities on earth. Inland waters are probably the most threatened types of ecosystems, with an estimated loss of 50% of the world's wetlands after 1900. Deforestation can also be a significant threat to catchment of water and the quantity and quality of fresh water available. With water use growing much more than double the population growth rate, it's vital for us to understand correctly and manage our ecosystems, the links between water, forests and wetlands. The interaction of water, woodlands and wetlands to produce healthy, productive ecosystems is continuous. For proper function and the wilderness of the hydrological cycle, for example, play an important role. A hydrological cycle exam reveals how the management and conservation of forests are directly associated with water basin health and water quality downstream. Forests also control and reduce soil erosion and pollution, settlement and desertification. The capacity of forests to support water capture and water storage can help mitigate floods during periods of heavy rain. In return, many forests rely on soil water to sustain their fauna and flora and depend on wetlands to support them. Wetlands can also be important for the preservation of many natural cycles and also the promotion of a broad range of biodiversity.

5.1.4. Plant and Animal Life and Sheltering

The forest is an important part of the environment because it acts as the earth's lungs, helping to cleanse the environment and the water. It also provides food and shelter to animals, microbes, and plants in living organisms. It supplies raw materials to industries such as wood, gums, resins, and others.

5.1.4.1. Forest as a Habitat

The forest is a habitat for a lot of plants and animals since it provides them with a good environment. For example, elephants, deer, boars, tiger, jackal, etc. The soil underneath the forest tree is the forest floor. It comprises a root, soil, decomposer and decomposing organic matter. Long woody trees provide refuge for climbers and crawlers in the forest. Millions live in woods, and they're reliant on the products we get from the wood for their livelihoods, including those living outside the forest. Eighty percent of terrestrial biodiversity is found in forests worldwide.

Forest animals are very specific to their habitats. Everyone needs the forest to survive. The species migrate to find a new home to adapt to climate change. When these animals move, they rely on the surrounding forest to get from one location to the next. Many species do not live elsewhere and require these specific habitats for their survival. This particular need is more common than ever. As the population grows, so does the demand for resources, resulting in deforestation. Wildlife habitats continue to deteriorate with each hectare of forest lost due to activities such as development, road construction, or timber production. The vast majority of animal, microorganism, and plant species find their natural habitat in forests. In tropical rainforests, this is common. As these rainforests are converted into farmland, biodiversity loss is having a significant impact on forest habitats. Forests contain a variety of plants.

5.2. HYDROLOGICAL PROCESSES AFFECTED BY FOREST LANDS

Water covers more than 70% of the world's surface area, with woodland accounting for nearly a third. Solar energy and gravity comprise the hydrologic cycle of the hydrosphere. The forest is critical to maintaining the dynamic balance of the physio-ecosystem as well as exchanging material and energy flows. For example, a forest participates in the hydrological cycle between the soil surface and the organism air through its ecological and physiological behavior, so it has become a participant in the hydrological cycle as well as a dedicated protector of the quantity balance of the various parts of the hydrological cycle.

Because of this, the study of forest hydrological effects is crucial as a result of the hydrological cycle transformation and protection and the cycle orientation for the gain of the human race. The primary portion of the vegetation consists of forests. One hectare of forest has a biological yield of 20–100 times that of one hectare of agricultural or perhaps grassland. The entire photosynthesizing leaf area of a tree is 5–10 times that of grassland plants.

6 $C0_2$ molecules and 6 H_2O molecules are absorbed in the photosynthesis of chlorophyll under solar energy. One glucose molecule is produced and 6 O_2 molecules are released. The technique may be expressed through the equation:

$$6\ CO_2 + 6\ H_2O + \text{light energy} = C_6H_{12}O_6 + 6\ O_2.$$

Because of the extended floor, large leaf areas, high biological yields and long stretches of growth, forests influence precipitation in a more substantial way than many other vegetation. Forests feed on soil moisture, like some other plants, to protect the lives of theirs and also to enhance their surviving environment. Forest root systems, trunks, leaves, etc., absorb a great deal of humidity and transform it into bio-water from soil, ground and surface. The forest then converts bio-water into gaseous waters by respiration, transpiration, etc., and releases it into the environment to form the soil-plants-atmosphere hydrological cycle and keep the mechanism of precipitation development (Figure 5.3). The forest is consumed with a great deal of solar energy when moisture is transpired and released to the atmosphere. Consequently, in the woodland area the temperature and humidity field is very low and humidity is high. Water vapor will probably saturate the atmosphere.

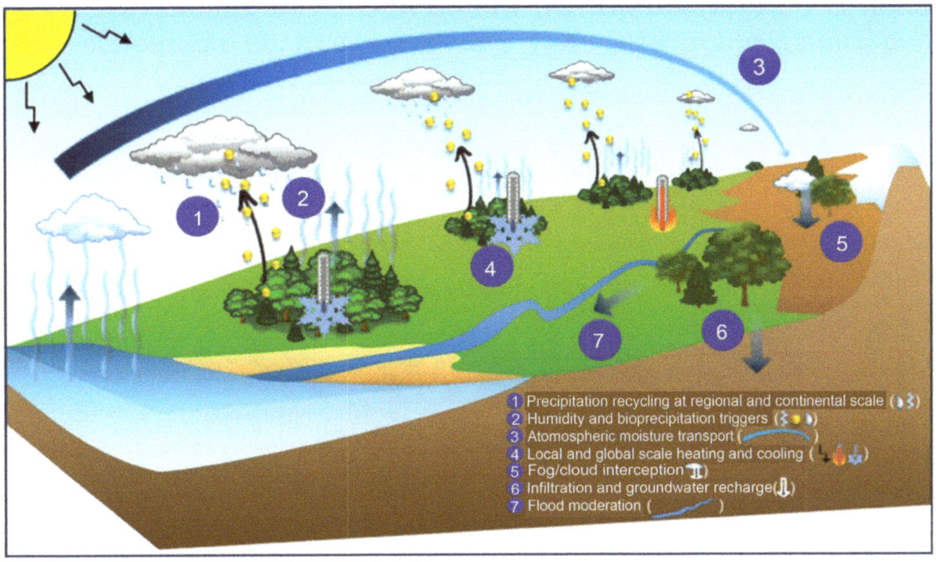

Figure 5.3. *Effect of forest on precipitation.*

Source: Ellison, D., Morris, C. E., Locatelli, B. D., Sheil, D., Cohen, J., Murdiyarso, D., Gutierrez, V., Van Noordwijk, M., Creed, I. F., Pokorny, J., Gaveau, D., Spracklen, D. V., Tobella, A. B., Ilstedt, U., Teuling, A. J., Gebrehiwot, S. G., Sands, D. C., Muys, B., Verbist, B., . . . Sullivan, C. A. (2017). Trees, forests and water: Cool insights for a hot world. Global Environmental Change, 43, 51–61. https://doi.org/10.1016/j.gloenvcha.2017.01.002.

In the forest area and surrounding areas, there's therefore a major difference between surface temperature and air temperature. A convergent local updrawing flow would therefore easily produce orographic rush. The data observed show that, whether or not the ocean is located at an equivalent latitude and spreads over the same location as the forest, it is able to transpire 50% more moisture than the ocean evaporates. The air humidity in the forest exceeds that of farmland by 10-20% For instance, 50% of the precipitation around the Amazon River basin is generated by vapor from the forest. While the Tianmu Mountains are closer to the ocean than the Huangshan Mountains, with vapor conditions, normal annual precipitation on Tianmu is 8.7% lower, since the forest coverage on the former is 16.2% less than that of the Huangshan Mountains. Huangshan Mountain and the mountain Tianmu Mountains are both at similar latitudes; the amount of rainfall is apparently interrelated with forest cover.

5.2.1. Effects of Forest on Annual Runoff

Forests are determined for an annual rushing by the area of forest, crown density, composition of age and forests of forest. When similar, the impacts of different forests would differ from geographical locations, weather and forest areas. To produce a compensation effect, as well as to affect transpiration and evaporation efficiency, the influence of forests on annual fluctuations is primarily a result of the changes in the concentrations and the yield, so the impact on compensation fluid efficiency is exceeded. The transpired humidity cannot develop local precipitation. In arid regions this phenomenon is quite apparent. On the contrary, the annual runoff would increase if the effect on transpiration efficiency was less than on compensation efficiency in humid regions.

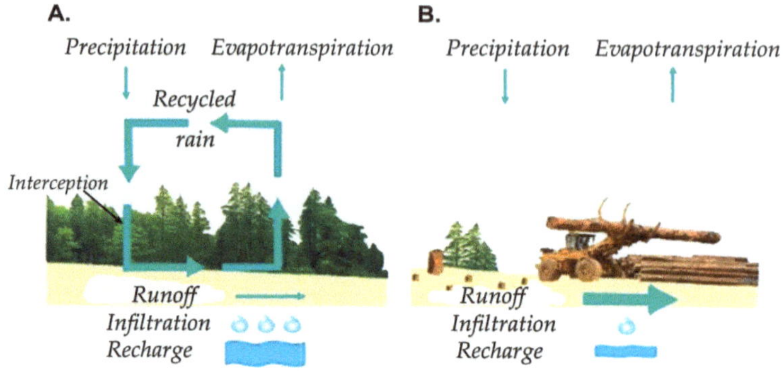

Figure 5.4. *Effect of forest on annual runoff.*

Source: Castruita-Esparza, L.U.; Gutiérrez, M.; Olivas-García, J.M.; Rubio-Arias, H.O. The Pine-Oak Forest of the Rio Conchos Basin, Mexico: Key to Rain Production and Soil Erosion Control. Environments 2023, 10, 37. https://doi.org/10.3390/environments10030037.

However, the annual flux and annual flux distributors in arid areas and wetlands are improving equally regulatory activities and forest compensation. It means that the fluxes between flood and the degree and drought years of flush distribution are reduced by the scope of the forest cover within one year (Figure 5.4).

5.2.2. Effects of Forests on Floods

When precipitation begins, the foliage of the crown canopy intercepts it. For example, a pine tree that has been growing for a hundred years can have a total needle length of approximately 250 kilometers. In a hardwood forest, the leaf surface area can be 75 to 100 times greater than the area that is shaded by the canopy. This significant leaf surface area plays a crucial role in intercepting rainfall, which affects water distribution and soil moisture levels in forested areas. The forest can therefore perfectly intercept rainfall. The outcome of the interception depends on crown density, crop composition, forest age and stage of growth. The bigger the amount of leaf storage per unit area is, the more precipitation it takes to moisten the leaves and the more interception. The intensity of precipitation will be inversely proportional to the quantity of the interception in forest plants of similar compositions and, where the intensity of the precipitation reaches a limited value, the interception amount also tends to be restricted. The earth is generally covered by a heavy layer of plant remains in the forest regions. Lose soils, good aeration and rich microbial activity create a great absorption of water and increase infiltration function. The capacity for infiltration varies with crown density, root osmotic range and forest litter thickness. A research study indicates that in the Poplar forest of David, the quantity of water absorbed by the soil is 3.16 times greater compared to the cover. It might be 2.2 times with the formed tabular pine wood. The first forest land infiltration capacity is 3.4 times grassland, 10 times soil, and 4 times farmland.

The rise in the surface roughness of forest land results in the extension of the flood concentration on slopes. The peak discharge is therefore reduced and the flood period is increased. The concentration velocity of the flood is determined by water surface gradient, hydraulic radius, surface roughness and other hydraulic factors, from the perspective of hydraulics, in cases of slope confluence and channel concentration. The forest's roughness is 10 to 100 times that in a non-forest area, which helps to delay the peak, reduces the peak discharge and prevents potential disasters from flooding.

5.2.3. Erosion and Sedimentation

Erosion is accountable for the Influence of the forest. The weathering of parent rocks generally keeps soil erosion in equilibrium with soil formation. The sediment of the river is caused primarily by human and natural erosion. A major factor in the erosion intensity and erudite that are determined by land use will be the amount of sediment produced. The degree of erosion is reduced and the durability of the earth is enhanced when land resources are rationally utilized. The forest contributes positively to the conservation of soil. For starters, it mitigates top soil erosion by wind, gravity, water, etc. Next, the surface area has a litter cover that strengthens the surface soil's erosion durability and reduces erodibility. This process confirms the following principles: rainfall removes soil and flow carries sediment along with the hydraulic laws. Based on the Law of the Falling Bodies, the rate of rainfall that results in the washing of the best soil is directly proportional to the vertical root of the precipitation. Only once the rains fell to the environment through the crown canopy is the washing force described as the corn canopy. The distance has thus fallen significantly. Forestry is able to delay erosion as well as improve the durability of soil erosion. This is because the origins of the forest and other plants and vegetation build up a system like a net under the ground, which provides soil protection and prevents soil from being washed away by precipitation. It's also due to this. The infiltration increases to postpone the time of the concentration. When the forest litters infiltrate, the sediment flow decreases the sediment content. Moreover, forest humus increases soil aeration by enhancing the microbial activity. All this is good for the preservation of water and soil.

5.2.4. Influence of the Forest on Water Quality

Water quality balance is dependent on the water cycle balance. Forests are generally located in mountainous regions from which rivers originate. Because of this, the forest is exerting nearly unique effects on the regulation and enhancement of water source quality: the forest regulates the balance of atmospheric 02 and maintains the oxygen cycle through photosynthesis, thereby strengthening auto purification of the atmosphere; the forest is needed in the hydrological cycle and also protects the water balance. The balance of quantity, heat, energy and quality isn't simply part of it but also regulated. Thus, to enhance the water balance in the hydrology cycle and increase the benefits to humans, it's incredibly crucial to protect the forest. It's not possible to change the role of forests with that of anyone other structure. We might achieve two times the result and have a half effort by weighing structural measures as crucial as biological measures in the development of water resources and regulation of cubicles.

5.3. HYDROLOGICAL EVALUATION OF LAND TREATMENT

Land use and treatment measures reduce the amount of direct storm runoff, either through increased infiltration or perhaps surface storage rates or perhaps both. All other factors which affect the amount of runoff are usually small. Increased interception isn't usually necessary for conservation of natural resources, for example, only in certain climatic or perhaps vegetative conditions.

5.3.1. Rapid Infiltration Land Treatment

Rapid infiltration (RI) is one of three major soil treatment methods that make use of the soil's ecosystem in the processing of wastewater. RI can also be known as a soil aquifer. In a much smaller territory than other land treatment concepts, the RI process can, however, treat a much higher volume of wastewater. Wastewater can be used in RI systems in the shallow basins of highly porous soil, built in permeable and deep deposits. Wastewater use may be by flooding or perhaps sprinklers from time to time. The wastewater passes in the soil matrix with the procedure of treatment including filtration, adsorption, and exchange of ions, precipitation, and microbial action. The soil retains phosphorus and most metals, while the degradation or perhaps adsorption of toxic organic and natural products. It may be collected or perhaps transferred to native surface water or perhaps groundwater aquifers as wastewater percolates through the soil. In the case of relatively shallow groundwater table, the use of sub drainage control permits groundwater mounding and water recovery. The water is recovered by wells in areas with deeper groundwater. This water could be useful for irrigation or perhaps industrial applications. The water that wasn't recovered can charge groundwater aquifers. This is referred to as "beneficial reuse." Figure 5.5 shows the typical hydraulic paths for RI-treated water. RI might be utilized in a range of climates and at sites that are different. Unlike some other land treatment and aquiculture concepts, there are no seasonal restrictions on RI systems, and in the north of the United States and in southern Canada they've been operating successfully over the winter months. In terms of location, RI is extremely versatile. Unless groundwater is usually to be recharged and reclaimed, the most desirable sites are situated alongside surface water to reduce impacts on the overall quality of the groundwater. There can easily be an under-drained system wherever suitable conditions exist on the ground and groundwater.

Advantages:

- The techniques of gravity distribution consume no energy.
- There's simply no need for chemicals.

- The RI treatment is easy and affordable.
- Seasonal changes don't restrict the process.
- The effluent is quite good.
- With sufficient rest periods, the procedure is quite reliable.
- RI provides groundwater recharge, groundwater level control, discharge and reuse of refurbished water to a specific surface water body, and temporary storage in an aquifer of refurbished water.
- The procedure is ideal for plants that are small with limited operator knowledge.

Disadvantages:

- As always, RI systems don't usually meet the high nitrogen level for drinking water discharges. The long term involvement associated with a substantial treatment area with minimum secondary benefit, like various other natural treatment systems (i.e., crop or perhaps forest production, enhancement of habitat, etc.,) needs to be achieved.
- Requires annual removal on basin infiltration surfaces of accumulated organic matter deposits.
- May have to possess the top several inches of soil removed and removed from at times for the exposure of clean material.
- Clogging can occur when significant lagoons and polishing ponds are received at high rates.

5.3.2. Evaluation by Infiltration Procedure

Probably the most common statistical methods for analyzing hydrological data are frequency analysis, time series tests, and regression analysis. Frequency analysis can be used to anticipate just how often certain values of a variable phenomenon is able to occur as well as to evaluate prediction reliability. It's a tool for the look and design of drainage planning and waste, particularly with regard to the required hydraulic capacity.

5.3.2.1. Multiple Regressions

A regression analysis is accustomed to identifying a connection of two or even more variables, one of which undergoes random variation, and also to determine whether a relationship is statistically significant, whether assumed or perhaps computed. It's one tool to detect hydrological parameters in locations that are different, hydrological model parameters, hydraulic parameters and ground parameters, crop growing and depth of water tables. Time series screening is accustomed to monitoring the coherence of data which are time-dependent, i.e., data collected over a length of time. This precaution is needed to stay away from wrong hydrological predictions (e.g., annual peak or precipitation rush rate). Thanks to an intricate hydrological natural phenomenon and having a bigger role in the style of the Hydraulic structures, the Runoff Data Model requires always extensive and complicated fieldwork. Temporal and spatial variability increase the complexities of measuring these parameters. A special tool for evaluating hydrological parameters in environmental science and engineering is the Rainfall Runoff Model (RRM). They're used to extend stream flow time series in time and space for the evaluation of management and/or catchment response strategies for climatic and/or land use variables for real-time flood forecasts. Because the rainfall-runoff is an extremely complex, dynamic and non-linear process, it's very hard to model. Statistical regression model for stream flow prediction from the concurring data on the rainfall and ruins of the drainage basin if measured is among the effective and many simple statistical tools for streaming (Pilgrim et. al., 1988).

Records are usually not offered at the relationship and the site should be derived from and transferred to the website from a measured basin of the exact same dimension in the region. For designing a domestic water supply or perhaps irrigation, either with or perhaps with no storage, annual or monthly return is generally needed. The water drainage area is typically fairly little in the arid zone. In semi-arid and arid regions, main water supplies and large storage areas with large drainage basins are seldom necessary. Furthermore, these losses should be considered when the basin is extensive enough to contain a couple of 100 meters of channel. Reversion relationships between outflow and reach inflow are possible approaches using only surface flow data. A typical assumption in frequency analysis is that random processes generate hydrological extremes (high precipitation) or flows. It means that the distribution of extreme events can't be changed by natural climate variability. To be able to test the random hypothesis over secular periods, a regional frequency analysis approach is proposed. Extreme events occur at regional levels compared to random events. Because of climate variability, significant retreats may indicate no randomness. By analyzing hydrological records, engineers and hydrologists estimate the probabilities of severe or large floods. To map the frequency of extreme events, a probability distribution is selected and is fit for historical details with the help of statistical methods. One significant assumption of this strategy is it's a random process. What this means is that extraordinary events are independent and distributed identically with a particular (unchanging) probability distribution. The treatment of hydrological ends as a random process doesn't affect the occurrence of extreme events by the natural climatic variability. Nevertheless, there's evidence that is strong in several regions that extreme hydrological conditions are responding to large scale forcing.

5.3.2.2. Regional Analysis

A common assumption in frequency analysis would be that the random process will be the generation of hydrological extremes (floods or perhaps heavy precipitation). This implies that the distribution of extreme events isn't altered by natural climate variability. To be able to test the random hypothesis over secular time scales a regional frequency analysis approach is proposed. The regionally observed occurrences of extreme events are compared to random events. Due to climate variability, considerable departures may indicate non-randomness. By analyzing hydrological records, engineers and hydrologists estimate probabilities of heavy rains and large floods. To model the frequency of extreme events, a distribution of probabilities is chosen and fits historical data with statistical techniques. It's a crucial assumption that the process is random. What this means is that extraordinary events are autonomous and sent out in equal measure with some probability distribution (unchanged). The random treatment of hydrologic extremities means that the occurrence of extreme events won't be impacted by natural climatic variability. But there's evidence that is clear for a few regions that extreme hydrologic reacts to large scale forcing.

5.3.2.3. Hydrographic Analysis

A hydrograph is a continuous record of immediate release over time. It is the result of a combination of conditions in a watershed, demonstrating the significant effects of climate, surface rivers, hydrological losses, interflow, and groundwater flows. Detailed hydrographic analyses are required to mitigate flood damage, forecast floods, and possibly create design flows for structures that transport floodwaters.

5.3.2.4. Drainage Basin Responses

The way a river responds to rainfall has significant consequences. In the short term, flash floods can cause an abrupt and extreme increase in discharge, leading to more frequent flooding events. The characteristics of the drainage basin determine how a river manages rainfall, and it's important to consider how water re-enters the river since the river's response may lag behind rainfall.

There are three primary pathways for water to enter the river: surface runoff, throughflow, and groundwater flow. Surface runoff is by far the fastest route for water to reach the river. Factors in the drainage basin that promote surface runoff while limiting infiltration often produce a "peak" hydrograph, with a rapid rise in water levels.

In summer, forests with dense leaf cover can increase transpiration rates by up to 40%, largely due to interception. This process prevents water from immediately reaching the ground surface, reducing surface runoff and allowing more time for water to infiltrate into the soil. Water that does make it through the canopy, known as throughfall, is more likely to infiltrate and percolate compared to surface flow.

Human activities that clear dense vegetation from forests are likely to increase surface runoff, as less water is intercepted by plants. Figure 5.5 illustrates the inverse relationship between plant density and surface runoff, highlighting how vegetation plays a key role in managing water flow within a drainage basin.

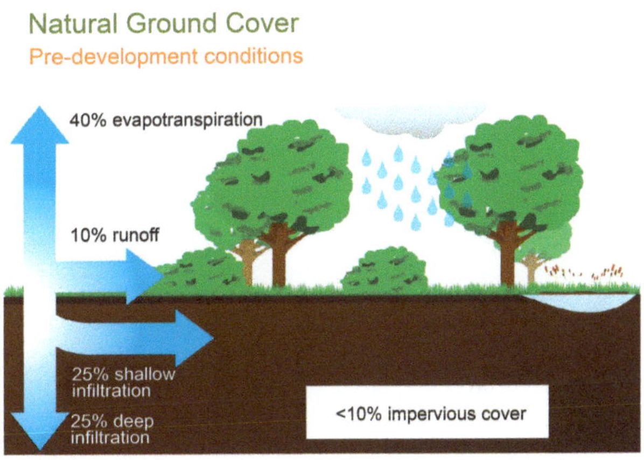

Figure 5.5. *Shows urbanization and surface run-off.*

Source: https://wiki.sustainabletechnologies.ca/wiki/Urbanization.

We can also see a strong correlation between increasing urbanization and imperishable surfaces, as well as increased surface fluctuations. We should take care that urbanization does not cause too many flash floods in rivers; after all, the town or city is only a small part of the drainage basin as whole. Construction of highways and parks naturally increases the downfall of roads, as does building in floodplains of businesses and homes. The look of drainage and roads can also help to reduce lag time by quickly transporting water into the river.

These effects, however, should be common at the local level, with the exception of major metropolitan areas. They are not solely responsible for the floods themselves. River management is commonly used to reduce the likelihood of flooding in built-up areas. Furthermore, urbanization frequently exacerbates downstream floods within the city.

However, it should be noted that the combined effect of growth and urbanization, which has removed the natural forest from its countryside, is primarily responsible for the widespread floods that appear to be more common today. Certain engineers, planners, and community leaders are only now beginning to understand the importance of forests and trees in storm water management and water quality protection.

Forests would be the best land use because of their ability to collect, filter, and retain water and air contamination from the air, thereby protecting water quality. Forests are also important for providing clean drinking water, carbon capture, flood control, wildlife habitat, and forest products for more than 10 million people living in watersheds, as well as providing valuable ecological and financial benefits.

CASE STUDY

HYDROLOGICAL PROCESSES IN FOREST ECOSYSTEMS

Forest ecosystems play a crucial role in regulating hydrological processes, influencing water quality, quantity, and availability. A notable case study is the effects of forest cover on watershed hydrology in the Amazon Rainforest, particularly in the Tapajós River Basin in Brazil. The region showcases how dense forest cover can enhance rainfall interception, soil infiltration, and groundwater recharge, leading to healthier streamflow patterns. Research has shown that forests contribute to maintaining the hydrological cycle by absorbing and redistributing precipitation, which mitigates flood risks and sustains dry-season flows.

Hydrological Impacts

The conversion of forests to agricultural land in the Amazon has led to significant alterations in hydrological dynamics. Studies indicate that deforestation decreases evapotranspiration rates, resulting in reduced precipitation and altered seasonal flow patterns. For instance, after the clearing of large forest areas, researchers observed increased runoff and sedimentation in rivers, negatively affecting water quality and aquatic habitats. Furthermore, changes in land cover can lead to more intense flooding events and extended dry periods, impacting both local communities and biodiversity. The interplay between deforestation, land-use changes, and hydrology emphasizes the importance of preserving forest ecosystems to sustain watershed health.

Conservation and Management Strategies

To address the challenges posed by deforestation, several conservation strategies have been implemented in the Amazon, including reforestation initiatives and sustainable land management practices. These strategies aim to restore hydrological functions by promoting forest regrowth and enhancing soil conservation. Additionally, integrated watershed management approaches that involve local communities can foster sustainable land-use practices while ensuring the preservation of critical hydrological functions. The case of the Tapajós River Basin highlights the need for a multifaceted approach to conservation that considers the intricate relationships between hydrology, forest ecosystems, and human activities.

Discussion Questions

- How can the restoration of forest ecosystems contribute to improving local water quality and availability in regions affected by deforestation?
- What are the potential challenges in implementing sustainable land management practices in forested watersheds, and how can these challenges be addressed?

SUMMARY

Hydrological processes in forest ecosystems encompass a complex interplay of precipitation, interception, throughfall, infiltration, and groundwater dynamics. Precipitation interacts with the forest canopy, where interception by foliage and stems affects the amount reaching the forest floor as throughfall. Infiltration rates vary with soil type, vegetation cover, and topography, influencing groundwater recharge and runoff patterns. Forest structure and species composition significantly impact these processes, altering water availability and regulating streamflow. Human activities like deforestation and land-use changes can disrupt these natural hydrological cycles, leading to erosion, reduced groundwater recharge, and altered microclimates. Understanding and managing these processes are crucial for sustainable forest management, watershed conservation, and mitigating the impacts of climate change on water resources.

REVIEW QUESTIONS

1. How do forests influence the hydrological cycle, and why is this important?
2. What role do roots play in the hydrological significance of forests?
3. Discuss the effects of forests on annual runoff and floods.
4. How do forests contribute to erosion control and sedimentation prevention?
5. What methods are used for evaluating rapid infiltration land treatments in forest ecosystems?

REFERENCES

1. Arnáez, J., Larrea, V., & Ortigosa, L. (2004). Surface runoff and soil erosion on unpaved forest roads from rainfall simulation tests in northeastern Spain. *Catena, 57*(1), 1–14. https://doi.org/10.1016/j.catena.2003.07.002.
2. Benda, L., James, C., Miller, D., & Andras, K. (2019). Road Erosion and Delivery Index (READI): A model for evaluating unpaved road erosion and stream sediment delivery. *Journal of the American Water Resources Association, 55*(2), 459–484. https://doi.org/10.1111/1752-1688.12714.
3. Bowling, L. C., & Lettenmaier, D. P. (2001). The effects of forest roads and harvest on catchment hydrology in a mountainous maritime environment. In M. S. Wigmosta & S. J. Burges (Eds.), *Land Use and Watersheds: Human Influence on Hydrology and Geomorphology in Urban and Forest Areas* (pp. 145–164). American Geophysical Union. https://doi.org/10.1029/WS002.
4. Buchanan, B. P., Falbo, K., Schneider, R. L., Easton, Z. M., & Walter, M. T. (2013). Hydrological impact of roadside ditches in an agricultural watershed in Central New York: Implications for non-point source pollutant transport. *Hydrological Processes, 27*(17), 2422–2437. https://doi.org/10.1002/hyp.9323.
5. Butzen, V., Seeger, M., Wirtz, S., Huemann, M., Mueller, C., Casper, M., & Ries, J.

B. (2014). Quantification of Hortonian overland flow generation and soil erosion in a Central European low mountain range using rainfall experiments. *Catena, 113*, 202–212. https://doi.org/10.1016/j.catena.2013.05.005.

6. Croke, J., & Mockler, S. (2001). Gully initiation and road-to-stream linkage in a forested catchment, southeastern Australia. *Earth Surface Processes and Landforms, 26*(2), 205–217. https://doi.org/10.1002/1096-9837(200102)26:2<205::AID-ESP168>3.0.CO;2-G.

7. Croke, J., Mockler, S., Fogarty, P., & Takken, I. (2005). Sediment concentration changes in runoff pathways from a forest road network and the resultant spatial pattern of catchment connectivity. *Geomorphology, 68*(3–4), 257–268. https://doi.org/10.1016/j.geomorph.2004.11.020.

8. Cuo, L., Giambelluca, T. W., Ziegler, A. D., & Nullet, M. A. (2008). The roles of roads and agricultural land use in altering hydrological processes in Nam Mae Rim watershed, northern Thailand. *Hydrological Processes, 22*(23), 4339–4354. https://doi.org/10.1002/hyp.7043.

9. Cuo, L., Giambelluca, T. W., Ziegler, A. D., & Nullet, M. A. (2006). Use of the distributed hydrology soil vegetation model to study road effects on hydrological processes in Pang Khum Experimental Watershed, northern Thailand. *Forest Ecology and Management, 224*(1–2), 81–94. https://doi.org/10.1016/j.foreco.2005.12.010.

10. Dutton, A. L., Loague, K., & Wemple, B. C. (2005). Simulated effect of a forest road on near-surface hydrologic response and slope stability. *Earth Surface Processes and Landforms, 30*(3), 325–338. https://doi.org/10.1002/esp.1140.

11. Dymond, S. F., Aust, W. M., Prisley, S. P., Eisenbies, M. H., & Vose, J. M. (2014). Application of a distributed process-based hydrologic model to estimate the effects of forest road density on stormflows in the Southern Appalachians. *Forest Science, 60*(6), 1213–1223. https://doi.org/10.5849/forsci.13-152.

12. Eastaugh, C. S., Rustomji, P. K., & Hairsine, P. B. (2008). Quantifying the altered hydrologic connectivity of forest roads resulting from decommissioning and relocation. *Hydrological Processes, 22*(14), 2438–2448. https://doi.org/10.1002/hyp.6821.

13. Foltz, R. B., & Burroughs, E. R., Jr. (1990). Sediment production from forest roads and wheel ruts. In *Proceedings of the Watershed Planning and Analysis in Action*, Durango, CO, USA, 9–11 July 1990; American Society of Civil Engineers (pp. 266–275). https://doi.org/10.1061/40695(2003)41.

14. Foltz, R. B., Copeland, N. S., & Elliot, W. J. (2009). Reopening abandoned forest roads in northern Idaho, USA: Quantification of runoff, sediment concentration, infiltration, and interrill erosion parameters. *Journal of Environmental Management, 90*(8), 2542–2550. https://doi.org/10.1016/j.jenvman.2009.01.014.

15. Forsyth, A. R., Bubb, K. A., & Cox, M. E. (2006). Runoff, sediment loss, and water quality from forest roads in a southeast Queensland coastal plain Pinus plantation. *Forest Ecology and Management, 221*(1–3), 194–206. https://doi.org/10.1016/j.foreco.2005.09.024.

16. Grace, J. M., III. (2002). Effectiveness of vegetation in erosion control from forest road side slopes. *Transactions of the ASAE, 45*(3), 681–685. https://doi.org/10.13031/2013.8846.

17. Grace, J. M., III. (2017). Predicting forest road surface erosion and storm runoff from high-elevation sites. *Transactions of the ASABE, 60*(3), 705–719. https://doi.org/10.13031/trans.11869.

18. Hairsine, P. B., Croke, J. C., Mathews, H., Fogarty, P., & Mockler, S. P. (2002). Modelling plumes of overland flow from logging tracks. *Hydrological Processes, 16*(11), 2311–2327. https://doi.org/10.1002/hyp.101.

19. Harden, C. P. (1992). Incorporating roads and footpaths in watershed-scale hydrologic and soil erosion models. *Physical Geography, 13*(4), 368–385. https://doi.org/10.1080/02723646.1992.10642464.

20. Hartanto, H., Prabhu, R., Widayat, A. S. E., & Asdak, C. (2003). Factors affecting runoff and soil erosion: Plot-level soil loss monitoring for assessing sustainability of forest management. *Forest Ecology and Management, 180*(1–3), 361–374. https://doi.org/10.1016/S0378-1127(02)00656-4.

21. Hawbaker, T. J., Radeloff, V. C., Hammer, R. B., & Clayton, M. K. (2005). Road density and landscape pattern in relation to housing density, land ownership, land cover, and soils. *Landscape Ecology, 20*(5), 609–625. https://doi.org/10.1007/s10980-004-5647-0.

22. Heede, B. H., & King, R. M. (1990). State-of-the-art timber harvest in an Arizona mixed conifer forest has minimal effect on overland flow and erosion/L'exploitation rationelle du bois d'oeuvre dans une forêt mixte de conifères a un effet minimal sur l'écoulement de surface et l'érosion. *Hydrological Sciences Journal, 35*(6), 623–635. https://doi.org/10.1080/02626669009492478.

23. Jones, J. A., & Grant, G. E. (1996). Peak flow responses to clearcutting and roads in small and large basins, Western Cascades, Oregon. *Water Resources Research, 32*(4), 959–974. https://doi.org/10.1029/95WR03493.

24. Jordán-López, A., Martínez-Zavala, L., & Bellinfante, N. (2009). Impact of different parts of unpaved forest roads on runoff and sediment yield in a Mediterranean area. *Science of The Total Environment, 407*(3), 937–944. https://doi.org/10.1016/j.scitotenv.2008.09.048.

25. Jordán, A., & Martínez-Zavala, L. (2008). Soil loss and runoff rates on unpaved forest roads in southern Spain after simulated rainfall. *Forest Ecology and Management, 255*(3–4), 913–919. https://doi.org/10.1016/j.foreco.2007.10.021.

26. Kastridis, A., & Kirkenidis, C., & Sapountzis, M. (2020). An integrated approach of flash flood analysis in ungauged Mediterranean watersheds using post-flood surveys and unmanned aerial vehicles. *Hydrological Processes, 34*(14), 1–20. https://doi.org/10.1002/hyp.13752.

27. Kastridis, A., & Stathis, D. (2012). Natural and anthropogenic flash flood generation in mountainous watersheds—The case of Apollonia torrent. In *Proceedings of the Protection and Restoration of the Environment XI*, Thessaloniki, Greece, 3–6 July 2012; Water Resources Management (pp. 126–135). https://doi.org/10.13140/RG.2.1.1204.8725.

28. Kastridis, A., & Stathis, D. (2020). Evaluation of hydrological and hydraulic models applied in typical Mediterranean ungauged watersheds using post-flash-

flood measurements. *Journal of Hydrology, 7*(3), 1–12. https://doi.org/10.1016/j.jhydrol.2020.125660.

29. King, J. G., & Tennyson, L. C. (1984). Alteration of streamflow characteristics following road construction in north-central Idaho. *Water Resources Research, 20*(8), 1159–1163. https://doi.org/10.1029/WR020i008p01159.
30. La Marche, J. L., & Lettenmaier, D. P. (2001). Effects of forest roads on flood flows in the Deschutes River, Washington. *Earth Surface Processes and Landforms, 26*(2), 115–134. https://doi.org/10.1002/esp.184.
31. Lane, P. N. J., & Sheridan, G. J. (2002). Impact of an unsealed forest road stream crossing: Water quality and sediment sources. *Hydrological Processes, 16*(13), 2599–2612. https://doi.org/10.1002/hyp.1050.
32. Lane, P. N. J., Hairsine, P. B., Croke, J. C., & Takken, I. (2006). Quantifying diffuse pathways for overland flow between the roads and streams of the mountain ash forests of central Victoria, Australia. *Hydrological Processes, 20*(9), 1875–1884. https://doi.org/10.1002/hyp.5956.
33. Laurance, W. F., Sayer, J., & Cassman, K. G. (2014). Agricultural expansion and its impacts on tropical nature. *Trends in Ecology & Evolution, 29*(2), 107–116. https://doi.org/10.1016/j.tree.2013.12.001.
34. Loague, K., & Kyriakidis, P. C. (1997). Spatial and temporal variability in the R-5 infiltration data set: Déjà vu and rainfall-runoff simulations. *Water Resources Research, 33*(12), 2883–2895. https://doi.org/10.1029/97WR02422.
35. Lotfalian, M., Babadi, T. Y., & Akbari, H. (2019). Impacts of soil stabilization treatments on reducing soil loss and runoff in cut slope of forest roads in Hyrcanian forests. *Catena, 172*, 158–162. https://doi.org/10.1016/j.catena.2018.08.031.
36. Luce, C. H. (1997). Effectiveness of road ripping in restoring infiltration capacity of forest roads. *Restoration Ecology, 5*(3), 265–270. https://doi.org/10.1046/j.1526-100X.1997.09731.x.
37. Luce, C. H. (2002). Hydrological processes and pathways affected by forest roads: What do we still need to learn? *Hydrological Processes, 16*(14), 2901–2904. https://doi.org/10.1002/hyp.5061.
38. Luce, C. H., & Cundy, T. W. (1994). Parameter identification for a runoff model for forest roads. *Water Resources Research, 30*(4), 1057–1069. https://doi.org/10.1029/93WR03298.
39. MacDonald, L. H., & Coe, D. B. R. (2008). Road sediment production and delivery: Processes and management. In *Proceedings of the First World Landslide Forum*, United Nations University, Tokyo, Japan, 18–21 November 2008 (pp. 710–713). https://doi.org/10.1201/9780203864201.ch79.
40. MacDonald, L. H., Sampson, R. W., & Anderson, D. M. (2001). Runoff and road erosion at the plot and road segment scales, St John, US Virgin Islands. *Earth Surface Processes and Landforms, 26*(3), 251–272. https://doi.org/10.1002/esp.184.
41. Megahan, W. F. (1983). Hydrologic effects of clearcutting and wildfire on steep granitic slopes in Idaho. *Water Resources Research, 19*(3), 811–819. https://doi.org/10.1029/WR019i003p00811.

42. Megahan, W. F., & Clayton, J. L. (1983). Tracing subsurface flow on roadcuts on steep, forested slopes. *Soil Science Society of America Journal, 47*(6), 1063–1067. https://doi.org/10.2136/sssaj1983.03615995004700060036x.

43. Mirus, B. B., Ebel, B. A., Loague, K., & Wemple, B. C. (2007). Simulated effect of a forest road on near-surface hydrologic response: Redux. *Earth Surface Processes and Landforms, 32*(1), 126–142. https://doi.org/10.1002/esp.1387.

44. Montgomery, D. R. (1994). Road surface drainage, channel initiation, and slope instability. *Water Resources Research, 30*(7), 1925–1932. https://doi.org/10.1029/94WR00538.

45. Negishi, J. N., Sidle, R. C., Ziegler, A. D., Noguchi, S., & Rahim, N. A. (2008). Contribution of intercepted subsurface flow to road runoff and sediment transport in a logging-disturbed tropical catchment. *Earth Surface Processes and Landforms, 33*(8), 1174–1191. https://doi.org/10.1002/esp.1591.

46. Orfánus, T., Jen⊠o, M., Bebej, J., & Benko, M. (2017). Simulation of the effects of forest roads on stormflow generation using GIS and 2D vadose zone hydrological model. *Ekológia (Bratislava), 36*(1), 25–39. https://doi.org/10.1515/eko-2017-0003.

47. Ramos-Scharrón, C. E., & Figueroa-Sánchez, Y. (2017). Plot-, farm-, and watershed-scale effects of coffee cultivation in runoff and sediment production in western Puerto Rico. *Journal of Environmental Management, 202*, 126–136. https://doi.org/10.1016/j.jenvman.2017.07.006.

48. Ramos-Scharrón, C. E., & LaFevor, M. C. (2016). The role of unpaved roads as active source areas of precipitation excess in small watersheds drained by ephemeral streams in the Northeastern Caribbean. *Journal of Hydrology, 533*, 168–179. https://doi.org/10.1016/j.jhydrol.2015.12.011.

49. Ramos-Scharrón, C. E., & LaFevor, M. C. (2018). Effects of forest roads on runoff initiation in low-order ephemeral streams. *Water Resources Research, 54*(10), 8613–8631. https://doi.org/10.1029/2017WR021961.

50. Ramos-Scharrón, C. E., & MacDonald, L. H. (2007). Runoff and suspended sediment yield from an unpaved road segment, St John, US Virgin Islands. *Hydrological Processes, 21*(1), 35–50. https://doi.org/10.1002/hyp.6181.

51. Reid, L. M., & Dunne, T. (1984). Sediment production from forest road surfaces. *Water Resources Research, 20*(11), 1753–1761. https://doi.org/10.1029/WR020i011p01753.

52. Rijsdijk, A., Sampurno-Bruijnzeel, L. A., & Sutoto, C. K. (2007). Runoff and sediment yield from rural roads, trails and settlements in the upper Konto catchment, East Java, Indonesia. *Geomorphology, 87*(1–2), 28–37. https://doi.org/10.1016/j.geomorph.2006.06.043.

53. Robichaud, P. R., Foltz, R. B., & Luce, C. H. (1993). Development of an on-site prediction model for forest roads and timber harvest areas. In R. F. Hadley & T. Mizuyama (Eds.), *Sediment Problems: Strategies for Monitoring, Prediction and Control* (Proceedings of the Yokohama Symposium, July 1993; pp. 135–140). IAHS Publication No. 217. International Association of Hydrological Sciences.

54. Robichaud, P. R., Wagenbrenner, J. W., & Brown, R. E. (2010). Rill erosion in natural

and disturbed forests: 1. Measurements. *Water Resources Research, 46*(4), 1–14. https://doi.org/10.1029/2009WR008314.

55. Sapountzis, M., & Stathis, D. (2014). Relationship between rainfall and run-off in the Stratoni region (N. Greece) after the storm of 10th February 2010. *Global NEST Journal, 16*(2), 420–431. https://doi.org/10.30955/gnj.001293.

56. Sheridan, G. J., Noske, P. J., Lane, P. N. J., & Sherwin, C. B. (2008). Using rainfall simulation and site measurements to predict annual interrill erodibility and phosphorus generation rates from unsealed forest roads: Validation against *in-situ* erosion measurements. *Catena, 73*(1), 49–62. https://doi.org/10.1016/j.catena.2007.09.007.

57. Sidle, R. C., & Ziegler, A. D. (2012). The dilemma of mountain roads. *Nature Geoscience, 5*(7), 437–438. https://doi.org/10.1038/ngeo1512.

58. Sidle, R. C., Sasaki, S., Otsuki, M., Noguchi, S., & Rahim, N. A. (2004). Sediment pathways in a tropical forest: Effects of logging roads and skid trails. *Hydrological Processes, 18*(4), 703–720. https://doi.org/10.1002/hyp.1364.

59. Sosa-Pérez, G., & MacDonald, L. H. (2017). Wildfire effects on road surface erosion, deposition, and road–Stream connectivity. *Earth Surface Processes and Landforms, 42*(5), 735–748. https://doi.org/10.1002/esp.4022.

60. Soulis, K. X., Dercas, N., & Papadaki, C. H. (2015). Effects of forest roads on the hydrological response of a small-scale mountain watershed in Greece. *Hydrological Processes, 29*(8), 1772–1782. https://doi.org/10.1002/hyp.10312.

61. Surfleet, C. G., Skaugset, A. E., III, & Meadows, M. W. (2011). Road runoff and sediment sampling for determining road sediment yield at the watershed scale. *Canadian Journal of Forest Research, 41*(9), 1970–1980. https://doi.org/10.1139/x11-110.

62. Tague, C., & Band, L. (2001). Simulating the impact of road construction and forest harvesting on hydrologic response. *Earth Surface Processes and Landforms, 26*(2), 135–151. https://doi.org/10.1002/1096-9837(200102)26:2<135:AID-ESP160>3.0.CO;2-R.

63. Takken, I., Croke, J., & Lane, P. (2008). A methodology to assess the delivery of road runoff in forestry environments. *Hydrological Processes, 22*(2), 254–264. https://doi.org/10.1002/hyp.6857.

64. Thomas, R. B., & Megahan, W. F. (1998). Peak flow responses to clear-cutting and roads in small and large basins, western Cascades, Oregon: A second opinion. *Water Resources Research, 34*(12), 3393–3403. https://doi.org/10.1029/98WR02500.

65. Thomaz, E. L., & Ramos-Scharrón, C. E. (2015). Rill length and plot-scale effects on the hydrogeomorphologic response of gravelly roadbeds. *Earth Surface Processes and Landforms, 40*(15), 2041–2048. https://doi.org/10.1002/esp.3773.

66. Thomaz, E., & Peretto, G. (2016). Hydrogeomorphic connectivity on roads crossing in rural headwaters and its effect on stream dynamics. *Science of the Total Environment, 550*, 547–555. https://doi.org/10.1016/j.scitotenv.2016.01.086.

67. United States Department of Agriculture Forest Service (USDA). (2004). *Forest Roads: A Synthesis of Scientific Information* (Gucinski, H., Furniss, M. J., Ziemer,

R. R., & Brookes, M. H., Eds.). General Technical Report PNW-GTR-509. United States Department of Agriculture Forest Service.

68. USDA Forest Service. (2001). *National Forest System Road Management Strategy*. United States Department of Agriculture Forest Service.

69. Wagenbrenner, J. W., Robichaud, P. R., & Elliot, W. J. (2010). Rill erosion in natural and disturbed forests: 2. Modeling approaches. *Water Resources Research, 46*(W10507), 1–14. https://doi.org/10.1029/2009WR008315.

70. Weiler, M., McDonnell, J. J., Tromp-van Meerveld, H. J., & Uchida, T. (2005). Subsurface stormflow. In M. G. Anderson (Ed.), *Encyclopedia of Hydrological Sciences* (pp. 1719–1732). Wiley and Sons. https://doi.org/10.1002/0470848944.hsa135.

71. Wemple, B. C., & Jones, J. A. (2003). Runoff production on forest roads in a steep, mountain catchment. *Water Resources Research, 39*(8), SWC81–SWC817. https://doi.org/10.1029/2002WR001744.

72. Wemple, B. C., Jones, J. A., & Grant, G. E. (1996). Channel network extension by logging roads in two basins, western Cascades, Oregon. *Water Resources Bulletin, 32*(6), 1195–1207. https://doi.org/10.1111/j.1752-1688.1996.tb03490.x.

73. Wemple, B. C., Swanson, F. J., & Jones, J. A. (2001). Forest roads and geomorphic process interactions, Cascade Range, Oregon. *Earth Surface Processes and Landforms, 26*(2), 191–204. https://doi.org/10.1002/1096-9837(200102)26:2<191:AID-ESP175>3.0.CO;2-U.

74. Wemple, B., Browning, T., Ziegler, A. D., Celi, J. E., Chun, K. P., Jaramillo, F., Leite, N. K., Ramchunder, S. J., Negishi, J. N., Palomeque, X., et al. (2018). Ecohydrological disturbances associated with roads: Current knowledge, research needs, and management concerns with reference to the tropics. *Ecohydrology, 11*(3), e1881. https://doi.org/10.1002/eco.1881.

75. Wigmosta, M. S., & Perkins, W. A. (2001). Simulating the effects of forest roads on watershed hydrology. In *Land Use and Watersheds: Human Influence on Hydrology and Geomorphology in Urban and Forest Areas*. American Geophysical Union.

76. Woldie, D. W., Sidle, R. C., & Gomi, T. (2009). Impact of road-generated storm runoff on a small catchment response. *Hydrological Processes, 23*(24), 3631–3638. https://doi.org/10.1002/hyp.7459.

77. Zemke, J. J. (2016). Runoff and soil erosion assessment on forest roads using a small-scale rainfall simulator. *Journal of Hydrology, 3*, 25–36. https://doi.org/10.1016/j.ejrh.2016.03.001.

78. Ziegler, A. D., & Giambelluca, T. W. (1997). Importance of rural roads as source areas for runoff in mountainous areas of northern Thailand. *Journal of Hydrology, 196*(1–4), 204–229. https://doi.org/10.1016/S0022-1694(96)03259-0.

79. Ziegler, A. D., Giambelluca, T. W., Sutherland, R. A., Vana, T. T., & Nullet, M. A. (2001). Horton overland flow contribution to runoff on unpaved mountain roads: A case study in northern Thailand. *Hydrological Processes, 15*(16), 3203–3208. https://doi.org/10.1002/hyp.1034.

80. Ziegler, A. D., Negishi, J. N., Sidle, R. C., Gomi, T., Noguchi, S., & Nik, A. R. (2007). Persistence of road runoff generation in a logged catchment in Peninsular Malaysia. *Earth Surface Processes and Landforms, 32*(13), 1947–1970. https://doi.org/10.1002/esp.1495.

81. Ziegler, A. D., Negishi, J., Sidle, R. C., Preechapanya, P., Sutherland, R. A., Giambelluca, T. W., & Jaiaree, S. (2006). Reduction of stream sediment concentration by a riparian buffer: Filtering of road runoff in disturbed headwater basins of Montane Mainland Southeast Asia. *Journal of Environmental Quality, 35*(1), 151–162. https://doi.org/10.2134/jeq2005.0114.

CHAPTER 6

The Environmental Value of Forests

LEARNING OBJECTIVES

After studying this chapter, you will be able to:

- Understand the components of the environment and their importance in forest ecosystems.
- Explain principles of environmental management and conservation relevant to forest protection.
- Evaluate the impacts of deforestation on climate change and other environmental factors.

KEY TERMS FROM THIS CHAPTER

Environment
Efficiency and Effectiveness
Principle of Precaution (PPr)
User Pays Principle (UPP)
Polluter Pays Principle (PPP)
Principle of Participation (PPa)
Principle of responsibility (PR)

INTRODUCTION

Forests play an important role in environmental protection. Forest protection in mountainous areas has a long history, and it helps to prevent soil erosion, landslides, and avalanches while also preserving the water quality of rivers that drain forested catchments. Special forestry methods should be used to keep forests in good condition indefinitely. Forests also help to protect the environment. Air pollution is one of the major issues, and it has been shown to have serious consequences for certain forests. Pollutants of concern for the air include sulfur dioxide, fluoride hydrogen, heavy metals, and ozone. Finally, reducing pollution benefits forests. Forests play an important role in maintaining the global carbon cycle. The conversion of forests to other land uses contributes to increased atmospheric carbon dioxide levels and is a major source of carbon dioxide emissions. Reforestation and afforestation may help reduce atmospheric carbon dioxide concentrations, while the use of biofuels may help reduce demand for fossil fuels. Forests have recognized a relationship with the environment for over a 1,000 years, but forestry practices continue to cause environmental damage such as soil erosion, water quality degradation, and other negative consequences. Today, however, forest protection is rapidly deteriorating due to a variety of factors, including the expansion of agriculture and timber plantations; other land uses such as pulp and paper plantations, urbanization, road construction, industry, and forests, which pose the greatest and most serious threat to environmentally damaged forests. As a result, raising public awareness is essential. In previous chapters, we discussed various strategies for forest conservation and raising public awareness, both of which play critical roles in maintaining a healthy environment. This chapter discusses various conservation strategies for forests, which play an important role in maintaining environmental balance. This chapter examines various aspects of forest-environmental protection relations to highlight the fact that forestry is a global issue.

6.1. ENVIRONMENT: COMPONENTS AND IMPORTANCE

The environment is a complex of many variables, both human and living organisms. Environmental studies describe the interrelationships between organisms, the environment, and the majority of the elements that influence earth life, such as feed chains, air, and water cycles. It is a fundamental science about our planet and how it works daily, so it is important to everyone. Almost everyone around us now uses the term "environment," including on television and in newspapers. All of them discuss environmental protection and pre-serration. Global summits on environmental issues take place regularly. Within the last century, geography, anthropology, and sociology have addressed the mutual relationship between the environment, cultural organization, and culture. All of this demonstrates the planet's increasing importance.

Additionally, life is linked to the planet. Social sciences borrow the concept of ecology from biology. Ecology is a biological industry that studies the relationship between living things and their surroundings. Biology has had a significant influence on sociology. Ecological research can also investigate the relationship between the environment and humans. Human Ecology in Sociology will conduct research on the relationship between man and his environment. The study of human ecology begins with Burgess and Park in sociology. There is a close relationship between humans and the planet. On the one hand, humans are present on the planet and coexist with it. Humans, on the other hand, grow tired of managing and changing their environment in response to their needs. As a result, people must understand their surroundings (Figure 6.1).

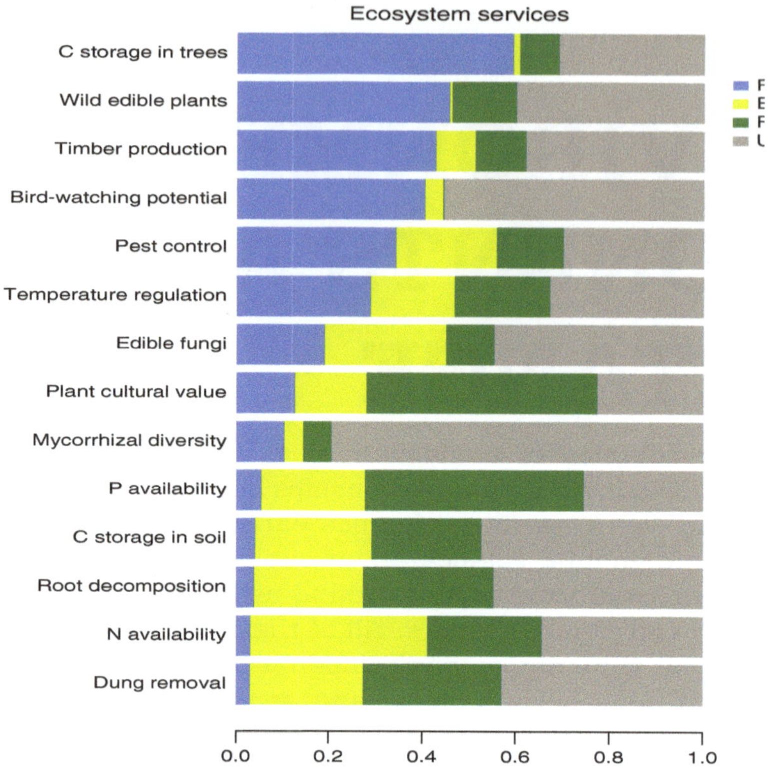

Figure 6.1. *Importance of forest attributes and environmental factors for ecosystem services. Bars show the variance exclusively explained by forest attributes and environmental factors, the shared variance between them and the unexplained variance.*

Source: Felipe-Lucia, M.R., Soliveres, S., Penone, C. et al. Multiple forest attributes underpin the supply of multiple ecosystem services. Nat Commun 9, 4839 (2018). https://doi.org/10.1038/s41467-018-07082-4.

6.1.1. Meaning and Definition

The term "environment" derives from the French word "Environia," which means to surround. It refers to both biotic and abiotic environments. The word environment refers to the surroundings in which creatures live.

The environment and organisms are two of nature's most powerful and complex elements. The environment has an impact on the lives of humans and organisms. People interact more forcefully with the environment than other living things.

The environment is defined broadly as the materials and forces that surround a living organism. The environment is the sum of the conditions we encounter in a specific location and time. It consists of interacting systems of physical, cultural, and biological elements that are both individually and collectively interconnected. The environment is unquestionably the sum of the circumstances that an organism must endure to survive or even exist. It has an impact on how living things grow and develop. The environment is defined as the space that surrounds and influences living things on all sides. It comprises the atmosphere, hydrosphere, biosphere, and lithosphere. Sun, water, air, bodies, and solar energy are all necessary elements. It gave us the strength to lead a comfortable life.

1. In terms of P. Gisbert "Environment is anything immediately surrounding an object and exerting a direct influence on it."
2. In terms of E. J. Ross "Environment is an exterior force which impacts us."

Therefore, environment refers to whatever is around an object promptly and has a strong impact on it. Our environment refers to those things or perhaps agencies that affect our activities or lives, albeit separate from us. Natural, social, artificial, psychological and biological factors could be the environment around which man is affected. The whole physical and natural system of humans and several factors and other entities affecting them are identified in the ecosystem. Factors include soil, light, water, air, temperature etc. These elements are called Abiotic. Aside from the abiotic factors, biotic factors, like plants, microorganisms, and animals significantly affect the planet. Hence human is also a component of the environment which is inseparable. Man and also the planet are very closely linked. Human social life is affected by the planet. This is why cultural and social activities of various kinds are performed worldwide. The hilly persons have various types of life than the normal humans. Furthermore, people worldwide differ in their food, festivals, clothes, and more. These factors are affected by him.

Three important components can be found in the environment. These are:

- Physical
- Biological
- Social

The physical constituents of the environment include soil, climate, air, water, light and temperature. These are sometimes known as abiotic environmental components. This particular component of the environment determines mainly the habitat type of the human population or perhaps its living conditions. Once again, this physical component is divided into 3 parts.

- Atmosphere (gas).
- Hydrosphere (liquid); and
- Lithosphere (mass).

These 3 parts represent the 3 key environmental states of matter. This physical part of the planet consists of only non-living items such as air, earth and water. All these nonliving things affect every living organism, such as man, to a great level. The main abiotic elements affecting living beings are temperature and water. Drinking water is accountable for a greater proportion of the body's weight. For their survival, all living organisms require water. Along with water, the main vital fluid for maintaining optimal body temperature is water. Every lifetime is activated in a particular temperature range. If the temperature is extreme, the living creatures will die. Air is the primary physical component providing breathing oxygen. All living creatures, which include animals and plants, require oxygen. By the respiration process, oxygen is taken into the body and derived from carbon dioxide. These plants, on the contrary, take carbon dioxide during photosynthesis and distribute oxygen to the planet for food preparation. Dirt is the central habitat for all living beings. It is the earth where plants grow and humans build houses. It is the soil's groundwater that provides for drinking activities and other agriculture.

The biological environmental component is also known as the green biotic component. All living things, including plants, animals, and small microorganisms like bacteria, make up the element. This component interacts with the environment's abiotic components.

In this component, two elements interact in a variety of ecosystems, including ponds, deserts, and marine ecosystems. The biosphere is defined as the Earth's self-sustaining large ecosystem. Most ecosystems contain at least three types of living organisms.

These three types are named as:

- Producers.
- Consumers.
- Decomposers.

Green plants and other photosynthetic bacteria use energy from water, light, and soil to produce a variety of natural substances such as carbohydrates, proteins, and more. Consumers rely on organic foods derived from green plants for nutrition. Decomposers decompose dead animals and plants, returning several essential minerals to biogeochemical cycles.

The social component of the environment consists primarily of diverse population groups of various living organisms, such as birds, animals, and others. Man is arguably the most intelligent living organism. Humans, like all other living beings, build houses, prepare food, and emit waste into the environment. Man, as the Greek philosopher Aristotle stated, is a social animal. He creates various policies and laws to ensure the proper functioning of society. The three environmental components combine to form four major areas. The air, hydrosphere, biosphere, and lithosphere are all the same. Between these four zones, there is constant interaction. These interactions involve the movement of various components, forms, and compounds of energy. The following sections explain these zones. The Earth's atmosphere, which is an intricate fluid system of gases and suspended particulates, did not form at the beginning of the world.

6.1.2. Atmosphere

The atmosphere that we have today is the result of chemical and biochemical reactions on Earth. While the fluid system forms a gaseous envelope around the earth, its boundaries are not easily defined. The interface between the Earth's atmosphere and space can be defined arbitrarily. The entire atmosphere is composed of gases such as nitrogen, argon, oxygen, carbon dioxide, and water vapor. Gaseous turbidity is the combination of suspended particles such as soot and dust, particularly in the troposphere. Nonetheless, the structure and composition of the atmosphere change over time and space. The vertical structure of the atmosphere is closely related to the absorption of radiant energy and is also known as the temperature variable (Figure 6.2).

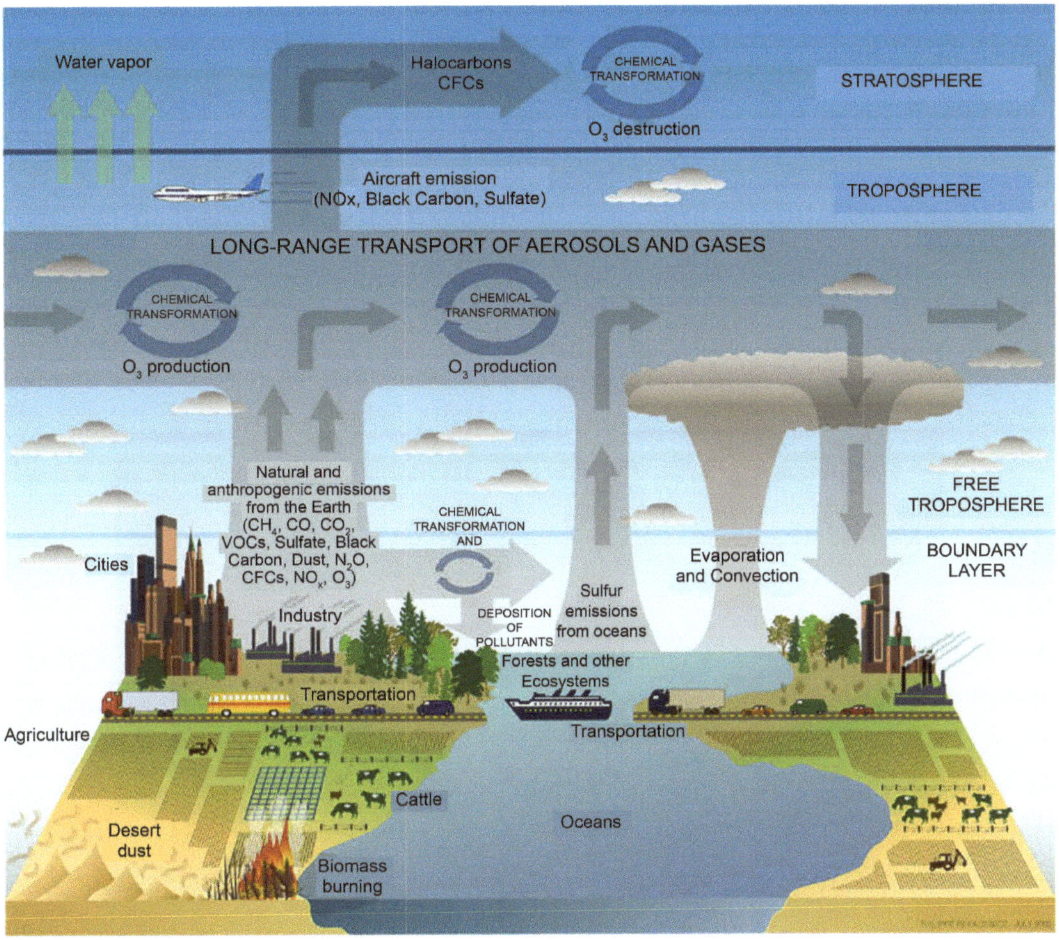

Figure 6.2. *Diagram showing the evolution/cycles of various elements in Earth's atmosphere.*

Source: Wikimedia commons.

Below 60 kilometers, the Earth's surface and the Ozone layer have two primary absorption zones. Radiation, conduction, and convection redistribute absorbed energy. As a result, there are two maximum temperatures: on the Earth's surface and 50 kilometers above it. Above each of these maxima, the mixing is primarily convectional. The temperature in those mixing layers decreases as one moves higher above the heat source. The bottom of these two areas is known as the top, and the troposphere as the mesosphere. The levels of small blending at which the atmosphere is likely to form a layered structure known as the stratosphere are distinguishable. The tropopause represents the near-upper limit of mixing in the lower atmosphere between the stratosphere and the ionosphere. It usually has an average height of 11 kilometers, but this varies around the world. It has an average height of 16 kilometers in tropical latitudes. It is only 10 kilometers in polar latitude. Another heating zone exists over the mesosphere at a height of over 90 km. Shortwave ultraviolet radiation is absorbed at this height by the earth's surface, where there are many oxygen molecules. This is known as the heat. Ionization occurs within this layer, resulting in free electrons and charged ions. Beyond the heat, about 700 kilometers high, is the exosphere, which has an extremely low atmospheric density. At this level, the concentration of ionization particles in bands known as Van Allen radiation belts is increasing. Nonetheless, it is possible to simplify this simple model of vertical structure to make the atmosphere a model of two concentric shells, the boundaries of which are defined by the stratopause at approximately 50 km. Over the surface area of the Earth and at approximately 80,000 km hypothetical atmospheric outer limit, 99% of the total atmospheric mass is below the stratopause, in the troposphere and stratosphere, and circulatory systems operate at this level.

It is a level of about 80,000 km beyond stratopause. Thick only contains 1% of the entire atmosphere mass and has high-energy ionization of short wavelength solar radiation. The atmospheric layer temperature profiles are shown in the Figure 6.3.

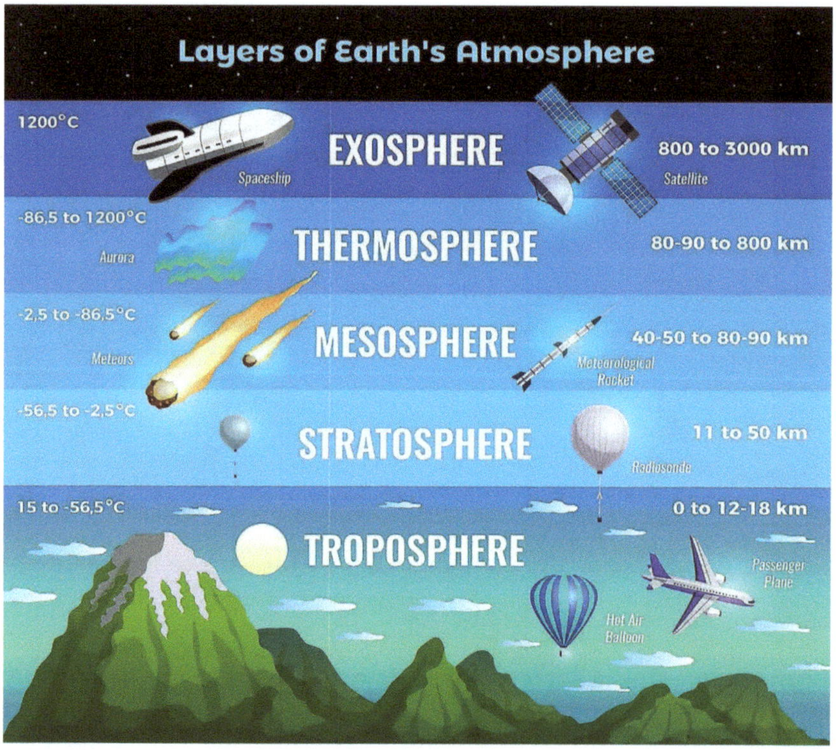

Figure 6.3. *The atmospheric layer temperature profiles.*

Source: https://www.worldatlas.com/articles/what-are-the-5-layers-of-the-earth-s-atmosphere.html.

6.1.3. Hydrosphere

The surface waters include the interface between them. It is crucial to survive for the molecule. Water has many physical and chemical characteristics that help the molecule to act as the very best medium for life. The movement of water through the hydrological cycle from the planet's surface to the atmosphere appears to be close. Water is probably the most abundant matter on the surface area of the Earth. The oceans cover approximately 71% of the planet's water, and glaciers and ice caps cover more areas. The streams and lakes, the soils as well as the sub-terranean reservoirs, the bodies and the air of all living organisms contain water. Consequently, we're well acquainted with water, in all its forms ice, fluid, vapor and water. We use water for home, industry, farming and leisure. The quality and quantities of the water they require differ widely in these applications. We use many sources of inland water, groundwater and oceanic water in one way or perhaps another. The need for worldwide water is growing day by day through the availability of clean natural water. This simply means that clean natural water and its fruitful storage and preservation must be used preciously (Figure 6.4).

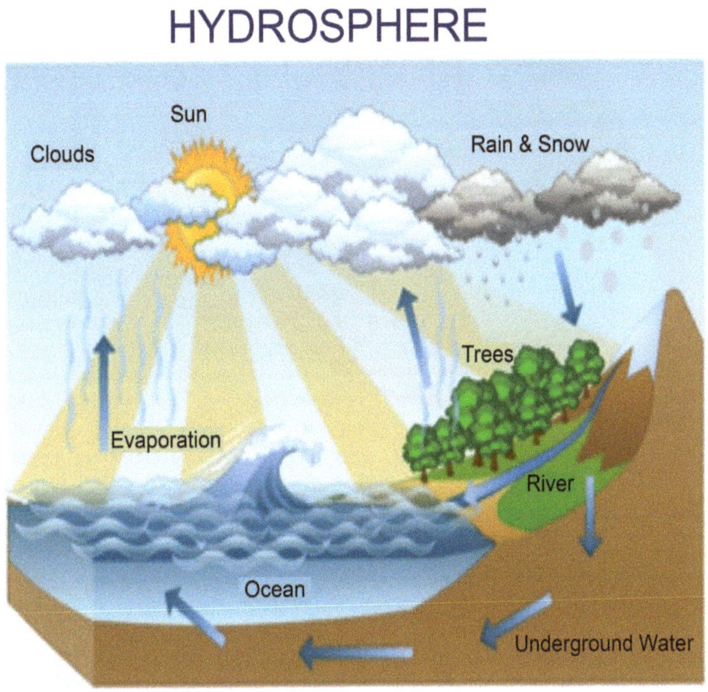

Figure 6.4. *A simplified outline of the hydrosphere.*

Source: https://www.teacharesources.com/product/grade-10-the-hydrosphere-in-animated-powerpoint/.

6.1.4. Lithosphere

It is the outer boundary layer of the discontinuity and solid earth within the mantle. The exterior frontier forms a complex interface with the atmosphere and hydrosphere and is where life has evolved. The inner limit of the rock is virtually its melting point and lets you move relative to the above-mentioned lithosphere. The lithosphere is basically nothing much more than a crustal system made up of many layers: core, crust and mantle (Figure 6.5). Different elements comprise another mixing proportion of the crustal layer. The ground curst consists generally of three main classes of rocks: igneous rocks, sedimentary rocks, and metamorphosed rocks (as classified by their mode of origin). There are two crusts-the continental crust consisting of silicon aluminum granite rocks with a mean density of 2.8, and the various other oceanic crusts consisting of much more basic minerals with a mean density of 3.0 which is basaltic in composition. The average earth density is 5.5 gm/c.c.

Interaction between the lithosphere, atmosphere and biosphere crustal system occurs where continental crust is exposed above the sea levels. At the interface between air and ground, crustal material is exposed to solar radiant energy, precipitation and air gas input. The consequences of the living systems of the biosphere often modify or perhaps operate these inputs. The effect of these inputs divides crustal rocks into weathering processes and transmits them to soils, and fine crustal layers.

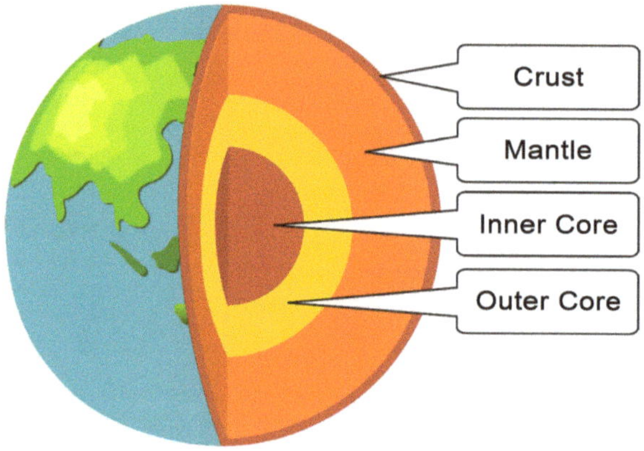

Figure 6.5. *Provides a contour of earth layers and the composition of crustal materials.*

Source: https://www.pinterest.es/pin/304485624782442372/.

6.1.5. Biosphere

The biosphere comprises all the areas of the world where life is present whole bio related sources. It has been developing on Earth through evolutionary processes for 4.5 billion years. Life of many types is present at the top part of the lithosphere, across the hydrosphere and in the lower atmosphere. The "biosphere" where mankind acts as probably the most evolved animal is the bio-resources and the surrounding area. The steps that lead to the beginnings of life right here on earth are extremely complicated and take several centuries. The details of the atmosphere, the processes involved as well as the sequence of certain events leading to the advancement of living cells form considerable uncertainty. The standard viewpoint was based on the heterotrophic prokaryotic bacteria being the first organism on the plant. Autotrophic eukaryotes and prokaryotes subsequently begin to show up as evolutionary changes step by step (Figure 6.6).

Water, a power source (several nutrients and sunlight) found in air, water, and soil are essential for life on the Earth. Please note: On the top of the atmosphere and beneath the ground, the right combinations of these essentials cannot be found. They're only sold in a narrow layer near the planet's surface. This particular level of biosphere stretches across the majority of the Planet's surface. The top layers as well as the thick layer of soil supporting the plant life of the earth are included in the earth. This living area covers about eight km as well. This zone of life also spreads about 8 km. up into the atmosphere and much as 8 km. down into the depths of the sea. But there are very few organisms that live on polar Ice caps and glaciers whereas there are numerous in tropical rainforests. The earth isn't evenly distributed to living organisms. The biosphere is home to a number of main regions that have particular types of ecosystems. These figures imply the natural world is very different on Earth.

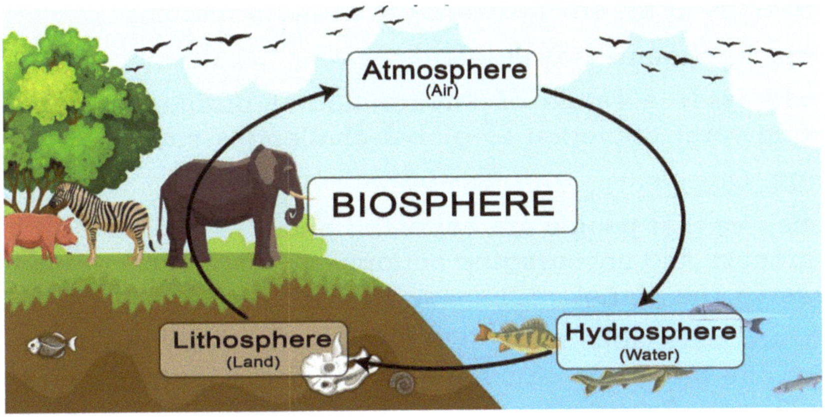

Figure 6.6. *An outline of the biosphere.*

Source: https://ar.inspiredpencil.com/pictures-2023/earths-biosphere-diagram.

6.1.6. Importance of the Subject

There are many environmental challenges facing the world, ranging from local issues like groundwater depletion to global problems such as climate change. These issues can only be effectively addressed if everyone is engaged in environmental stewardship and understands the causes, consequences, and remedial measures for various environmental issues. Green science is promoted and taught at various educational levels to achieve this goal. This field is crucial because it aims to protect the integrity of Earth, a unique planet that supports life. The significance of green science can be described in terms of the following seven key objectives:

- Understanding Environmental Impacts:
  - It helps us understand how our daily and developmental activities impact the planet and how environmental changes affect us.
- Pollution Prevention and Control:
  - It promotes the adoption of techniques for preventing and controlling pollution, aiming to create a pollution-free environment with clean air, water, soil, and food.
- Resource Conservation and Recycling:
  - It guides the efficient use of natural resources such as water, forests, minerals, and fossil fuels through conservation and recycling practices to minimize waste and maximize utility.

- Eco-Friendly Lifestyles:
  - It educates the public on understanding the effects of activities on the environment, pollution prevention and management, and the effective use of resources, leading to more sustainable and eco-friendly lifestyles.
- Sustainable Industrial Practices:
  - It encourages industries to adopt clean, efficient pollution control technologies to operate in an environmentally friendly manner.
- Addressing Environmental Problems:
  - It addresses a range of environmental problems, from local issues like groundwater depletion to global challenges such as climate change.
- Educational Outreach:
  - It ensures that people are educated about environmental issues, promoting awareness and encouraging actions to mitigate environmental damage and preserve the planet.

These issues are only able to be fixed whether everyone is worried about the environment; everyone has to be educated on the causes, effects and remedial actions of different green issues. Green science is promoted as well as taught at various levels of education to achieve this goal. The topic is very important, as it seeks to protect the integrity of the earth, a unique world that sustains life, and the livelihood environment. In terms of the different objectives it meets to save the planet, the value of the topic can be described.

6.2. PRINCIPLES OF ENVIRONMENTAL MANAGEMENT AND CONSERVATION

Environmental management principles refer to activities that enhance beneficial links and minimize adverse links between resource systems and their environments and that are aimed at attaining the desired environment. Particularly, the management of resources is often a special objective whereas environmental management is always multi-purpose, as resource management focuses only on its resource system and environmental management focuses both on the resource system and on its surroundings. Additionally, the needs and values of the community are essential to setting the environmental goal. These are some environmentally friendly management guiding principles. These are helpful in decision-making on the environment.

6.2.1. Polluter Pays Principle (PPP)

For the past two decades, many economists have recommended that companies that discharge polluting environmental effluents somehow pay the cost of the harm they cause to the planet. The OECD proposed PPP as a common timeframe for environmentally friendly policies. It states that polluters should bear the cost of pollution-reduction measures.

As defined by the OECD Council, "the so-called PPP is the concept to be used to allocate pollution prevention and control measures to encourage the rational use of scarce environmental resources and also to stay away from distortions of global trade and investment."

According to the Supreme Court of India, the principle for polluted countries means that total responsibility for harm to the planet includes not only compensating pollution victims but also treating the cost of environmental damage. As a result, it includes both direct and environmental costs for properties and individuals. The recovery of the damaged environment is an important part of the sustainable development process, and the polluter is likely to help both the prices of the damaged ecology and the costs of reversing the polluter's damage. The application of this principle is contingent on interpretations, cases, and specific circumstances. During the 1992 Rio Earth Summit, this principle sparked much more debate. To combat environmental degradation, the South has requested financial assistance from the North. In particular, liability and the use of financial tools have practical implications for assigning economic obligations to environmental degradation.

6.2.2. The User Pays Principle (UPP)

It is considered a part of the PPP. The principle states that all resource users should bear the full and long-term marginal costs of using related services and resources, including any associated treatment costs. Its purpose is to use and consume resources.

6.2.3. The Principle of Precaution (PP)

The primary goal of the precautionary principle is to ensure that, even if there is no conclusive scientific evidence linking a specific activity or substance to environmental damage, a substance or activity posing a risk to the environment is avoided; "activity" and "substance" are the result of human intervention. The Rio Declaration's Principle Fifteen emphasizes the concept of irreversible or serious damage threats. There is no reason to put off cost-effective environmental degradation measures due to a lack of complete scientific certainty. This process is therefore essential for implementing energy resources in the areas of distribution and production to protect the environment and human health.

6.2.4. Principle of Efficiency and Effectiveness

The use of policy tools that provide incentives to reduce destructive use could help to increase material efficiency. Furthermore, it applies streamlining procedures and processes to various issues of environmentally friendly governance to reduce environmental costs.

6.2.5. The Principle of Responsibility

All humans, businesses, and governments are responsible for maintaining ecological processes. Access to green resources comes with a responsibility to use them in an environmentally sustainable, economically efficient, and socially just manner.

6.2.6. The Principle of Participation

Everyone has a responsibility to participate in global decision-making. Several fields of participation concern the use of plants, soils, minerals, wildlife, or fish for food and materials, as well as recreation. The next issue concerns solid waste, which includes waste, demolition and construction materials, chemical waste, etc. The final issue concerns pollution-producing activities.

6.3. IMPACT OF DEFORESTATION

Forests produce vital oxygen in the aftermath of deforestation while also providing habitat for animals and humans. Many of the world's most endangered animals live in forests, and 1.6 billion people rely on forest products such as clothing, fresh water, food, shelter, and medicine. However, deforestation threatens the world's forests, putting these benefits at risk. Deforestation can take many forms, including fires, farming, ranching, and development reductions, unsustainable logging, and climate change degradation. This has an impact on human livelihoods and poses a threat to many animal and plant species. Every year, we lose 18.7 million acres of forest, equivalent to 27 soccer fields per minute. Forests play an important role as a CO_2 plinth, which traps carbon dioxide that would otherwise be free in the atmosphere and contributes to ongoing climate change. Deforestation undermines the important carbon sink function. According to estimates, deforestation accounts for 15% of total greenhouse gas emissions. Deforestation in tropical rainforests is a major concern because of the forests' global biodiversity. In the Amazon, for example, approximately 17% of the forest has been lost over the last 50 years, primarily due to conversion for cattle farming. Deforestation is especially rapid in this area, near more populous areas, rivers, and highways, but even remote locations containing valuable mahogany, gold, and petroleum have been entrusted.

The degradation of forests has a number of consequences, particularly for farmers' livelihoods and the community as a whole. Deforestation converts forests to permanent, alternative land uses such as agriculture, urban development, and pasture. Deforestation is a major issue in developing countries because tropical forests are shrinking, resulting in biodiversity loss and an increase in greenhouse gas emissions. The FAO considers tree planting primarily established for forest wood production to be deforestation and thus does not classify the conversion of natural forests to planting as such. Nonetheless, despite classifying rubber plantations as forests, the FAO does not consider tree plantations that produce forest products. Forest breakdown occurs when forest ecosystem functions deteriorate but the area remains mostly forested. This section discusses the causes, solutions, and effects of deforestation.

The most significant effects of forest degradation would be climate change, soil erosion, water loss, decreased biodiversity, loss of conflict and habitat, economic loss, and a fuelwood deficit. Agriculture expansion, logging, small farming, overgrazing, small-scale mining and associated pollution, over-harvesting of wood and non-wood products, and extreme drought are all indicators of forest degradation.

Smallholder farmers move through the forest to burn and slash for farming. Various scientific studies show that agricultural expansion is a major cause of forest degradation in the country. There has been debate about the impact of population density on forest degradation. According to international agencies such as intergovernmental bodies and the Food and Agriculture Organization of the United Nations, poverty and overpopulation are the primary causes of forest loss. These organizations are thought to address the issue by promoting development and attempting to reduce population growth. "An annual destruction rate appears to be stepping up even more, and could increase over the next decade," says Professor Norman Myers, the leading authority on tropical deforestation rates. The subtropical and temperate regions have been mostly deforested.

Deforestation is minimal in advanced temperate countries, and many temperate countries are expanding their forest areas. Developed countries are typically located in moderate areas, while tropical developing countries are in tropical areas. However, deforestation in tropical humid leeway forests decreased significantly from 1980–1990 to 1990–2000, but the FAO overestimated tropical leeway forest deforestation by 23% using satellite images. However, it is still debatable what is and is not. Rainforests receive the most attention; however, moist leafy and dry forests accounted for 60% of total deforestation in tropical forests between 1990 and 2010. Nonetheless, a somewhat modern event that gained momentum in the twentieth century, especially in the last half of the twentieth century, is extensive tropical deforestation. FAO FRA reports from 2001 and 2010 show significant global deforestation between 1990 and 2010, but almost entirely in tropical areas (Anon. 2001a; 2010). The fight to save other forests and rainforests continues, and global concerns are rising.

To save forests, we must first understand why they are being destroyed. To identify the primary causes of deforestation, the distinction between deforestation agents and causes is critical. Deforestation operators include burning producers, slashers, and farmers, ranchers, loggers, firewood collectors, and infrastructure developers who clear forests. Forces that motivate agents to clear forests are the root causes of deforestation. However, the vast majority of existing literature distinguishes between two types of deforestation causes: indirect and direct. Direct causes and agents of deforestation, also known as sources of deforestation, are relatively simple to detect; however, the most discrepancies and difficult to quantify are likely the most indirect causes of deforestation.

6.3.1. Climate Change

Distinguishing the effects of forests on climate, especially the outcome of tropical deforestation on the climate, among the microclimates, global and regional climate is essential. Deforestation is able to change the global energy shift by increasing the concentration of carbon dioxide in the atmosphere as carbon dioxide absorbs thermal infrared radiation in the atmosphere. Additionally, deforestation may result in an increase in land surface albedo and therefore affect the region's radiation budget. The effect on wind, solar energy and water vapor absorption is therefore clear in local and global climates. Deforestation affects wind flows. Lowland deforestation moves precipitation and cloud formation up to higher levels (Lawton et al., 2001). Deforestation interrupts normal weather patterns, which generates dry and hotter weather, therefore increasing desertification and drought, crop failures, polar ice cap melting, coastal flooding, and displaced vegetation regimes. Land degradation is a more serious problem for dry forest areas, leading to extreme desertification cases (Dregne, 1983). Desertification is the result of severe climate changes and unsustainable land use practices including forest overcutting. Global change or global warming includes anthropogenic climate and ecological issues, including recent seeming changes in precipitation and climate in some places, the rise of the sea level, the depletion of stratospheric ozone, atmospheric pollution and the fall of forests. Tropical forestry is declining by approximately 5% each decade, with the forests logged down and cleared to provide wood, livestock, agricultural products, and biofuels for local, global and national industries (Anon., 2007, 2010). The effect of deforestation on a global scale is among the most crucial ramifications. Deforestation helps global warming that occurs as greenhouse gasses (GHG) increase in the atmosphere, leading to a total increase in the average global temperature because forests are the primary terrestrial carbon sink. This disrupts the global carbon cycle by increasing the amount of carbon dioxide in the atmosphere. The emissions of approximately 2 billion tons of carbon into the atmosphere each year are attributable to tropical deforestation (Houghton, 2005). The global deforestation release of carbon dioxide is believed to be 25% of combustion emissions from fossil fuels (Asdrasko, 1990).

6.3.2. Losses and Flooding of Water and Land Resources

The global water cycle also undermines deforestation. The field cannot hold as much water to trigger a drier climate when part of the forest is removed. Including drinking water resources, fisheries and water habitats, flood/drought controls, waterways, and barriers, water recreation, crop damage and erosion-and turbidity irrigation systems, are impacted by deforestation. The protection of urban water is perhaps among the biggest forest services. It is expensive to filter and treat water. Forests are able to reduce their costs by filtering outflows passively or actively by replacing housing or perhaps runoff farms. Watersheds that are able to not anymore sustain and regulate flows of water from streams and rivers can lead also to deforestation. After they've passed, way too much water could lead to downstream flooding, which in most areas of the world have resulted in disasters. This downstream flow causes soil erosion so that waterways, lakes and dams are silted. For two reasons, deforestation increases flooding. For starters, the dirt is much more prone to be completely saturated with water with a smaller "tree fountain" effect. In the wet season the' sponge' fills up, bringing about more risks and precipitation to the floods. Second, deforestation often results in soil compaction which cannot absorb rain. This leads to local flux response to precipitation and hence the potential flash inundation to be faster. Additionally, dry season flows are reduced when deforested. Deforestation can have a major long-term effect on the soil resource. The deletion of slash-and-burn farming vegetative cover exposes the earth to the intensity of torrential rain and tropical sun. The forest floors are easy for heavy rainfall with the leaf litter and porous soils. The effect of deforestation on water availability, flash floods and dry seasonal flows depends on the consequences of these countervailing influences sponge versus fountain. The proportion of the bowl subject to erosion and therefore contributes to the long-term siltation has increased as an outcome of deforestation along with other land use changes. The river bed has increased the threat of flooding, especially in China's Yangtze River Basin, the main river wetland basins in East Asia, and in the Amazon basin.

6.3.3. Reduced Biodiversity, Habitat Loss, and Conflicts

Forests, especially the tropics serve as a biodiversity conservation facility, which consequently destroys the whole biodiversity and habitat for, including, but still some to be cataloged, migrant species. Two-thirds of all known species is supported by tropical forests, and 65% of the world's 10,000 species are endangered. The preservation of forest area biodiversity is as the maintenance of a kind of capital until research on several animal and plant species are able to create the relative value of it. At least partly traditional medicine depends on primary health care, based on the World Health Organization, approximately 80% of the world's population. The loss of biodiversity and the associated major forest cover changes may lead to strong, harmful and irreversible changes. Local climate change including feedback that could theoretically shift rainforests towards savannas and also the development of completely new pathogens, as the growing trade of bush meat increases human-animal contact. Another negative effect of deforestation is increasing incidents of conflict between humans and animals which are harmful to conservation success in a method in that alienates the participation of individuals in conservation. The elephant habitat in north-western Bengal in India is an element of a highly fragmented Eastern Himalayan biodiversity hotspot. The heavy breakups of this habitat resulted in a rigorous conflict between the human-elephant and the lives and the crops of elephants and people.

6.3.4. Economic and Social Consequences

The annual destruction of tropical forests is an amount of 45 billion dollars of forestry capital loss (Hansen, 1997). All possible future revenues and future jobs derived from sustainable timber and non-wood products are lost by the destruction of forests.

Quite simply, deforestation is a manifestation of social injustice. Deforestation has multiple social consequences, often with long-term damaging effects. The appearance of civilizations for indigenous communities typically means the destruction/change of the traditional life and their style as well as the disintegration of their social institutions, particularly by moving from their ancestral territory. The intrusion of outsiders destroys traditional lifestyles, practices and faith that intensify the structural development of roads, usually leading to the expansion of frontiers with social conflicts and land. With the loss of ecological services in forests, the immediate social impact of deforestation takes place locally. The forests provide important services to humans, like erosion prevention, flood control, water treatment, pollination and fishing, which are particularly essential for probably the poorest individuals in the world who make use of natural resources to survive on a regular basis. Through forest destruction, we're endangering our quality of life, putting local weather stability and climate at risk and undermining other species as well as the helpful services of biodiversity.

6.4. FOREST FIRES AND VARIOUS HUMAN ACTIVITIES

Fire is a vital tool used to clear wood for shifting, permanent farming and for the development of pastures. Fire is great but has a bad master. Fire used responsibly may be a useful tool in the management of forestry and agriculture, but abuse could result in deforestation. Typically, 19.8 million hectares or perhaps 1% of forests are reported to be significantly damaged every year by forest fires based on data that are available from 118 countries representing 65% of the global forest area (Anon., 2010). Mining is very intense and destructive.

6.4.1. Mining

The spot of the land is quite small and never regarded as a significant cause. Mining is a profitable business that promotes growth booms that can attract population growth with the resulting deforestation. Based on an evaluation by the World Wildlife Fund Guianas, the deforestation rate for mining activity in Guyana increased 2.77 times between 2000 and 2008. In the Philippines, mining and logging were the same among the key sources of forest loss: a decrease of just 3 million hectares in 2003 or perhaps just more than 82% from a total of 17 million hectares in 1934. Almost 2,000 ha of rainforests are being totally damaged by mining iron minerals planned by the Italy Argentine mining company TERNIUM (Anonymous, 2008) in the Municipality of Coahuayana in the State of Michoacan (Southwest Mexico). The Niyamgiri Mountain Range in Orissa India is under threat from an initiative of Vedanta Aluminum Corporation to begin mining bauxite. Uncontrolled and massive mining in Jharkhand, India of coal, bauxite and iron ore caused wide deforestation and created enormous water scarcity (Anon., 2011b). The agreements between the Jharkhand State Government, India as well as the mining companies in return for US $3.8 billion in investments will end up in massively acquired land which will deforest at least 57,000 ha of forest, and replace 9,615 families, a lot of whom reside in legally-protect scheduled areas reserved for indigenous populations in the state. Additionally, mining roads will open up the area for shifting farmers, ranchers, permanent farmers, land speculators and developers of infrastructure. For instance, infrastructure development projects including access roads to borders, large water and mining areas reservoirs had been at the heart of Brazil's development strategies on the Amazon (Mahar et al., 2004). Roads, railways, bridges and airports open the farmland to development and bring a lot more humans to the woodland border. If timber is used as a combustible fuel in mining and is a source for plantations, the region might be severely deforested. On the contrary, mining can be extensive in labor and can get rid of work from forest clearing. Cities and cities expansion requires land for building infrastructure required to support the growing population, through clearing of forests.

6.4.2. Urbanization/Industrialization and Infrastructure

Tropical forests are a key target for oil, logging, and hydroelectric dam construction infrastructure developments which inevitably promote the expansion of the road network and road construction in pristine regions. Building roads, railways, airports and bridges are going to open the area to development and take increasingly more individuals to the forest border. Those settlers have generally settle in the forest by logging new roads or roads for a chance to access the forest for subsistence land, whether supported or perhaps not by government programs. Rainforest clearing accounts for approximately 20% of anthropogenic CO_2 emissions which destroy significant around the world carbon sinks, and about 21% of the worldwide tropical forest has been lost after 1980 (Bawa, 2004).

1. **Industrializing Countries:** The wealthy or former colonial governments that have a debt in their very own natural resources mainly retain the resources, which are usually full of natural resources of the financially poorer countries. 20% of the world's population uses 80% of the world's resources. Sad to say, the rulers of these poor rich resources have often adopted the same syndrome of growth as their western neighbors or perhaps former colonial masters, emphasizing the maximization of exports, income and also the untenable exploitation of the wealth of organic sources for short-term benefits. Additionally, corruption is popular in government, economic and military strength. The issue is further compounded by the lower cost of most international third-world exports. The relations between development and deforestation are complex and dynamic.
2. **Exploitation by Industrialized Countries:** Rich countries are generally supported by resources of the financially weaker ones generally natural resources, or perhaps those formerly colonial powers with a deficiency in their very own natural resources. 20% of the world's population uses 80% of the world's resources. The governments of the resource-rich countries, also, have unfortunately often adopted a similar growth syndrome as their West neighbors or perhaps former colonial masters, with a focus on the maximization of short-term gains of exports, incomes and also the exploitation of the rich natural resources. Additionally, corruption is popular in government, economic and military powers. The issue is even more exacerbated by the lower cost of most international exports from the third world (Lohmann & Colchester, 1993).
3. **Debt Burden:** The financially poorer countries now are experiencing heavy international debt in pursuit of a guided development agenda and so feel the desire to pay off these enormous debts as a rise in interest rates. This forces the poorer countries to exploit the rich natural resources, including the forests, to make exchange payments for servicing their debt. For instance, in some Southeast Asian countries, the building of logging roads was funded with Japanese aid, which allowed forestry to be used by Japanese wood companies.

These timber companies exploited the forests profitably as Japan's money for the building of its roads remained owing in Southeast Asian countries (Lohmann & Colchester, 1993). The job of the population in deforestation is an issue of controversy.

6.4.3. Overpopulation and Poverty

There has been controversy about the effect of population density on deforestation. Based on international agencies such as intergovernmental bodies and FAO, poverty and overpopulation are the primary causes of forest loss. These organizations are usually thought that by fostering development as well as trying to reduce population growth they are able to fix the issue. The World Rainforest Movement as well as numerous other NGOs are, on another hand, directly accountable for the most forest losses due to uncontrolled development and too much consumption practices in the wealthy industrialized countries. Nevertheless, it is clear that fast population growth is a major indirect cause of deforestation. A lot more people want much more food and additional space for housing and agriculture. In turn, this leads to more forest clearing. The biggest struggle for sustainable management of human life support systems is most likely to create a growing population, and also the control of population growth is maybe the greatest thing to perform to be able to promote sustainability. Exclusively for third countries, overpopulation is not really a problem. In an industrialized country, a person will probably consume so many resources as a human being in a poor country in the order of 60 times that of the planet. Thus, the increasing population in rich industrialized countries has a big share in earth exploitation and a definite link exists between too much consumption in rich countries and also tropical deforestation. Inextricably linked are overpopulation and poverty. Poverty has mostly been the outcome of the greed of rich industrialized nations and the third world elites, which seek to copy them, but is the unmistakable cause of a lot of the harm to the rainforests.

Development is typically viewed as a solution to global poverty and rarely helps many in need. So it is usually the cause of poverty instead of the cure. Many aid and government agencies use overpopulation as an excuse for inaction as the cause of deforestation. In countries that are tropical, human settlement pressure is a lot more caused by unequal distribution of land than by population pressure. Most of the land is frequently run by little but powerful elites, who move poor farmers to rainforest areas. So long as these elites maintain power, it is going to be tough to achieve a long-lasting agricultural reform and deforestation will continue. Poverty is therefore popular as a serious underlying cause for the conversion of forests by small farmers as well as naturally wooded areas are usually linked to excessive levels of poverty. In the population also, investment in maintaining increasing yield or soil quality on existing cleared land is usually lacking. The uneven distribution of wealth is largely affected by deforestation. Shifting forest border farmers are among the most poverty-stricken and marginalized segments of the population. Usually, they do not own land and they've little money. Consequently, they've no option but to clear the virgin forest. For the livelihoods of farmers living in forest areas, deforestation including clearing for agriculture is often the only available option (Angelsen, 1999).

6.4.4. Logging and Fuel Wood

Logging and fuelwood don't always lead to deforestation. But logging can degrade forests seriously. Southeast Asian logging is intensifying and may be extremely devastating. The logging system however gives access to roads and log scales are able to help finance the cost of clearing remaining trees and also the planning of land for cultivation or perhaps pasture planting. This catalyzes deforestation through logging. Fuelwood collection in the tropically dry forests and degraded forest areas is often concentrated. The primary cause of deforestation in humid tropics isn't usually fuelwood, though in some populations with reduced forest areas including the Philippines, Parts and Thailand of Central America it is able to occur. The primary cause of deforestation and forest degradation in El Salvador was regarded as the fuelwood collection. Fuelwood collection can be a significant cause of depletion and deforestation in the drier areas of the tropics.

6.4.5. Overgrazing

Overgrazing is much more frequent in tropical drylands where soil erosion occurs in pastures that are degraded by overgrazing. Rubbing trees to feed animals may be an issue in some tropical dry areas, but might not be a significant cause of deforestation. Clear cuts and overgrazing made large areas of the province of Qinghai a desert in China. Overgrazing causes widespread areas of the grassland north of Desertification and Beijing in Inner Qinghai and Mongolia Province. One male who lived in a village on the eastern edge of the sandy Qinghai Tibet Plateau told the NY Times, "The pastures were here as environmentally friendly and as weighty as they had been, and so the grass has disappeared and the sand has gone." Great flocks of sheep and goats are stripping the land of vegetation.

The population of livestock, for example, in Xilingol Prefecture, Inner Mongolia, rose from two million in 1977 to 18 million in 2000, making one third of the grassland desert. The entire prefecture could be uninhabitable by 2020 unless something is done. Sociological phenomena, known as' the tragic events of the common' exacerbate the overgrazing, in which humans share land, but bring animals in concert to enrich them almost as they are able to. This results in even more animals than can be supported by the farmland.

6.4.6. Economic Causes – Developments/Lands Conversion Value, Fiscal Policies, Consumerism, and Markets

The development is going to increase the productivity of land and therefore decrease the demand for forest clearance to meet food needs. On the other hand, development is going to generate additional capital and stimulate expansion as well as forest clearing. The first might be the situation if a fixed food demand is restricted. Lastly, due to the continued exportation market along with an increasing domestic population with rising consumption, food demand may be unmet. Deforestation benefits vary, based on location, technology and land use systems, from under one dollar to 1,000 dollars per hectare. Workers also argue that rich farmers could finance deforestation better, while a poor farmer can't afford to clear a huge amount of forest. Conversely, transfers are able to increase deforestation and revenue by small landowners, by strengthening credit markets and much better opportunities for off season employment. Land with higher rents also promotes faster deforestation. Higher crop prices minimizing agricultural input prices also spur deforestation faster (Chomitz et al., 2007). Increased salaries also can boost deforestation. Techno-new developments make agriculture much more successful, either driving forest farmer's expansion or perhaps attracting new farmers at forest borders. Although the price increase is temporary, expectations about future prices often rise, thereby increasing the expected likelihood of land clearance and transformation to agriculture. Many policies in the area of development have failed since they support the advancement of the land, power, political influence and influence unwittingly or wisely. This alienates rural poor humans and reinvigorates the forests. The environmentally intensive effects of their actions are little encouraged in poor farmhouses or perhaps business loggers. Unreported costs such as neighborhood market failures, global appropriation failures and policies lead to economic failures. The marketplace doesn't achieve an optimum result due to an unregulated market economy. The prices generated from these markets don't reflect the true social benefits and costs of the resource use as well as convey misleading info about resource shortages, which provides inadequate management incentives, effective use and enhancement of natural resources. Policy breakdowns or perhaps distortions of the market are caused by misguided intervention as well as failure to mitigate failure that results in worse results (Panayotou, 1990).

Lack of compliance with traditional land rights, for instance, makes forest land rights uncertain and can encourage short-term exploitation of forests rather than sustainable long-term use. Additionally, as in the case of exotic woods, the global appropriation failures aren't shown in market allocation and in the benefits of preserving biodiversity and also the importance of genetic pool in developing new medicines, plants and pest control agents. For instance, much better conditions of trade for the export of farming and forestry products and higher real exchange rates make converting forests into other uses more profitable. The first empirical analysis by Scrieciu (2003), shows that the movement to maximize profits within the agricultural sector is also responsible for tropical deforestation. The developed countries have often claimed rampant consumerism to be a significant cause for tropical deforestation. The opening to world commodity markets of tropical countries has accelerated deforestation. Beef products, cotton, banana, sugar, and coffee are included in Central America and oil palm, wood and rubber in Southeast Asia. Underestimating forests is useful only in case they're cleared by improvement to get legal title (Mather, 1991; Sands, 2005). It has been recommended that the extraction of forest products is not really a method of adding value to the forest, but isn't economic in comparison with clearing options. If some actions may be devised to purchase the owners of forests and for agents who benefit from the green values, the choice of not clarifying them will be more competitive. Otherwise, it might use a tax or perhaps disincentives to clear up if the national governments value the green benefits. Whilst environmental services are vital to sustained economic growth, deforestation nations generally have much more immediate goals and aren't ready to do it. Sustainability isn't an option, it is imperative.

6.5. ROLE OF TREES AND FORESTS IN ENVIRONMENTAL CONSERVATION

We want clean air, nutritious food, pure water, and a proper environment and greenery around us to dwell in a greater world. Without having the deterioration of sustainability environmental conditions and economic decline, poor health, pollution, poverty, political unrest and turmoil will feed one another. It is not a self-sufficient concern for the planet. It covers all development sectors.

A major global concern has been expressed by rapid increases in the number of greenhouse gases in the air, land loss, growing droughts and inundations and the deepening desert conditions of fragile ecosystems, deforestation, environmental pollution and biodiversity loss. The general effect of this phenomenon will probably lead to ozone depletion, climate change, sea level rise, loss of natural resources, productiveness decrease, which eventually leads to an ecological crisis that will affect the development of livelihood options and overall quality of living deterioration. Our environmental assets have been seriously impacted by development according to the use of natural resources, population pressure and the growing poverty and demand of the people of ours. While natural assets have declined, demands have grown, leading to unsustainable overdraws. We have to boost our rate of economic development, provide an extensive segment of the population of ours with basic minimum life support services and address the problems of unemployment and poverty. While doing so, we should be careful to protect the natural resources as well as improve the environmental status. To be able to guarantee environmental and economic sustainability, we will have to address environmental degradation in a holistic way. For the country, particularly the planners of ours and policymakers today, this's probably the most demanding task. In economic and environmental sustainability, forestry plays a crucial role. They supply many services and goods and maintain lifestyle systems that are essential to life on Earth. Several of these methods which are of great financial and environmental value to support life are:

i. The provision of forestry products, fuelwood, and timber, in addition to a broad range of non-wood products.

ii. Natural habitat for genetic reserves and biodiversity.

iii. Recreation and ecotourism opportunities; and

iv. Integral watershed management, soil preservation, and co-regulations.

Despite significant resource flows and national concerns, forests have been not able to fully realize the potential to reduce poverty, achieve economic development and add to the global and local environment. A mix of market and institutional failures have led forests to contribute as significantly as is possible, under great financial and technical management, to rural revenue and economic growth and poverty alleviation.

The demand for effective forest conservation is very apparent with a lot of atmospheric carbon dioxide that affects and disrupts the climate of the planet. Trees have been made for the regulation of carbon dioxide levels through gas absorption, the release and carbon storage of oxygen as a crucial component in fighting global climate change. When the trees are cut down, the forest carbon is released into the environment along with the carbon dioxide and greenhouse effects are intensified.

Forests play a significant role in the defense of the planet. Forest protection in mountainous areas has much history of helping prevent soil erosion, avalanches and landslides and of maintaining water quality for rivers draining forest catchments. To make sure that these forests are maintained indefinitely, special forestry methods are necessary. Forests also react to the defense of the planet.

6.5.1. Role of Forests in Maintaining Air Quality

Air pollution for forests is a crucial issue in terms of human health and welfare, the environment, crops, ecosystem health, visibility and human materials. The role of forests is essential in maintaining air pollution. Sulfur dioxide, heavy metals, hydrogen fluoride, and ozone are all the air pollutants of concern. Lastly, forests benefit from the control of these pollutants. Forests play a significant part in protecting the world's carbon cycle. It constitutes a significant sink of atmospheric carbon dioxide. One cause of an increase in atmospheric carbon dioxide concentrations lies in converting forests into other land uses.

Foraging and reforestation could contribute to the decrease in concentrations of atmospheric carbon dioxide, and biofuels are able to bring about reducing fossil fuel demand. Forests and trees, like air pollution, vary from one tree to the next worldwide and are made up of various species. Trees affect air quality through the immediate removal of air pollutants, altering local energy and microclimates use of the emission and building compounds of VOCs that can add to the formation of O_3 and PM 2.5.

Trees have a bad impact on air quality. Nevertheless, integrative scientific studies show that trees can be practical in reducing urban O_3 levels, especially low VOC emitting species. Trees mainly remove gas air pollution by taking up leafy stomachs, though some gasses are removed by the surface area of the plant. Most pollution is removed through leaf stomata for O_3, NO_2, and SO_2. Gasses spread into intercellular spaces once inside the leaf and also can be absorbed into acids or perhaps react with internal leaf surfaces by the water films. Trees directly influence atmospheric particulate matter by intercepting particulate matter, emitting particulate matter (e.g., pollen), and rehabilitating particulate matter from the plant. Although the majority of the particles intercepted on the plant surface can be absorbed in the tree. The partitions are usually suspended into the atmosphere, cleaned up by rain and thrown right down to the ground with twigs and leaves. Particles are intercepted and re-suspended continuously during dry times, depending in part on wind speed. Photosynthesis is usually affected by the accumulation of particles on the leaves (e.g., Darley, 1971) and therefore could possibly have an effect on trees removing contamination. Particles may be washed away and dissolved or perhaps transferred into the earth during precipitation. Vegetation is therefore only a temporary retention site for lots of atmospheric particles, which eventually move particles back to the atmosphere or perhaps moves them to the earth.

6.5.2. Role of Forests in Water and Soil Management and Conservation

The importance of forests in water management and conservation is vital to life. Forests serve as supply systems, filtration, and our collection of natural waters, delivering and gathering snow and rain throughout the year into streams, water wetlands and aquifers. Drinking water flows into rivers from the forests, which provide the reservoirs, farm canals and water tables. Woodlands can also be crucial for controlling floods, absorbing and retaining vast quantities of water from major rainfall events as climate change, and also for releasing water a lot more slowly and cleaner than otherwise. Over the 1800s and 1900s, the tremendous wealth of old growth was diluted by a' gold rush' for wood products from the Pacific forests, with many different natural systems house holding wildlife, mitigating climatic change and dispensing water. Drinking water has become important "liquid gold" in the 21st century, as the constant threat of climate change and record droughts in the West has been constantly threatened. Forests are clearly influenced by availability and, particularly, water quality. Forested catchments provide high-value water and surface areas. The functions of the planet's natural systems are determined by tropical rainforests. The forests regulate global and local weather by absorbing and generating precipitation and exchanging atmospheric gasses. The preservation of ecological equilibrium and improved livelihoods in arid regions involve shrubs and trees. Clear drops of rainforests change the Earth's reflectivity, affecting global climate by changing the patterns of rainfall and wind and ocean currents patterns. The global weather patterns will be a lot more unstable and intense if forests continue to be destroyed. Water quality and water availability in numerous areas of the world are increasingly endangered by overuse, pollution and misuse and also the reality that both are clearly affected by the forests is increasingly recognized. Moreover, the availability of water resources by forest is altered by climate change. Therefore, a crucial issue must be given higher priority on the relationship between water and forest. Forest catchments provide large quantities of water in both the downstream and upstream areas for domestic, farming, ecological and industrial needs. Here we emphasize the benefits of trees and forests in influencing precipitation and water supplies to preserve forests and promote tree planting.

6.5.3. Role of Trees and Shrubs in Rainfall and Water Supply

Much of the world's drinking water originates from C. An obvious connection seems to exist between the forests and the water quality of a catchment, the significantly larger sporadic connection between forest and also the quantity of water available, along with a variable link between forests based on type, constancy and age of flow. Forests therefore often form the basis of integrated water resources management, though precise effects vary from one spot to another and therefore are susceptible to hydrological dispute. Knowledge about the type and age of trees, soil conditions and user needs are able to make it possible to set the most beneficial type of forest management policy. Loss of forest protection and transformation to other land use can adversely impact the availability of fresh water, threaten the survival of countless persons and hurt the environment (FAO, 2003).

6.5.4. Roles of Forest in Maintaining Water Cycle

The role of rainforests in the water cycle is to bring water to the atmosphere through transpiration (in which plants release water from their leaves during photosynthesis) as well as to form rain clouds that return the water to the forest. 50-80% of humidity in the Amazon remains in the water cycle of the environment. If the forests are reduced, less humidity enters the atmosphere, with decreases in drought and rainfall. Deforestation, and humidity produced by rough forests traveling all over the world, have made them worse. Based on Janet and Clare (2013), isolated rainforests are thus vital that you male all over. Forests are an important component of the global water cycle because the evaporation rate is high and also contributes to the circulation of the atmosphere. Aragão (2012) reported that forest canopy is much more successful in recycling water through evapotranspiration than surfaces with sparse vegetation such as fields. The mixture of sweating and evaporation from the vegetation to the atmosphere is evapotranspiration.

The local hydrological cycle is weakened by deforestation, and the changed land cover gives a unique pattern of thermal release. The deforestation of vegetation may reduce evapotranspiration. This reduces the circulation of moisture and lowers precipitation. Deforestation also affects local thermodynamics which result in a reduction in atmospheric heat. This affects the air circulation and its corresponding precipitation. Changes in the plumage pattern may have damaging effects on agriculture and also the availability of water, particularly during the dry season. Well-managed nature forest is typically renewable water and falls from the skies unbroken and regularly throughout much of the world as a superior quality drinking water benefits urban populations. But because of carelessness, depletion, speed of population growth and increasing per capita water demands, sufficient safe provision of water now is a significant source of concern, cost as well as international animosity. What this means is that water supplies are safe and adequate. Additionally, the need for water, along with growing pressure on water from pollution, urbanization and water use, creates a freshwater biodiversity crisis. Well-managed natural forests typically provide water of better quality than water coming from many other catchments, with much less sediment and less pollutant pollution. Some natural forests increase the general water flow, though this's untrue sometimes and also can decrease net water flows among exotic plantations and young forests. Forests have much less impact on the security of supply and flood mitigation, but forests could reduce flooding on the local headwaters. These various advantages ensure that natural forests are protected to make sure that cities are maintained with high-quality water supply. Only the vegetation is in the river channel and along the banks of the river, as it is strictly defined. The phrase was used much more extensively not too long ago to include the adjacent landscape of a stream which specifically affects the river and lake borders and associated aquatic ecosystems and water. Riparian habitats are the interface of terrestrial and water ecosystems within the landscape. They're used for controlling water, nutrients and sediments, as conduits, barriers or filters. A crucial element of good management of natural resources is ensuring riparian ecological functions like filtering polluted surface land and subsurface flows, stabilization of stream banks and control of instream habitats (Hayakawa & Mander, 2005).

6.5.5. Forests Control Floods

The belief that forests covering the upper shores of the river will prevent floods along the mainstream and upstream rivers remains quite prevalent (Hamilton, 1991). Forests continue to be observed in excess of countless times (the event of the storm) and after that release them gradually, to increase dry season flow during the after storms or perhaps after monsoon seasons. Thus, cutting the forest usually increases dry season flow, according to Nigel and Sue (2003), but also increases flood flows somewhat. Storage of water in the ground depends on the soil depth, soil penetration capacity (lack of compaction), texture, degree and structure of previous water saturation, and may be stashed away to reduce flooding in a storm. Some features such as for instance anterior soil humidity, infiltration capacity and structure are affected by the forests. These generally represent the very best hydrological bet to reduce the amount of storm flows, reduce the peak fluxes and delay watercourse peaks emitted by the water border (Bruijnzeel & Hamilton, 1997). The vegetation of forests also protects the exposed soil from the breaking of soil aggregates. Individual soil particles are then easily eroded and washed into soil pores, preventing and obstructing the absorption of rainwater. Water tends to run over the surface area of the ground if such conditions are present, increasing the risk of erosion. In the forest, nonetheless, rain is intercepted with a wooden canopy, with leaves of small trees or shrubs in the forested area along with an organic level of litter covering the forest ground reducing the rain falling on the ground. Pores of dirt continue to be blocked and infiltration is permitted (Vincent, 2008).

6.5.6. Forests Protect Soils and Reduce Forest Erosion Rates

Forest protects the soils and reduces the rates of sedimentation and erosion in rivers. Sedimentary losses may rise, but best management practices are able to control that risk, drainage, including cultivation, forestry operations, road construction and timber collection. Forest planting is able to reduce, and intercept, sediments in erosion susceptible soils and runoff tracks (IUFRO, 2007). Woody vegetation protects the earth much better compared to annual plants and lasts longer. Its roots deepen and enhance the soil and its shade facilitates the metabolism of the environment. These functions are necessary for the stability of the continuity and the soil of farm work. By intercepting the rain, the result of heavy rainfall on the forest floor decreases the forest canopy and reduces soil disturbance. Leaves and organic waste on the ground is able to slow down the rate of rushing water and trap soil from nearby areas. Tree roots are able to hold ground and stabilize banks in the stream. Furthermore, coastal forests and forested wetlands safeguard coastlines through the absorption of several erosion-related effects, impacts, and the energy of storms (TVA, 2003.). Trees reduce the erosion rate by protecting soils against the rain impact, transpiring a lot of water that counteracts highly wet soil and ties soil and roots to sloping soil. A tampon strip planted along lakes and rivers can contribute to the filtering out of pollutants before entering waterways. Studies indicate that forest buffers take in and use huge amounts of water to grow, therefore consuming stormy water, by up to 90%. Forest buffers are able to take in and consume tremens of fertilizers, pesticides and sediment rushes. In a mature tree, more than 151,416 liters of water may be consumed (transpired) a season. Evapotranspiration cools and modifies the summer temperatures surrounding it. The evaporation decreases to 35.56 cm when the forest is removed or perhaps harvested, so the flux of the stream increases to 66.04 cm of precipitation annually. Therefore, the sole impact on streams in the waterfront is only able to be the removal of forests. The soil aggregation (and consequently more space for the pores) also contributes by moving and mixing soils to microorganisms, small livestock, insects, and cultivating tree roots. This has brought soil particles into contact, increasing the probability that soil particles will clump together, resulting in big pores that are very easily drained by water. The littered layer, consisting of wood leaves and pieces of wood, can be used to have a proper soil organism population during different stages of decay in the forest floor. Liter layer also protects the soil against elements that produce a much less hostile and stabilizing environment for soil organisms (Vincent, 2008).

6.5.7. Trees and Forest Aid to Regulate Stormwater

Runoff is the rain or even melts water that sweeps into the planet Earth. Stormwater is produced by rain on roads, car parks, entrances, rooftops along with other paved surfaces that don't permit water to penetrate the soil. Runoff of stormwater will be the top cause of urban stream damage. The amount of pollutants and stormwater rush reaching local waters are able to decrease the presence of trees in roadside, neighborhoods and communities. Trees reduce the runoff of stormwater by capturing, releasing and storing rainfall to the atmosphere. Tree roots and leaf litter create soil conditions that encourage rainwater to get into the soil. Trees help temporarily and slowly reduce pollutants and store runoff through their roots by absorbing water and soils in other pollutants and nutrients. Trees, acting as mini ponds, control the cause of storm water.

Recommendations for sustainable water and rainfall:

1. **Natural Regeneration by Land Protection:** Natural trees and forests are improving the water cycle in order to decrease rivers and to enhance the regeneration of the water table. The most apparent method of restoring vegetable cover is to defend it against degradation causes: mainly exploitation through harvesting, grazing, and fires. Even on bare lands, vegetation is able to spread easily, however the procedure is usually slow. It is not simply easy to protect, as it should be carefully maintained over quite a while. The procedure is going to be expedited by planting trees, grass and bushes. The restored lands must then be managed sustainably.

2. **Aggressive Afforestation:** Modern rainfall patterns and surface temperatures are the major indirect global impacts of deforestation.

   Probably, the lowest average rainfall in African forests occurs out of all forests, and also the change in precipitation because of deforestation is expected to be very affected (Janete & Clare, 2013). Boasting could be the right instrument for environmental restoration through tree plantation. The plantation of trees was often proposed as a means to boost precipitation. 60% of the rainfall in the moist, evergreen Amazon forest has been estimated to be triggered by evapotranspiration from the forest itself. Planting trees produces tangible results only when huge areas are converted into forest, as rainfall is increasing in neighboring areas. The afforestation of a higher land environment like the development area is likely influenced by surface water runoff rates, both through plowing and planting effects and the following forest canopy impacts.

3. **Sound Management and Forest Conservation:** Forests provide a number of water supply options based on the type, the location, the needs and the age of its users. A number of forest management options could be picked from cities, including protection, sustainable management and restructuring. Sound sustainable land management, vegetable protection, biodiversity and water resources will basically be used to maintain and sustain the capital and its productive capacity by making use of the renewable portion, i.e., its efficient production. The participation of local communities and people and their traditional practices and knowledge are of key importance. For sound land management, clear land use rights are essential. Strengthening people's control of resources as well as ensuring a healthy engagement to conserve resources.

4. **Policy Formation and Implementation:** The management of forests to preserve water, given that forests influence water quantity and, quality, in particular, is a suitable objective. The forest decision makers have to foster forest management integration into river, watershed and groundwater management national and regional strategies, programs and plans; and also to work with international and national organizations and institutes to increase the understanding of forest hydrological and environmental services. The prevention of forest degradation should be given priority including restricting pasture on forest lands and promoting reforestation and afforestation (including trees outside the forest), for the defense of the environment and local timber supply, especially fuelwood. Along with wood production, various other forest economic functions have to be stressed because of the critical state of the forest and some dependence on it for their livelihoods; Non-wood Forest Products (Ecotourism and NWFPs) must be promoted via appropriate policies. The transfer of duties for local forest management is able to provide communities with

incentives and rights for sustainable management and utilization of forest resources. Countries should prepare catchment reforestation source protection plans, including measures for controlling possible pollution sources, like septic systems and fuel tanks, and identifying alternative drinking water sources in the event of contamination. This will involve mapping and identifying suitable rapidly growing species of the refueling area of the water sources. Forests and trees are important instruments to enhance precipitation, reload soil waters and prevent flooding and erosion. As changes in land cover, which include deforestation, can result in lower evapotranspiration and enhanced surface temperatures, disrupt the hydrological cycle and reduce moisture circulation, efforts must thus be focused on aggressive foresters, natural forest conservation by means of effective policy formulation and implementation to guarantee a ready source of water for increasing usefulness.

6.5.8. Role of Forest in Noise Pollution

Solid obstacles like fences or perhaps earth mounds are usually used as sound obstacles, but space-enabled shrubs and forests to reduce noise can be also effective. Shrubs and trees can be utilized to get the very best of all scenarios together with good obstacles. The sound is closest to the source and decreases by distance. Sound travels on various surfaces differently. Hard surfaces reflect almost every sound of the incident, while soft surfaces absorb sound. Through a visual barrier between the hearer and the source, trees reduce noise perception. Everyone is much less aware of the noise when they can't see the source. It was suggested. Trees and shrubs mask the sounds of birds, other animals and insects by rustling leaves, and moving branches in the wind.

Sound is generally attributed to reflectance, deflection and absorption processes by vegetation. Foliage appears to be the best part of a tree to disperse large leaves and sound seems to be more efficient compared to leaves that are small. Lower shrubs and/or hedges along the edge of the trees, particularly on the edge nearest to the good spring, could improve the good reduction. With tree height up to 10–12 meters, noise reduction is much more apt, after which it tends to decrease. This most likely is because lower branches die and the sound is much easier to travel. The reduction of noise corresponds to the breadth of the tree belt, i.e., the larger the racket is, the larger the reduction. The length extended by a tree and an arbutus belt affects its noise reduction efficiency. Nevertheless, current prescriptions are hard since they depend on the sound source's dimensions. A screen located relatively near a noise source is more efficient than one situated near the secured area. Nevertheless, the noise reduction is significantly less at halfway between the receiver and the source. The efficient reductions of noise in summer and spring can be of great flowered feeding stocks, but evergreen species provide a much better attenuation throughout the entire year (Baggethun et al., 2010).

CASE STUDY

DEFORESTATION IN AFRICA

Deforestation in Africa has emerged as a critical environmental challenge, significantly impacting biodiversity, climate, and the livelihoods of millions. One of the most affected regions is the Congo Basin, home to the second-largest rainforest in the world. Over the past few decades, the rate of deforestation in this area has accelerated due to logging, agriculture, and infrastructure development. According to the World Wildlife Fund (WWF), the region has lost approximately 30% of its forest cover since the 1970s. The main drivers of this deforestation include illegal logging, which undermines local ecosystems and contributes to habitat loss, as well as the expansion of agriculture, particularly for palm oil, soy, and other cash crops, which has led to further encroachment into forested areas.

The consequences of deforestation are profound, affecting not only the environment but also the social and economic fabric of local communities. The loss of forests disrupts ecosystems and threatens the species that depend on them. Moreover, many indigenous and rural communities rely on forests for their livelihoods, using forest resources for food, medicine, and shelter. As deforestation continues, these communities face increased vulnerability, as their access to essential resources diminishes. Additionally, deforestation contributes to climate change by releasing stored carbon dioxide into the atmosphere, exacerbating global warming and its associated impacts on weather patterns and food security across the continent.

Efforts to combat deforestation in Africa have gained traction, with initiatives aimed at sustainable forest management and conservation. Various NGOs and government programs have been implemented to promote reforestation and afforestation projects, raise awareness about the importance of forests, and involve local communities in conservation efforts. For example, the African Forest Landscape Restoration Initiative (AFR100) aims to restore 100 million hectares of deforested and degraded landscapes across the continent by 2030. While progress has been made, significant challenges remain, including political instability, corruption, and the ongoing demand for land and resources. Addressing these issues requires coordinated action at local, national, and international levels, involving various stakeholders in the pursuit of sustainable development and environmental protection.

Discussion Questions

- What role do local communities play in both the causes and solutions to deforestation in Africa, and how can their knowledge and practices be integrated into conservation efforts?
- How can international organizations and governments effectively balance economic development and environmental conservation to mitigate the impacts of deforestation in Africa?

SUMMARY

Forests support diverse ecosystems, providing habitats for countless species and contributing to global biodiversity. They act as carbon sinks, sequestering atmospheric carbon dioxide and mitigating climate change impacts by stabilizing global temperatures and rainfall patterns. Additionally, forests play a vital role in regulating water cycles, influencing local and regional hydrological processes such as precipitation, runoff, and groundwater recharge. Their preservation is essential for maintaining soil stability, preventing erosion, and filtering pollutants from water bodies. Beyond ecological benefits, forests also contribute to human well-being by offering recreational opportunities, cultural significance, and economic resources through timber and non-timber forest products. Therefore, safeguarding forest ecosystems is critical for sustainable development, ensuring their continued provision of environmental, economic, and social benefits to present and future generations.

REVIEW QUESTIONS

1. What are the main components of the environment, and why are they important in the context of forests?
2. Explain the Polluter pays principle (PPP) and how it applies to forest conservation.
3. Discuss the impact of deforestation on climate change according to Chapter 6.
4. Why is the Principle of precaution (PP) important in environmental management?
5. How does the Principle of Responsibility influence conservation efforts in forest ecosystems?

REFERENCES

1. Aarikka-Stenroos, L., Ritala, P., & Thomas, L. D. W. (2021). Circular economy ecosystems: A typology, definitions, and implications. In S. Teerikangas, T. Onkila, K. Koistinen, & M. Mäkelä (Eds.), *Handbook of Sustainability Agency* (pp. 1–20). Edgar.
2. Azul, A. M., Brandli, L., Salvia, A. L., & Wall, T. (2020). Life on land. In *Encyclopedia of the UN Sustainable Development Goals* (pp. 1–11). Springer. https://doi.org/10.1007/978-3-319-95726-5_51-1.
3. Baggethun, B., Martín-López, M. V., Armsworth, P., Christie, M., Cornelissen, H., Eppink, F., Farley, J., Loomis, J., Pearson, L., & Perrings, C. (2010). The economics of valuing ecosystem services and biodiversity. In P. Kumar (Ed.), *The Economics of Ecosystems and Biodiversity: Ecological and Economic Foundations* (pp. 183–255). Earthscan.
4. Balian, E. V., Drius, L., Eggermont, H., Livoreil, B., Vandewalle, M., Vandewoestjine, S., Wittmer, H., & Young, J. (2016). Supporting evidence-based policy on biodiversity and ecosystem services: Recommendations for effective policy briefs. *Evidence & Policy, 12*(3), 431–451. https://doi.org/10.1332/174426415X14377319715027.

5. Belmontes, J. A., Pintor, A. L., Rodríguez, M. A., & Gómez-Sal, A. (1997). On the usefulness of ecosystem services evaluations. *Pirineos, 149–150*(1), 145–152. https://doi.org/10.3989/pirineos.1997.v149.183.
6. Bernetti, I., Sottini, V. A., Marinelli, N., & Marone, E. (2013). Quantification of the total economic value of forest systems: Spatial analysis application to the region of Tuscany (Italy). *Aestimum, 2013*(63), 27–44. https://doi.org/10.13128/Aestimum-12871.
7. Bjørn, A., Molin, C., Hauschild, M. Z., & Owsianiak, M. (2017). LCA history. In M. Hauschild, R. K. Rosenbaum, & S. Olsen (Eds.), *Life Cycle Assessment: Theory and Practice* (pp. 17–30). Springer. https://doi.org/10.1007/978-3-319-56475-3_2.
8. Bolger, M., Schweitzer, J. P., Arditi, S., & Lutter, S. (2019). A circular economy within ecological limits. In *Friends of the Earth Europe; European Environmental Bureau, Institute for Ecological Economics* (pp. 1–25). WU.
9. Brady, E. (2006). Aesthetics in practice: Valuing the natural world. *Environmental Values, 15*(3), 277–291. https://doi.org/10.3197/096327106778226293.
10. Caetano, M. (2007). *LULC Applications, Advanced Training Course on Land Remote Sensing*. ESA. https://earth.esa.int/landtraining07/D3L1-Caetano.pdf (accessed on 29 August 2024).
11. Carpenter, S. R., Mooney, H. A., Agard, J., Capistrano, D., DeFries, R. S., Diaz, S., Dietz, T., Duraiappah, A. K., Oteng-Yeboah, A., & Pereira, H. M. (2009). Science for managing ecosystem services: Beyond the millennium ecosystem assessment. *Proceedings of the National Academy of Sciences, 106*(5), 1305–1312. https://doi.org/10.1073/pnas.0808772106.
12. Chan, K. M. A., Guerry, A. D., Balvanera, P., Klain, S., Satterfield, T., Basurto, X., Bostrom, A., Chuenpagdee, R., Gould, R. K., Halpern, B. S., et al. (2012). Where are cultural and social in ecosystem services? A framework for constructive engagement. *BioScience, 62*(8), 744–756. https://doi.org/10.1525/bio.2012.62.8.7.
13. Chaplin-Kramer, R., Sim, S., Hamel, P., Bryant, B., Noe, R., Mueller, C., Rigarlsford, G., Kulak, M., Kowal, V., & Sharp, R. (2017). Life cycle assessment needs predictive spatial modelling for biodiversity and ecosystem services. *Nature Communications, 8*(1), 15065. https://doi.org/10.1038/ncomms15065.
14. Coscieme, L., & Stout, J. C. (2019). Ecosystem services evaluation. In B. Fath (Ed.), *Encyclopedia of Ecology* (2nd ed., pp. 288–293). Elsevier. https://doi.org/10.1016/B978-0-12-409548-9.10840-5.
15. Costanza, R., De Groot, R., Braat, L., Kubiszewski, I., Fioramonti, L., Sutton, P., Farber, S., & Grasso, M. (2017). Twenty years of ecosystem services: How far have we come and how far do we still need to go? *Ecosystem Services, 28*(B), 1–16. https://doi.org/10.1016/j.ecoser.2017.09.008.
16. Costanza, R., de Groot, R., Sutton, P., van der Ploeg, S., Anderson, S. J., Kubiszewski, I., Farber, S., & Turner, R. K. (2014). Changes in the global value of ecosystem services. *Global Environmental Change, 26*, 152–158. https://doi.org/10.1016/j.gloenvcha.2014.04.002.
17. Crook, S., Levine, A., & Lopez-Carr, D. (2021). Perceptions and application of the

ecosystem services approach among Pacific Northwest national forest managers. *Sustainability, 13*(3), 1259. https://doi.org/10.3390/su13031259.

18. Curran, M. A. (2013). Life cycle assessment: A review of the methodology and its application to sustainability. *Current Opinion in Chemical Engineering, 2*(3), 273–277. https://doi.org/10.1016/j.coche.2013.02.002.
19. Czúcz, B., Haines-Young, R., Kiss, M., Bereczki, K., Kertész, M., Vária, A., Potschin-Young, M., & Arany, I. (2020). Ecosystem service indicators along the cascade: How do assessment and mapping studies position their indicators? *Ecological Indicators, 118*, 106729. https://doi.org/10.1016/j.ecolind.2020.106729.
20. D'Amato, D., Bartkowski, B., & Droste, N. (2020). Reviewing the interface of bioeconomy and ecosystem service research. *Ambio, 49*(12), 1878–1896. https://doi.org/10.1007/s13280-020-01374-5.
21. D'Amato, D., Korhonen, J., & Toppinen, A. (2019). Circular, green, and bio economy: How do companies in land-use intensive sectors align with sustainability concepts? *Ecological Economics, 158*, 116–133. https://doi.org/10.1016/j.ecolecon.2018.12.026.
22. Daily, G. C., Polasky, S., Goldstein, J., Kareiva, P. M., Mooney, H. A., Pejchar, L., Ricketts, T. H., Salzman, J., & Shallenberger, R. (2009). Ecosystem services in decision making: Time to deliver. *Frontiers in Ecology and the Environment, 7*(1), 21–28. https://doi.org/10.1890/080025.
23. Daniel, T. C., Muhar, A., Arnberger, A., Aznar, O., Boyd, J. W., Chan, K. M. A., Costanza, R., Elmqvist, T., Flint, C. G., Gobster, P. H., et al. (2012). Contributions of cultural services to the ecosystem services agenda. *Proceedings of the National Academy of Sciences, 109*(23), 8812–8819. https://doi.org/10.1073/pnas.1114773109.
24. David, W., Pearce, R., & Turner, K. (1991). *Economics of Natural Resources and the Environment*. American Journal of Agricultural Economics.
25. De Groot, R. S., Wilson, M. A., & Boumans, R. M. J. (2002). A typology for the classification, description, and valuation of ecosystem functions, goods, and services. *Ecological Economics, 41*(3), 393–408. https://doi.org/10.1016/S0921-8009(02)00089-7.
26. de Groot, R., Brander, L., van der Ploeg, S., Costanza, R., Bernard, F., Braat, L., Christie, M., Crossman, N., Ghermandi, A., Hussain, L. H. S., et al. (2012). Global estimates of the value of ecosystems and their services in monetary units. *Ecosystem Services, 1*(1), 50–61. https://doi.org/10.1016/j.ecoser.2012.07.005.
27. Dick, J., Al-Assaf, A., Andrews, C., Díaz-Delgado, R., Groner, E., Halada, ⊠., Izakovi⊠ová, Z., Kertész, M., Khoury, F., Kraši⊠, D., et al. (2014). Ecosystem services: A rapid assessment method tested at 35 sites of the LTER-Europe network. *Ekológia, 33*(3), 217–231. https://doi.org/10.2478/eko-2014-0021.
28. Dietz, T., Börner, J., Förster, J. J., & Von Braun, J. (2018). Governance of the bioeconomy: A global comparative study of national bioeconomy strategies. *Sustainability, 10*(9), 3190. https://doi.org/10.3390/su10093190.
29. Droste, N., & Bartkowski, B. (2018). Ecosystem service valuation for national accounting: A reply to Obst, Hein and Edens (2016). *Environmental and Resource Economics, 71*(1), 205–215. https://doi.org/10.1007/s10640-017-0127-7.

30. El-Chichakli, B., Von Braun, J., Lang, C., Barben, D., & Philp, J. (2016). Policy: Five cornerstones of a global bioeconomy. *Nature, 535*(7611), 221–223. https://doi.org/10.1038/535221a.

31. European Commission. (2019). *EU Guidance on Integrating Ecosystems and Their Services into Decision-Making*. Commission Staff Working Document SWD (2019) 305 Final PART 1/3.

32. Fang, X., Zhou, B., Tu, X., Ma, Q., & Wu, J. (2018). What kind of a science is sustainability science? An evidence-based reexamination. *Sustainability, 10*(5), 1478. https://doi.org/10.3390/su10051478.

33. Fisher, B., Turner, R. K., & Morling, P. (2009). Defining and classifying ecosystem services for decision making. *Ecological Economics, 68*(3), 643–653. https://doi.org/10.1016/j.ecolecon.2008.09.014.

34. Fitzgerald, J., & Lindner, M. (2016). Adaptive challenges for European forests. In *Adapting Forests to Climate Change Symposium* (pp. 1–15). International Conference on Ecological Sciences, Marseille, France.

35. Freeman, A. M. III, Joseph, A., & Catherine, L. K. (2002). *The Measurement of Environmental and Resource Values: Theory and Methods* (3rd ed.). RFF Press.

36. Furman, E., Häyhä, T., & Hirvilammi, T. (2018). *A Future the Planet Can Accommodate*. Finnish Environment Institute. ISBN 978-952-11-4938-2.

37. Ghilarov, A. M. (2000). Ecosystem functioning and intrinsic value of biodiversity. *Oikos, 90*(2), 408–412. https://doi.org/10.1034/j.1600-0706.2000.900224.x.

38. Gibbons, S., Mourato, S., & Resende, G. M. (2014). The amenity value of English nature: A hedonic price approach. *Environmental and Resource Economics, 57*(2), 175–196. https://doi.org/10.1007/s10640-013-9664-9.

39. Giurgiu, V., & Badea, O. (2015). *Pădurile și schimbările de mediu în Romania*. Editura Academiei Române.

40. Gómez-Baggethun, E., Martín-López, B., Barton, D., Braat, L., Saarikoski, H., Kelemen, M., García-Llorente, E., van den Bergh, P. J., Arias, P., Berry, L., et al. (2014). *State-of-the-Art Report on Integrated Valuation of Ecosystem Services* (Deliverable D.4.1/WP4). European Commission.

41. Grainger, D., & Stoeckl, N. (2019). The importance of social learning for non-market valuation. *Ecological Economics, 164*, 106339. https://doi.org/10.1016/j.ecolecon.2019.106339.

42. Heckwolf, M. J., Peterson, A., Jänes, H., Horne, P., Künne, J., Liversage, K., Sajeva, M., Reusch, T. B., & Kotta, J. (2021). From ecosystems to socio-economic benefits: A systematic review of coastal ecosystem services in the Baltic Sea. *Science of The Total Environment, 755*, 142565. https://doi.org/10.1016/j.scitotenv.2020.142565.

43. Hernández-Blanco, M., Costanza, R., Anderson, S., Kubiszewski, I., & Sutton, P. (2020). Future scenarios for the value of ecosystem services in Latin America and the Caribbean to 2050. *Current Research in Environmental Sustainability, 2*, 100008. https://doi.org/10.1016/j.crsust.2020.100008.

44. Hove, S. V. D. (2007). A rationale for science-policy interfaces. *Futures, 39*(7), 807–826. https://doi.org/10.1016/j.futures.2006.12.004.

45. Jacobs, S., Dendoncker, N., Martín-López, B., Barton, D. N., Gomez-Baggethun, E., Boeraeve, F., McGrath, F. L., Vierikko, K., Geneletti, D., Sevecke, K. J., et al. (2016). A new valuation school: Integrating diverse values of nature in resource and land use decisions. *Ecosystem Services, 22*, 213–220. https://doi.org/10.1016/j.ecoser.2016.11.007.

46. Jamie, A., Carr, G., Petrokofsky, D., Spracklen, V., Simon, L., Lewis, D., Nicholas, R., Trull, A., Vidal, S., Wicander, J., et al. (2021). Anticipated impacts of achieving SDG targets on forests—A review. *Forest Policy and Economics, 126*, 102423. https://doi.org/10.1016/j.forpol.2021.102423.

47. Joachim, H., & Spangenberg, J. S. (2016). Value pluralism and economic valuation—Defendable if well done. *Ecosystem Services, 18*, 100–109. https://doi.org/10.1016/j.ecoser.2016.02.008.

48. Karen, E., & Rebecca, M. A. (2016). Moving beyond the exchange value in the non-market valuation of ecosystem services. *Ecosystem Services, 18*, 78–86. https://doi.org/10.1016/j.ecoser.2016.02.022.

49. Kates, R. W. (2011). What kind of a science is sustainability science? *Proceedings of the National Academy of Sciences of the United States of America, 108*(49), 19449–19450. https://doi.org/10.1073/pnas.1116097108.

50. Kim, M. S., & Stepchenkova, S. (2020). Altruistic values and environmental knowledge as triggers of pro-environmental behavior among tourists. *Current Issues in Tourism, 23*(13), 1575–1580. https://doi.org/10.1080/13683500.2019.1628188.

51. Kindu, M., Schneider, T., Teketay, D., & Knoke, T. (2016). Changes of ecosystem service values in response to land use/land cover dynamics in Munessa–Shashemene landscape of the Ethiopian highlands. *Science of The Total Environment, 547*, 137–147. https://doi.org/10.1016/j.scitotenv.2015.12.127.

52. Kontogianni, A., Tourkolias, C., Machleras, A., & Skourtos, M. (2012). Service providing units, existence values and the valuation of endangered species: A methodological test. *Ecological Economics, 79*, 97–104. https://doi.org/10.1016/j.ecolecon.2012.04.004.

53. Kopperoinen, L., Itkonen, P., & Niemelä, J. (2014). Using expert knowledge in combining green infrastructure and ecosystem services in land use planning: An insight into a new place-based methodology. *Landscape Ecology, 29*(8), 1361–1375. https://doi.org/10.1007/s10980-014-0014-2.

54. Kornatowska, B., & Sienkiewicz, J. (2018). Forest ecosystem services—Assessment methods. *Forestry Letters, 60*, 248–260. https://doi.org/10.2478/frp-2018-0025.

55. Ligate, E. J., Chen, C., & Wu, C. (2018). Evaluation of tropical coastal land cover and land use changes and their impacts on ecosystem service values. *Ecosystem Health and Sustainability, 4*(9), 188–204. https://doi.org/10.1080/20964129.2018.1534330.

56. Liu, X., Bakshi, B. R., Rugani, B., de Souza, D. M., Bare, J., Johnston, J. M., Laurent, A., & Verones, F. (2020). Quantification and valuation of ecosystem services

in life cycle assessment: Application of the cascade framework to rice farming systems. *Science of The Total Environment, 747*, 141278. https://doi.org/10.1016/j.scitotenv.2020.141278.

57. Loomis, J. J., Knaus, M., & Dziedzic, M. (2019). Integrated quantification of forest total economic value. *Land Use Policy, 84*, 335–346. https://doi.org/10.1016/j.landusepol.2019.03.016.

58. Mandle, L., Douglass, J., Lozano, J. S., Sharp, R. P., Vogl, A. L., Denu, D., Walschburger, T., & Tallis, H. (2016). OPAL: An open-source software tool for integrating biodiversity and ecosystem services into impact assessment and mitigation decisions. *Environmental Modelling & Software, 84*, 121–133. https://doi.org/10.1016/j.envsoft.2016.06.001.

59. Marre, J. B., Brander, L., Thebaud, O., Boncoeur, J., Pascoe, S., Coglan, L., & Pascal, N. (2015). Non-market use and non-use values for preserving ecosystem services over time: A choice experiment application to coral reef ecosystems in New Caledonia. *Ocean & Coastal Management, 105*, 1–14. https://doi.org/10.1016/j.ocecoaman.2014.12.009.

60. Martin, C., & Czellar, S. (2017). Where do biospheric values come from? A connectedness to nature perspective. *Journal of Environmental Psychology, 52*, 56–68. https://doi.org/10.1016/j.jenvp.2017.05.004.

61. Millennium Ecosystem Assessment. (2005). *Ecosystems and Human Well-Being: Synthesis*. Island Press.

62. Morelli, F., Jiguet, F., Sabatier, R., Dross, C., Princé, K., Tryjanowski, P., & Tichita, M. (2017). Spatial covariance between ecosystem services and biodiversity pattern at a national scale (France). *Ecological Indicators, 82*, 574–586. https://doi.org/10.1016/j.ecolind.2017.07.017.

63. Nabuurs, G. J., Verkerk, P. J., Schelhaas, M. J., Olabarria, J. R. G., Trasobares, A., & Cienciala, E. (2018). Climate-smart forestry: Mitigation impacts in three European regions. *Science & Policy, 13*, 456–465.

64. Naidoo, R., & Ricketts, T. H. (2006). Mapping the economic costs and benefits of conservation. *PLoS Biology, 4*(11), e360. https://doi.org/10.1371/journal.pbio.0040360.

65. Neill, A. M., O'Donoghue, C., & Stout, J. C. (2020). A natural capital lens for a sustainable bioeconomy: Determining the unrealized and unrecognized services from nature. *Sustainability, 12*(19), 8033. https://doi.org/10.3390/su12198033.

66. Nkonya, E. M., Phillip, D., Mogues, T., Pender, J., & Kato, E. (2012). Impacts of community-driven development programs on income and asset acquisition in Africa: The case of Nigeria. *World Development, 40*(9), 1824–1838. https://doi.org/10.1016/j.worlddev.2012.04.021.

67. Nunes, P. A., & Schokkaert, E. (2003). Identifying the warm glow effect in contingent valuation. *Journal of Environmental Economics and Management, 45*(2), 231–245. https://doi.org/10.1016/S0095-0696(02)00052-9.

68. Obst, C., Hein, L., & Edens, B. (2016). National accounting and the valuation of ecosystem assets and their services. *Environmental & Resource Economics, 64*(1), 1–23. https://doi.org/10.1007/s10640-015-9921-1.

69. Ojea, E., & Loureiro, M. L. (2007). Altruistic, egoistic and biospheric values in willingness to pay (WTP) for wildlife. *Ecological Economics, 63*(4), 807–814. https://doi.org/10.1016/j.ecolecon.2007.02.003.

70. Pagiola, S., Bishop, J., & Landell-Mills, N. (2002). *Selling Forest Environmental Services: Market-Based Mechanisms for Conservation and Development* (1st ed.). Routledge.

71. Parida, V., Burström, T., Visnjic, I., & Wincent, J. (2019). Orchestrating industrial ecosystem in circular economy: A two-stage transformation model for large manufacturing companies. *Journal of Business Research, 101*, 715–725. https://doi.org/10.1016/j.jbusres.2019.01.006.

72. Pascual, U., Muradian, R., Brander, L., Gómez-Baggethun, E., Paracchini, M. P., Zulian, G., & Alkemade, R. (2012). Synergies and trade-offs between ecosystem service supply, biodiversity, and habitat conservation status in Europe. *Biological Conservation, 155*, 1–12. https://doi.org/10.1016/j.biocon.2012.06.016.

73. Pearson, T. R. H. (2007). *Measurement Guidelines for Forest Carbon Sequestration*. U.S. Department of Agriculture, Forest Service, Northern Research Station.

74. Plato, L., & Meskin, A. (2014). Aesthetic value. In A. C. Michalos (Ed.), *Encyclopedia of Quality of Life and Well-Being Research* (pp. 76–78). Springer.

75. Popa, B., Coman, C., Borz, S. A., Nita, D. M., Codreanu, C., Ignea, G., Marinescu, V., Iora⊠, F., & Ionescu, O. (2013). Total economic value of natural capital—A case study of Piatra Craiului National Park. *Notulae Botanicae Horti Agrobotanici Cluj-Napoca, 41*(2), 608–612. https://doi.org/10.15835/nbha4129355.

76. Popa, B., Pascu, M., Nita, D. M., & Codreanu, C. (2013). The value of forest ecosystem services in Romanian protected areas—A comparative analysis of management scenarios. *Bulletin of the Transilvania University of Bra⊠ov, Series II: Forestry, Wood Industry, Agricultural Food Engineering, 6*(55) No. 2, 41–50.

77. Quintas-Soriano, C., Castro, A. J., Castro, H., & García-Llorente, M. (2016). Impacts of land use change on ecosystem services and implications for human well-being in Spanish drylands. *Land Use Policy, 54*, 534–548. https://doi.org/10.1016/j.landusepol.2016.03.011.

78. Sabbadin, D. (2021). *The elephant is in the room. Why it makes sense giving priority to circular economy measures in the building industry in the 2020s. EEB.*

79. Salles, J. M. (2011). Valuing biodiversity and ecosystem services: Why put economic valuation on nature? *Comptes Rendus Biologies, 334*(5–6), 469–482. https://doi.org/10.1016/j.crvi.2011.03.008.

80. Schaefer, M., Goldman, E., Bartuska, A. M., Sutton-Grier, A., & Lubchenco, J. (2015). Nature as capital: Advancing and incorporating ecosystem services in United States federal policies and programs. *Proceedings of the National Academy of Sciences of the United States of America, 112*(24), 7383–7389. https://doi.org/10.1073/pnas.1420500112.

81. Schmidt, K., Sachse, R., & Walz, A. (2016). Current role of social benefits in ecosystem service assessments. *Landscape and Urban Planning, 149*, 49–64. https://doi.org/10.1016/j.landurbplan.2016.01.005.

82. Simpson, R. D. (2016). Ecosystem services: What are the public policy implications? *PERC Policy Series, No. 55*. Available online: https://www.perc.org/wp-content/uploads/old/pdfs/PS-55-EcosystemServices_Simpson_PERC.pdf (accessed on 29 August 2024).

83. Sousa-Silva, R., Verbist, B., Ângela, L., Valent, P., Suškevi's, M., Picard, O., Hoogstra-Klein, M. A., Cosofret, V. C., Bouriaud, L., & Ponette, Q. (2018). Adapting forest management to climate change in Europe: Linking perceptions to adaptive responses. *Forest Policy and Economics, 90*, 22–30. https://doi.org/10.1016/j.forpol.2018.02.007.

84. Turner, B. L., Kasperson, R. E., Matson, P. A., McCarthy, J. J., Corell, R. W., Christensen, L., Eckley, N., Kasperson, J. X., Luers, A., Martello, M. L., et al. (2003). A framework for vulnerability analysis in sustainability science. *Proceedings of the National Academy of Sciences of the United States of America, 100*(14), 8074–8079. https://doi.org/10.1073/pnas.1231335100.

85. Turner, K., Badura, T., & Ferrini, S. (2019). Natural capital accounting perspectives: A pragmatic way forward. *Ecosystem Health and Sustainability, 5*(1), 237–241. https://doi.org/10.1080/20964129.2019.1660281.

86. United Nations Framework Convention on Climate Change. (1992). *UNFCCC*. New York, NY, USA.

87. United Nations. (1992). Agenda 21: Programme of action for sustainable development; Rio Declaration on Environment and Development. Statement of Forest Principles. In *Proceedings of the United Nations Conference on Environment and Development, Rio de Janeiro, Brazil*, 3–14 June 1992 (Vol. 3).

88. Vallecillo, S., La Notte, A., Ferrini, S., & Maes, J. (2019). How ecosystem services are changing: An accounting application at the EU level. *Ecosystem Services, 40*, 101044. https://doi.org/10.1016/j.ecoser.2019.101044.

89. Varul, M. Z. (2011). Value: Exchange and use value. In D. Southerton (Ed.), *Encyclopedia of Consumer Culture* (pp. 1502–1504). Sage Publications.

90. Vasiljevic, N., & Gavrilovic, S. (2021). Cultural ecosystem services. In W. Leal Filho, A. M. Azul, L. Brandli, A. Lange Salvia, & T. Wall (Eds.), *Life on Land* (pp. 209–218). Springer International Publishing. https://doi.org/10.1007/978-3-319-95981-8_128.

91. Verkerk, P. J., Costanza, R., Hetemäki, L., Kubiszewski, I., Leskinen, P., Nabuurs, G. J., Poto⊠nik, J., & Palahí, M. (2020). Climate-smart forestry: The missing link. *Forest Policy and Economics, 115*, 102164. https://doi.org/10.1016/j.forpol.2020.102164.

92. Villegas-Palacio, C., Berrouet, L., López, C., Ruiz, A., & Upegui, A. (2016). Lessons from the integrated valuation of ecosystem services in a developing country: Three case studies on ecological, socio-cultural and economic valuation. *Ecosystem Services, 22*, 297–308. https://doi.org/10.1016/j.ecoser.2016.10.017.

93. Wegner, G., & Pascual, U. (2011). Cost-benefit analysis in the context of ecosystem services for human well-being: A multidisciplinary critique. *Global Environmental Change, 21*(2), 492–504. https://doi.org/10.1016/j.gloenvcha.2010.12.008.

94. Wijewardana, D. (2008). Criteria and indicators for sustainable forest management: The road travelled and the way ahead. *Ecological Indicators, 8*(2), 115–122. https://doi.org/10.1016/j.ecolind.2006.11.003.

95. Xu, L., Saatchi, S. S., Shapiro, A., Meyer, V., Ferraz, A., Yang, Y., Bastin, J. F., Banks, N., Boeckx, P., & Verbeeck, H. (2017). Spatial distribution of carbon stored in forests of the Democratic Republic of Congo. *Scientific Reports,* 7(1), 1–12. https://doi.org/10.1038/s41598-017-00107-8.

96. Zhang, P., Zhou, C., Wang, P., Gao, B. J., Zhu, X., & Guo, L. (2014). E-Tree: An efficient indexing structure for ensemble models on data streams. *IEEE Transactions on Knowledge and Data Engineering, 27*(2), 461–474. https://doi.org/10.1109/TKDE.2014.2322592.

97. Zulian, G., Polce, C., & Maes, J. (2014). ESTIMAP: A GIS-based model to map ecosystem services in the European Union. *Annals of Botany, 4*, 1–7. https://doi.org/10.4467/23920601EB.19.01.11496.

98. Zulian, Z., Paracchini, M. L., Maes, J., & Liquete, C. (2013). ESTIMAP: Ecosystem services mapping at European scale. *EUR 26474*. Publications Office of the European Union. https://doi.org/10.2788/64370.

CHAPTER 7

Regeneration of Forests

LEARNING OBJECTIVES

After studying this chapter, you will be able to:

- Understand the objectives and ecological principles of forest regeneration.
- Differentiate between natural and artificial regeneration methods.
- Identify factors influencing the choice between natural and artificial regeneration.
- Learn the process and management practices involved in tree planting.
- Assess the advantages and disadvantages of natural and artificial regeneration methods.

KEY TERMS FROM THIS CHAPTER

Artificial regeneration
Forest regeneration
Tree planting
Factors affecting regeneration
Natural regeneration
Vegetative method

INTRODUCTION

Forestry is concerned with the preservation and enhancement of forests. Forests serve a variety of important non-wood production functions for nature and society. It creates channels and slows the flow of water on the land, protects the soil from erosion, maintains air humidity, reduces temperature fluctuations, protects the environment, preserves natural resources, improves the land, and houses animals and plants.

People benefit from the forest's renewable raw materials, which include forest fruit, fungi, medicines, and timber. Financial earnings from commercial forest activities, such as wood sales, help to fund the construction and maintenance of forest roads, the treatment of small streams, and the provision of structures for the entertainment and leisure of forest visitors.

Thus, in terms of sustainable forest management, a forester's primary role and responsibility is to ensure the continuous availability of renewable raw materials, as well as to maintain a balance between tree harvesting and planting with the goal of preserving forests in a sustainable manner for future generations.

This chapter covers all natural plant regeneration processes in forest ecosystems, including pollination, seed dispersal, predation, and seedling development. It also provides information on the methods and techniques of artificial regeneration, also known as tree planting.

7.1. FORESTS REGENERATION: OBJECTIVES AND ECOLOGY

Regenerating forest is the process of planting new trees after harvesting or the death of forest trees due to fire, insecticides, or disease (Figure 7.1). Regeneration is a critical activity for sustainable forestry, and it can be achieved using the following basic approaches:

- Deer were kept on higher ground in this coniferous forest in the United Kingdom to allow new trees to grow well.
- Natural regeneration takes place when the new seedlings or sprouts are established by the trees left on or near the place.
- Artificial regeneration is a common practice of tree planting.

Figure 7.1. *Forest regeneration.*

Source: Photo: Russel Wills, (CC BY-SA 2.0).

Timing is a significant factor in the process of regeneration. Prompt reforestation can be crucial if the objective of a landowner is to produce timber or a habitat in forests. Every year a forest is understood to mean a loss of growth in timber. Furthermore, the delayed regeneration of the plant can lead to the invasion of brush, grass or other undesirable vegetation making it difficult and costly to establish desired trees. In addition, many countries need reforestation within a few years of wood harvest in forest practice laws.

To depart from the country where a forest is located, the degree to which a forest is renewed through its natural regeneration or tree planting. Land owners usually choose not to plant specific trees but to establish new wood stalls by promoting natural regeneration. Existing trees in forests reproduce themselves and become a natural community on the base of site conditions through natural regeneration.

The owners of rural properties may create conditions for the natural regeneration of the selected species. It is often a more practical approach to create such conditions than to develop an artificial hardwood planting. Naturally regenerated trees, which are established to rise in the natural soil, are normally less expensive to start and can occur at high densities producing straight, high-quality stems. Goals forests will be subject to varying degrees of impacts on climate change according to current climate change projections.

Temperatures will generally increase, but precipitation will change in various ways in Europe. The rises risk of drought and/or flooding, augmented frequency as well as concentration of acute weather events and consequently rise fire risk, insect, and pest damage can be caused, depending on regional conditions.

An increase as well in the damaged forest area is expected to regenerate the forest area under less favorable conditions. As a result, improving forest resilience against the abiotic and biotic hazards is an important objective of forest reproduction and requires a re-evaluation of existing regeneration strategies and techniques, besides the traditional targets of stand efficiency and quality. The decisions in use during the phase of forest regeneration can generally save the considerable cost. Consequently, in times of increasing risk, the careful choice of regeneration strategies, methods and materials becomes especially important.

Natural regeneration for forestry ecosystems is a significant ecological process. It includes a cycle of life stages from the seed phase to plants for adults and promotes the succession and restoration of forest communities after disorder. It is significant to measure the population growth and large quantity, and also to establish the patterns of species turnover at the community level, which are affected under interspecific success or the timing of recruits, under various conditions. It is important for the population level. Plant regeneration is therefore essential for the community dynamics and for the maintenance of the biodiversity of forests and ecosystems. Robustness to small disturbances is a feature of healthy ecosystems. After major (lethal) disturbances such as an outbreak or a fires outbreak in a forest, it will not be possible to return immediately to a previous dynamic balance.

The newly opened habitat will instead be occupied by pioneering species, compete in space and establish itself. New growth is known as the ecologic regeneration of seedlings and the community assembly process. As the ecological succession enters into force a forest will slowly regenerate into its former state (climax or any intermediate stage), as long as all outer parameters (climatic condition, availability of nutrients for soil fertility, trajectories for animal migration, air pollution and absence, etc.) remain unchanged.

Natural fires are necessary for a cyclically stable ecosystem with cyclical regeneration in certain regions such as Australia. It outlines the life cycle of forest woodlands and the process of regeneration (natural regeneration). There is insight into how the suitability of a forest stand can be evaluated for natural regeneration and for the planning of large-scale natural regeneration. The forestry systems are described in the context of cutting processes, initiation and establishment of natural regeneration, technical limitations and applicability, and effects on the renewing forest properties and characteristics. The regenerative systems are also described in this chapter. The main control factors for natural regeneration are analyzed, including availability of seeds and seed predation and seed growth and survival.

7.2. NATURAL FOREST REGENERATION

The natural renovation of the forest is the process by which young plants are established from seeds (seedlings). Forest trees experience different phases throughout their life cycle: an intensive competitive and high mortality seedling stage; a young, intensive growth in height determining how fast trees grow up to the overstory canopy layer; and a maturity stage characterized by canopy prevailing, as well as the beginning of processes in breeding and recruiting. Only when a sufficient number of "crowding space" for seed germination and subsequent seedling growth is available can natural regeneration be successful (Figure 7.2). The understory lighting regime is determined by canopy trees and tends to reduce the growing space for recruiting young trees into the canopy and therefore consolidates the dominance of these trees.

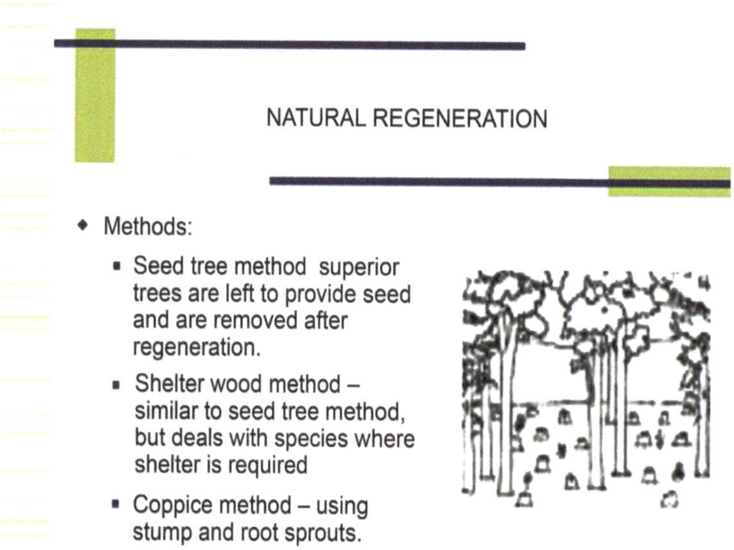

Figure 7.2. *Illustrates the natural regeneration methods.*

Source: https://www.slideserve.com/lavina/introduction-to-forestry.

The beginning of regeneration processes in natural forests depends on the development of open areas (gaps) due to the death and degradation of mature trees; the deficiencies become preferential sites for natural regeneration, and forest characteristics are largely dependent on the dynamics of gap recruitment. The beginning of natural regeneration in managed forests depends on the use of appropriate cuttings for regeneration, reducing the canopy cover and therefore creating the required area to recruit the new generation of trees. In a certain way, regeneration cuttings should be aimed at reproducing patch disturbances in natural forests.

7.2.1. Process of Natural Regeneration

- Seed supply.
- Production.

Almost every year, some seed is produced at almost all locations, although the timing of large seed crops varies. The production periods extend from 1 year to 2–11 year interval between large Douglas-fir seeds and lodge pole pine seed. While no relationships were established between cause and effect, weather patterns for certain species were linked to seed production.

Abundant Douglas-fir plants usually precede a cool summer two years before seed falls, followed by a cool clear winter, a wet spring and last hot January. Similar combinations of climate factors are likely to influence seed production in Gran Tan. Ponderosa seed is best produced at higher summer temperatures in the year in which it is produced (Daubenmire, 1960).

Douglas-fir cone plants are usually larger in poor locations than at good locations. Cone size decreases and the seed weight increases at ocean distance. Seeds are frequently produced only by cones and insects, in some places almost all seeds of certain species of the tree are destroyed.

A large amount of the seeds are produced by stalks in good seed years but not always reflects the ratio of each seed species in a mixed plant. For example, Port-Orford-cedar and Western Redcedar are generally over-represented in the Western hemlock seedfalls and Douglas-fir and white fir are generally under-represented. Furthermore, environmental conditions can almost lead to exclusion of other species (Dunning, 1923). When forestry treatments create a conducive environment for natural regeneration, the problem becomes sufficient seed source.

In general, seedbeds supply more seed on a renewable site for an adjacent clearcut than uncut trees. Seeds from seedbeds are also broader than seeds from neighboring uncut stalls, as their distribution in distance is restricted. However, lonely wind-filled trees (Garman, 1951) are wind sensitive.

When seedbed or shelterwood systems are to achieve natural regeneration, large, vigorous trees should be preserved with signs of wind force and past wind production. Excessive spacing of the seed tree is essential, as most Douglas-fir pollen falls in close proximity to the parent tree (Wreight, 1953). Shelterwoods are produced with less windthrows than seed trees by area. The shelterwood trees produce more uncut grain than the trees for the trees.

Even when cone plants are poor, some seeds are usually produced every year, so the cutting of seed cannot occur until seedbeds remain receptive. Sadly, seedbeds are not quite responsive in many places. Seed production can be stimulated at these sites in seed-source trees. Fertilization, root cutting, injection of gibberellins and girdling have been employed. In mature Douglas-fir, fertilization is ineffective, but young stands could produce more seeds if it is fertilized with nitrate. Spring applications are best, but a site-based approach is required. Low nutrient levels are better than sufficient sites to produce seed nutrients (Edwards, 1986). Root cutting and gibberellin injection stimulates the initiation of strobilus in young trees and can increase the size of poor plants in seed gardens. Nevertheless, Girdling seems to be more practical in the field and encourages students to start with poor grain years. The ribbon is to be performed by removing or cutting a bark band on the other side of the stem on two sides with a screw around half the rim. Girdling should take place approximately 1 month before the break of vegetative buds.

The girdle cures slowly without serious tree damage and continues to boost strobilus initiation for several years. Dilution increases the size of a good cone crop on young stalls, but does not promote seed production in poor years. The release of a shelter wood would appear to boost seed production. After a shelter wood overstory in the third summer, the development of bumper cone cultures often followed.

7.2.1.1. Seed Dispersal

The distribution of seeds is local and annual. There are also variations because of the spread of the seed trees, the local topography, the prevailing winds and each other. Exposed slopes and steep slopes from the valleys prevented effective scattering of seeds (Kos, 1947); small, isolated stalls could not move enough to be effective.

The majority of seeds fall in the tree's height (McDonald, 1980). As the number of seeds is reduced at distance, the exponential distribution of the timbered edge. Coniferous seeds are spread by gravity and wind. How far they travel depends on their weight and wing size, their speed and direction, their tree height and even their slope towards the downwind.

The first conical flight is the main factor in the distribution of seeds. Seeds often move later from water and from animals. The weight and feasibility of the seeds are dispersed naturally as the distance from the seed source increases. Because of the slower and slight fall and farther than the sound seeds, the distance between the hollow seeds increases, since they are unfathomable and hollow.

Seed dispersal usually lasts for a number of months and the maximum distribution for each trees varies widely (Garman, 1951). However, sound seeds usually fall in the first week of the seed and the peak of feasible seed dispersion earlier this season than hollow seeds. Wind and moisture influence the full dispersal of the wind. Cones open or disintegrate in warm, dry winds, and slowly in cold and wet winds.

The dormancy seeds are resistant to extreme temperature and humidity conditions, whilst seeds of a majority of coniferous species enjoy cold, wet stratification throughout the winter. Seed dispersal distance is most important for the process of natural regeneration if it is considered as an' efficient seed distance.'

7.2.1.2. Seed Survival

Dispersed, dormant seeds resist extreme temperatures and humidity conditions, with cool moist stratification during the winter months benefitting from seeds from the majority of coniferous species. Most seeds do not, however, survive. Meteorological factors appear not to directly affect survival, but indirectly by influencing the biological factors that contribute to the majority of this mortality. In rain water, for instance, seeds are often carried to holes and cranes, which are concealed from seed foods while protecting microsites with mineral soil. Winter snowpacks also cover seeds from seedlings, which may have been accountable for improved spring survival in high altitudes by Minore (1986c). The seed dispersal timing can be important compared to snowfall. The only effective seed is that falls before the soil is covered with snow for genuine high-level firs. The most seeds in a seed crop are consumed by animals. White-footed deer mice do the most harm, but shrews, chipmunks and birds are also serious predators. In 3 months after logging (Tev, 1956b), populations of moose can return to pre-harvests levels, but even small populations may cause a great deal of damage during seed fall. Where there is an adequate source of food for after-harvest vegetation these rodents population grow much larger in the first two or three years after harvest, and by further increasing their use for the increasing availability of food (Hooven, 1976), they react to large seed crops. The predators of seed prefer to others certain coniferous species. When plants are small, the consumption of seeds is usually most harmful. In the light and medium years of light Douglas fir seed crops, rodents and birds often consume all of the seeds. However, if a small crop is pre-germinated, its size could decrease so much further that rodents and birds no longer seek it.

7.2.1.2.1. Germination

Many important factors are affecting the critical transition from sleeping seeds in the soil to growing seedlings. The germination factors were classified as endogenous and exogenic by Zlobin in 1977. Seed maturity, dormancy, size and weight, nutrient reserves, genotypes and inherent growth capacities are endogenous factors.

Moisture, the temperature, light, aeration, soils and external biological influences are exogenous factors. The seed must soak moisture out of its immediate environment to germinate. For the seed to absorb water, the close touch between the seed and its substrate is crucial, but its water content must not be continuous. In fact, the seed strength improves with a brief redrying period before the seed germinates (DeMatos et al., 1985). The air temperatures affect germination relatively directly (Fowells & Stark, 1965). When there is enough moisture (Haig et al., 1941) then the temperature of the substrate probably becomes critical for stimulating germination.

The optimal temperature for germination of most coniferous seeds is 20–25° C, but slow germination occurs at 5–14° C. The primary biological factors for birds and rodents directly affect the success of germination, as seed and seedlings are consumed during germination. Other biological influences indirectly affect germination through microenvironmental modification. The same gophers that destroy seedlings, for instance, (Tevis, 1956a), produce mineral soil seedlings, and vegetation that subsequently compete for soil moisture may provide a live shade that benefits germination.

7.2.2. Requirement for Natural Regeneration

The seedbed quality is a primary factor in natural regeneration as a surface on the surface of a seed lands determines the humidity, temperature and nutrient regimes under which germination occurs. The most common seedbeds are duff, rotten wood and mineral soil. Sufficiently decayed root wood, which is well distributed, is rarely sufficient to cover a large part of the reconstructed area. Dead wood also burns often before it breaks down sufficiently to make a seedbed. The resulting charcoals stimulate early germination, but their higher surface temperatures can be lethal to tender plants later in the season. In most conditions, mineral soil is the best seedbed for most species. The temperature of the soil under mineral soils is more extreme than under the organic soil, but the surface temperatures of mineral soil are less extreme (Hermann, 1963). Germination tends to be the best place for moist and shaded mineral soil. The physical features of a soil affect the survival and growth of seedlings directly.

For example, root penetration on Sandy Soils (Eis, 1965) tends to be greatest, but exposed B2 horizons limit growth due to their high bulk density and low fertility. Many land features influence survival and growth indirectly, by altering the present amount and type of planting. Competitor vegetation in productive soils is often a major factor (Minore, 1984), with sometimes palate vegetation encouraging sufficient animal seed harm to cause mortality (Cooper, 1961).

Survival and development can benefit from soil scarification by creating favorable root growth requirements (Valentine, 1975) and reducing competition between the plants. But where there is checked plant competition, the piling and burning of slash scarification and soil displacement seem to damage plant growth. Moisture is the most vital factor limiting seed survival and growth, but it is also important to maintain sufficient light and favorable temperatures.

Forestry treatments affect all three factors. In the first year after germination, most mortality of naturally regenerated seedlings happens in the early weeks, when germs are most likely to cause moisture stress and heat damage. This mortality can be reduced by shading through an overstory canopy.

7.2.3. Importance and Potential of Natural Regeneration

Alongside the regeneration type and regeneration technique, it would be an important error, once and for all, that the choice of the regeneration type and the regeneration technique should be clearly linked to the basic characteristics of the managerial type and the underlying decomposition practices. The reasonable practice of forestry, aimed at continuously achieving a maximum economic performance over a period of time, always profoundly changes the original forestry situation. These changes vary from the minor altering of stand structure to the composition of non-native species after conscious destruction of the original undesirable stands, and from the significant trend of structural unifying to the creation, by planting, of pure or mixed stands with non-native ones, or in areas in which the forest has been lost a long time ago. It is possible to differentiate between extreme situations due to the large extent of forestry influences, but, on the other hand, it is not easy to establish a clear line between manufactured forests and natural sites. The fact is that natural regeneration can be promoted and achieved in a number of cases, while artificial regeneration must not be debarred from stands, wherein the original combination is well-maintenance.

The situation, in fact, is even more complicated because the choice between natural and artificial regeneration is fundamentally a matter of bio-ecological, economic, and technical opportunities, strongly influenced by the ever-moving conditions of production. In a future evolution, the elimination of all doctrinal prejudices, which until now have been only the cause of a useless division of the forest world, is increasingly intermixed by the two systems of stand regeneration. The only way to achieve a greater degree of freedom in silvicultural action and intervention can be natural logic and more basic knowledge of growing phenomena in a way that makes them more adaptable to different situations. Natural regeneration is the main process by which a forester can use the dynamics of woodlands and the genetic properties of local population of tree species to be preserved. The most successful method to move from equally old artificial stands conquered by coniferous species to higher species diversity mixed standards is to promote natural regeneration of the late-succession tree species. Costly plantings after logging may be avoided or reduced at least as a consequence of natural regeneration, the operational budget can be increased in commercial forests.

The need to move towards near-natural forestry systems and multiple forest management objectives that take into consideration major public opinion issues is met in the circumstance of natural regeneration. Natural regeneration in most cases is better-quality to artificial regeneration because seedlings from local seeds are often better tailored to adverse environmental conditions; this is predominantly important at sites where trees have been exposed for their entire lifetime to natural hazards, such as the Subalpine Forestry or the Mediterranean region, where harsh weather conditions are present.

Natural regeneration is an important process that we can rely on to restore the forest after severe disruption across large areas. In the tropical areas, there are examples of the secondary rainforests derived after shifting cultivation from the natural recovery, as well as intensive settlement of pastures from the seeds dispersed from the neighboring forest. As one of the most obvious features of global landscape changes in temperate regions, secondary succession by natural regeneration of trees on abandonment of farms or pastures is becoming increasingly important. The best option for plant breeding in the burnt regions in the Mediterranean region is the reliance on natural regeneration, e.g., after large fires in the Aleppo pine woodlands in southern Italy, as large quantity of seed are rapidly isolated from cones on the crown of burnt trees (crown seeds "banks"); In general, natural regeneration forest practices are considered a valuable means of progress to forest types that are more competent at providing valuable products for the forest and simultaneously assuring multipurpose functions for forest (water conservation, recreation, site protection, and habitat preservation in endangered plant and wildlife species). Substantial international efforts are considered to be developing sustainable management criteria based on natural regeneration.

7.2.4. Natural Regeneration Planning and Implementation

Natural regeneration planning requires good ecosystems and dynamism and highly experienced staff of forestry. Increased efforts in research are aimed at filling gaps in knowledge about the ecological factors that control regeneration processes and the development of decision making criteria to select the best forest option for natural regeneration standards. Goal criteria can help with natural regeneration planning.

To assess the suitability of the forest stand for the natural regeneration, regeneration indices have been proposed; indices based primarily on parameters such as seedling density (i.e., number of seeds per m2), seedling height, or a mixture of the height and density. A forest may be well thought-out suited for natural regeneration if, at a certain standing age (for instance in the normal age of first thinning in uniform ages), the regeneration index reaches a certain "threshold" based on both ecological and economic considerations.

On the basis of multivariate site analyzes, such as regional variables in climate (temperature, moisture, frost, and radiation) and ground conditions (topography of soil characteristic, water availability, and vegetation), or stochastic methods for predicting forest dyna, efforts to assess the suitability of natural regenerations can be made; Examples of significant results can be found in these methods for explaining and forecasting the suitability of Scot pine forestry in the boreal region to natural regeneration and long-term fluctuations in oak stalls in the temperate area.

Vegetation analysis on the basis of the assignment to understory plant species of the ecological indicator can also be useful in the prediction of ecologically significant conditions for the process of natural regeneration; the applicability of this method to the Scots pine stalls under the Mediterranean conditions is an example. The establishment of scenario maps of the natural regeneration, useful tools for the management of forests, can extensively contribute to geographical information systems (GIS). When planning natural regeneration, foresters should be conscious of local socio-economic conditions.

For example, in areas such as the Mediterranean, where open herbivore-grasping is still an important traditional activity, they should be made aware of possible problems with shepherds, and the temporary exclusion from the forest of grazing animals necessary for recruiting tree may lead to conflict. The application of a regeneration system is necessary to implement natural regeneration.

In forestry, a system of recovery is defined as a cutting process leading to the creation of a new generation of trees and the creation of a new forest stand. Different systems can provide favorable conditions for the initiation of reproduction and natural regeneration. The choice of a given system depends on the forest structure and the aims of the management plan.

This choice will have a significant influence in terms of biodiversity, structure, site protection, growth rate and wood quality on the characteristics of renewed forests. In scientific and technical literature, qualitative and quantitative information is more and more available on the impact of different forest regeneration methods and regimes, for example, in recent years a significant number of information has been produced about the impact of various regeneration systems on the genetic diversity of forest recruitment systems.

Managing the environment by harvesting methods and cultural treatments uses seeds produced in and near the area to create a new forest stand. This method provides a flexible, practical, cheap alternative to planting in certain industrial and domestic forestry lands if a proper seed source is available. It is especially suitable for non-industrial private land. For the major southern pine species, a number of reproductive cutting methods are used. Methods of clearcutting, seed-tree, and shelterwood establish equal-aged stands, while selection cutting creates or maintains uneven-aged stands.

7.2.5. Disadvantages of Natural Regeneration

The natural regeneration failure is common in clear cuts along hot, dry southern hillsides, at hazardous frost sites and at sites where there is insufficient seed source. If these conditions exist, successful natural regeneration will require mitigation, usually by leaving the tree cover. Tree cover does not guarantee the survival of plants, but tender germinant require more cover protection than robust, tree-by-tree survival (Helgerson et al., 1982), than natural plants under a Southwest Oregon shelter wood. When this overstory is not available and cutting is dictated by stand conditions or other considerations, the only possible means is artificial regeneration. The time it takes to plant an area is costly (Horton, 1985), but only costly if the regeneration effort fails. The regeneration is not a problem. The time it takes is not expensive. Regeneration methods are often less cost-effective than clear-cutting (Streeby, 1977). The repeated entries compact the soil and damage regeneration and the residual trees when seed trees, shelterwoods or selection methods are used.

Under an overstory, controlling competition can be harder. To manage a removal of wilderness, which produces an ideal combination of the desired products and values, extra time, care and money is required. The removal of overstory also requires understory seedlings of a good quality and quantity (Laacke & Fiddler, 1986).

In simple terms, protecting mature trees and new growth during forest management involves a more complex process. This includes careful planning and management, as well as dealing with slash (debris left after logging). Creating suitable seedbeds for regeneration might require specialized machinery and techniques such as underburning, growing, and tractor scarification, which are not typically used in artificial regeneration.

Prepared sowing beds over time decrease and investment in preparation could be wasted if the preparation is not in sequence with a seed plant (Oliver et al., 1983). In most species, good seed crops are rare and irregular and vary according to location as well as the time and nature of manipulation of the stands. The pines are warning at least for a year when seed crops will occur, but for other species to produce mature seeds it takes less than 2 years. The lead time for the preparation of seeds and harvest is inadequate to several months for these species. Even if cone counts show that a constructive crop is produced, rodents or birds often destroy the majority of the seed. Under these tentative conditions, operations involving people and machinery are difficult to plan and coordinate.

Prompt planting can give an immediate after-harvest lead for one or two years (Shearer, 1982) and successful regeneration cannot always take place immediately. The entire inventory in subalpine environments may require between five and 15 years (Minore & Dubrasich, 1981). At lower heights, similar delays are not unusual. It depends on how often the plantation must be planted and how capitalized plantings affect the economics of the stand before it is harvested whether the time of delay is more costly than a failed artificial regeneration attempt.

When full stock is obtained, there is often overabundant natural regeneration in places and must be diluted at a later date. The species to be managed are only those that grow on or adjacent to the site when natural regeneration is used. Shade-tolerant species tend to prevail in a new stand on the site of a relatively long period of natural regeneration overstory protection. The presence of a host specific disease or parasite can sometimes make the change of species desirable; even if dwarf-mistletoe infection can be avoided in the true firs through the removal of overhead infection sources before a 1 meter high natural regeneration, some diseases are not so easy to control forests. Thus natural regeneration on such an area would not achieve management goals.

In essence, after establishment artificial and natural regeneration requires the same conditions. Both grow best without an overstory and both suffer from vegetative competition; release is often required. At times, planter inventories retain and increase their initial size advantage in the field, when they are sufficient for rapid early growth with the first shoot-to-root relationship (Smith et al., 1968). Performance of genetic improvement schemes that enhance survival and growth is perhaps the clearest drawback of natural regeneration. An undertaking to regenerate the natural is a commitment to forget about the benefits of planting GM stocks.

7.3. ARTIFICIAL REGENERATION

Artificial Regeneration refers to the planting of tree seedlings after a timber harvest but not depending on the stump sprouts, seeds, and other types of natural regeneration. Planting is usually the last phase of preparation for other premises aimed at enhancing plantability, controlling competition from the native vegetation, improving growth and increasing the survival of seedlings. The deforestation continues, with a large amount of forest wealth (Figure 7.3).

Figure 7.3. *Artificial regeneration in a forest.*

Source: https://www.azolifesciences.com/article/What-is-Regenerative-Forestry.aspx.

7.3.1. Regeneration by Artificial Methods

This affects not only the forests but also animals and the entire ecosystem. Alarming deforestation (1.5 million hectares each year) is ongoing. There are some preliminary considerations that are urgently needed to perform artificial regeneration. Basic steps in artificial regeneration is based on:

- Presence of seed orchards or pits.
- presence of a seedling industry (nurseries).
- Suitable labor force availability.
- Site preparation.

Preparation of seed and plant materials on the forest premises (in most cases) permits a preliminary selection of reproductive material in relation to certain seeds. Such a seed is the most important prerequisite for the successful establishment of artificial regeneration. Improving tree growth, quality and/or resilience to pests, illnesses and drought is the most important objective. Plant breeding and genetic improvement is based on two categories of selection criteria: physiological and genetic characteristics. Based on its physiological properties, a proportional wider root system and a high density of small active roots are selected for healthy, vital plants with a single leader. The maintenance of seeds of grafted clones of productive, high quality, and healthy trees ensures genetic quality.

This can enrich the adaptation potential of trees to predictable climate change and associated risks by continually improving the genetic and physiological quality of plant material. The use of artificial regeneration is particularly recommended for (re) introducing additional tree species, for sites with rapid development of competing ground vegetation or for the restoration of severely damaged or burned sites.

1. Choice of Species:

i. **Objective of raising the trees:** The choice of species is very important in artificial regeneration. Thus, before selecting the tree species, the objective of raising the trees must be specified.

ii. **Climate and Microclimate:** The selection of species is dependent upon the prevailing micro-climatic and climatic conditions.

iii. **Soil Requirements:**

a. **Wet Soils:** Salix species, Populus species, etc.

b. **Water:** logged soils, *Eucalyptus robusta*, *E. saligna* etc.

c. **Sandy Loam:** *Albizia procera*, *Acacia nilotica*, *Dalbergia sissoo*, etc.

iv. **Market Facilities:** The availability of market facilities for the sale and processing of tree products should be considered to ensure economic viability.

v. **Growth Rate:** Fast developing tree species - *Acacia nilotica* (Babul), *Leucaena leucophala* (Subabool), *Melia azedarach* (Bakain), *P. deltoides* (Poplar), Salix species (Willow).

vi. **Availability of Exotics:** In easy which means pertains to, not indigenous to the spot of question.

The exotic could be referred to as "an organism in a location which isn't native of the spot but has its origin in various other regions. For instance, *Robinia pseudacacia*, *Leucaena leucocephala*, Eucalyptus species, and *Populus deltoides*, etc.

vii. **Base of Establishment:**

viii. **Management Goals.** The artificial regeneration depends upon the goals of management.

ix. **Site Conditions:** The site is definitely the complex of biological and physical aspects of a location which figure out what forest of some other vegetation might vary.

x. **Succession:** The succession is the easy replacement of a single area by an alternate in the improvement of vegetation towards a climax.

xi. **Cost of Growing:** This is additionally really crucial element impacting the option of species.

xii. **Availability of Seed/ Propagation Substance:** The seed supply must be sound.

2. **Choice of Method:** The achievements of unnatural regeneration depend to a fantastic level in the option of approach. Right now, there are generally 2 strategies of this particular regeneration, viz. (i) Sowing and (ii) Planting.

   i. **Sowing:** Sowing, in probably the simplest words, is the procedure of scattering the seeds in a specific location, e.g., nursery bed, area, etc.

   Ø **Benefits of Sowing:**

   - It's the least expensive procedure and costs much less.
   - Sowing is an immediate procedure and not one other problems.
   - It requires a shorter time and hence the tasks are accomplished before long.
   - In the sowing technique, there's simply no doubting of disturbances of origins.
   - Sometimes, planting is completed straight in the area (in forests), and therefore it doesn't involve some nursery.
   - The planting becoming the easy technique, is meant to be much less cumbersome.

   ii. **Planting:** Planting is one other approach to artificial regeneration. Nevertheless, growing is referred to as the transferring nursery stock to the growing location as contrasted with transplanting in the greenhouse.

   Ø **Benefits of Planting:**

   - More Success.
   - Less seed required.
   - No harm.
   - Cheaper weeding.

   Ø **Disadvantages of Planting:**

   - Need of nursery.
   - Disturbance of roots.
   - Time consuming.
   - Need for proficient labor.
   - Incurred huge costs.

3. **Site Selection:** The choice of site is additionally an important element in artificial regeneration. There are many factors which affect the preparation of site.

Below are the elements that are vital to performing the preparation:

1. **Ground Cover:** Evaluate the existing vegetation and ground cover to determine how it will affect the planting and establishment of new trees. This includes managing competing vegetation and preparing the soil for planting.
2. Physical Factors:
   (a) Topography.
   (b) Exposure.
   (c) Soil type.
   (d) Erosion hazards.
   (e) Size of therapy area; and
   (f) Access.
3. **Preparation Requirement:** To develop a good setting for establishment of attractive species.
4. **Man Power and Equipment:** The site choice and preparation techniques demand great ability as well as helpful equipment.

5. External Constraints:
   (a) Legal duties.
   (b) Smoke control recommendations.
   (c) Proximity to very sensitive areas; and
   (d) The mindset of adjacent farmers or land owners.
6. **Spatial Arrangement:** This is additionally called as Spacing.

7.3.2. Artificial Regeneration by Vegetative Methods

Planting material besides seeds, e.g., bare root seedlings, bulbs, offsets, suckers, rhizomes, layering, cuttings, containerized seedlings, and corms can also be employed for vegetative propagation material.

Propagation by Cuttings: Cuttings are of two kinds.

- Stem cutting; and
- Root cutting

1. **Stem Cutting:** Few species respond best because of this method. The species, which are not hard to root are ideal for this method of growing. Especially dicotyledons, having a proactive cambium layer e.g., shisam, nimbara drumstick, mulberry inga dulsis, pimpal, pangara, dhaman, and banyan tree, etc.

   Stumps for growing are utilized in only a couple of species, for instance. shivan, teak, shisam, *Cassia spp*. Stumps will be moved very easily, require less space and may be moved over long distances. These are ready in the time of growing or right before planting. The simple transplantation is of new uprooted seedlings. The percentage of 20% stem as well as 80% taproot throughout stump planning is to be maintained. To prevent any damage, the stem part shall be cut from 5 to 6 cm above the collar region and the remaining portion or root shall be cut to prepare stumps.

   The fine-edge knife or equipment shall be used. The stumps are packed in bundles, holding the stumps on one side and the roots on the other. They should be immediately transported. For transport stumps, moist gunny bag tissue is used to prevent desiccation. Within 2–3 days without much loss you can transport them.

   These stumps are planted after three to four rain showers at the beginning of the monsoon when the land gets humid enough and the temperatures on the soil warm. Stumps are planted with the help of crowbar by preparing small troughs in a layer so that the new shoot rises up straight. Then the stumps are inserted inside and the soil is firmly pressed so that the hole is keen to stagnate. The cooler area is located near the surface of the ground.

2. **Root Suckers:** Root suckers for seeding purpose may be used as well, for example, garnets, cocum, anjan, shisam, nimbara, pangara, and erythrona, etc. Potting material can be prepared using layers, grafted plants, magnetic plants. All of them are only used to preserve the superior genotype in forestry. It is used for commercial planting because it is short-lived and stretched and does not develop long straight (trunk) timber.

7.3.3. Scope and Objectives of Artificial Regeneration

- Artificial regeneration is carried out to replenish the naked areas in the forests, where natural regeneration never occurs. (The spp. is well developed and economical and requires less care.)
- To introduce valuable exotic spp., an artificial regeneration is also made.
- Artificial regeneration to replace less precious spp. with more precious spp.
- It is also made for the purpose of afforestation.
- The density of the forest is also increased.
- The natural regeneration is also replaced.
- It is also made to improve the environment and esthetically.

7.3.4. Merits and Demerits of Artificial Regeneration

The following are the merits of artificial regeneration:

- Faster results because all the work is systematically performed.
- Full storage and more evenly arranged.
- facility to control the composition of species.
- Forest Work concentration reduces costs and facilitates supervision.
- Faster early growth and thus less risk of weeds.
- Better revenues, even if the actual operation is less expensive.

The following are the demerits of artificial regeneration:

- Labor intensive.
- Temporary root development disrupted.
- Less adapted to microsites.
- Cost-intensive.

7.4. FACTORS AFFECTING THE CHOICE OF A METHOD OF REGENERATING FORESTS

7.4.1. Factors Affecting Artificial Regeneration

It is very important to choose a method for site preparation. Seven factors that have a vital role to play in selecting a method are the following:

1. Nature of the existing terrain cover.
2. Physical site factor.
3. Site preparation requirement.
4. Labor and equipment available.
5. External limitations.
6. Environmental impacts; and
7. Cost incurred in the operation.

7.4.2. Factors Affecting Natural Regeneration

In forest dynamics and the recovery of ecosystem functions, natural regeneration is an essential component. For ecological studies, therefore, it is important to understand the status of regeneration and how abiotic and biotic factors affect it. Both abiotic (environmental) and biotic factors could affect the natural regeneration, and the safeguarding of sufficient forest regeneration is a major challenge for forest management. Reducing precipitation, especially drought, may limit plants 'potential for regeneration and composition, since almost every species is highly sensitive to seedling environmental factors. Forests are generally very wide with a large number of species of various sizes and qualities in terms of land area. A continuous and up-to-date record of the height, shape factor, age, quality, etc. is required for each tree in the forests.

If this effort is considered throughout the forest, it is necessary to create a huge amount of funding and to carry out a single forest inventory. The flash or electric current passes through the roots of the trunk of the tree. The tree may decrease and die without major over ground damage to the root from electricity. The lightning in rough bark trees causes severe over ground damage. The leaves may turn into smooth bark trees, and the tree may die, and trees may recover their leaves in the following growing season. Land change is an all-embracing phenomenon caused by both human and natural landscape alterations. It is important to study change in land cover because it helps explain; the types of modifications that occur, the rates at which they occur and the location of changes in land cover in the landscape.

Understanding the spatial, temporal and thematic dynamics of land transformation makes hypothesizing the main drivers and effects of change researched and developed, help identify future situation scenarios, and is useful when understanding impacts on other ecosystem resources. Understanding the forest dynamics can help improve productivity, preserve the composition of species, financial input and develop forestry prescriptions. Continuous forest monitoring on a long-term basis is helpful in documenting satisfactory vegetation dynamics. The term 'disruption' refers to events that change the forest structure relatively rapidly to release space and resources and change the microclimate. Difference in the dynamic of disorders has profound effects of diversity in natural forests through the production of a variety of habitat types that have developed and adapted to the native species. Forests are key food, wood, and habitat sources. More than 80% of energy needs are supplied from fuelwood.

Food for cattle, non-wood products like sweet wilderness, gum, tubers, roots and wild animals, aromatic and medicinal plants also is provided by forests. In addition, forests are contributing to the control of desertification, soil fertilization, protection of watersheds and mitigation of climate change. Increased pressure on forests because of high forest dependency, population growth, population degradation and desertification problems, a drop in agricultural productivity, limited government budgets, and other environmental issues all contribute to urgent research on tree dynamics and human disturbance.

Natural forests face a wide range of problems from overgrazing and over-exploitation, especially as the forest products are the main source of income for the local population. These activities have directly altered the natural regeneration, the distribution of species of trees and eventually their natural dynamics.

7.4.3. The Hypothetical Contradiction Between Natural and Artificial Regeneration

A rich literature has repeatedly emphasized the many advantages of natural regeneration. A more critical approach towards those arguments shows that an important section of the papers available is based on preconceived ideas on regeneration or on specific situations that lead to further generalizations. On the other hand, there is little comparative trial carried out over a long period of time to prevent it from being inspired by some form of natural philosophy or the controversial viewpoint of the forest as an agency, in many cases. This mindset was responsible for regrettable exaggerations, which made sometimes natural regeneration an end in itself. At the same time, the extreme rejection of artificial regeneration led to more flexible forestry treatments, which allowed individual trees to make maximum use of their growing potential. The deeper rooting of seedlings and their rapid and much better fixation into the mineral soil is undoubtedly an almost exclusive advantage in the natural regeneration. This observation applies especially to poorer podzolic soils, where there is a significant difference between the development of root crops and natural seedlings. In particular, the aeration and permeability of water and gasses exert a positive influence on soil conditions.

This enables the potential productivity of the site to be further or maintained. On the other hand, well-rooted seedlings have a greater degree of vitality and resistance to outdoor biotic and abiotic attack. This is very important in the youthful stages of growth, whereas plantations have to go through several critical periods (collection, transplantation, final planting on the site) before they survive. Natural recovery therefore has a positive effect on the actual and ultimate stability of the structural stand.

Natural regeneration will also be better adapted to normal stand development and community development dynamics. It fulfills its function more successfully of connecting successive generations and stages of development or in any case, as can normally be expected, even if the highest precautions are taken, in a more fluent and complete way.

Natural regeneration will result directly in the conservation of native tree species, and typical species combinations, thus enhancing the stability of the forest and maintaining the site in good condition. However, since it is easy to manipulate forest treatment and the selection of the species and their mixing patterns and depends heavily on a range of decisions that can be made at any stage of stand development, it is sufficiently important to maintain the presumption of ecologic and local races to have more attention.

The real potential and objective value for preserving local types is, in fact, little known: the problem has an ecological, climate and genetic aspect. The former seems moral to be the latter. The local population is the survival product, already reduced by previous actions, by the selective influence of the total environment on a broader population or on a population. 6 It does not have constant or permanent internal stability, since at the beginning of every new generation it tends to enlarge again.

The population mainly dwindled during the initial construction, under the influence of soil conditions almost invariably or extremely slowly changing; the same factors continuously react to the extended subpopulation which develops after each regeneration. Action on the complex of air factors, which are not entirely independent of the soil conditions, accentuates the main trend of development.

In conjunction with regular action in almost invariable climatic conditions, an annual or at least regular manifestation of extreme weather conditions leads to a loss of collective and individual vitality and eventually to the endless disappearance of less durable and poorly adapted elements. A smaller local population is thus built over a very long time in successive phases and first of all as a selection product of the abiotic environment.

It should be understood, however, that such local populations are by no means invariable: at the start of every new generation, they tend to expand without reaching, in most cases, their original composition; at the end of each generation, the population has reduction, in part under the influence of acting factors of the biotic and abiotic environment, and in part allotted to them; There is no direct significance to forestry practice as the dynamics of formation of smaller or local populations are found in the real acting factors of selection in environmental influences which cannot be deeply influenced by forestry management. It should also be considered that forestry and even forestry management are designed on a different time scale to ensure that a population is effectively restricted. The study of the time factor in building natural populations has amply demonstrated this. Although ecological selection of this type does not affect forestry treatment directly, its consequences are important enough to be studied and approached realistically.

Indeed, the surviving elements are better adapted to local conditions and site factors, thereby creating more stability within the forest. Similarly, and while the selection is based on the abiotic environment and results primarily in the organic and physiological similarity between survivors, not necessarily all morphological changes are beneficial: its significance and desirability must be assessed based on the technological quality of the timber and its economic value. Practical forestry has nevertheless repeatedly shown an important tendency to overrate ecotypes and local races' morphological and technological supremacy. A very optimistic generalization of the positive effects of ecological research and its economic consequences has often been deduced from the enthusiastic study of specific situations and especially successful site selections.

The result is also that apparently good populations are being established in new sites, with eventually a few poor and even catastrophic results, because population dynamics have not been clearly understood and there has been no thorough analysis of the equilibrium of the initiation site with the origin site. The following concept is still more doubtful that, even in the single rotation or management period, the combination of the environmental and the positive forestry selection can lead to a real and stabilized improvement in quality of the stand after a limited number of trees. After each new regeneration, the inevitable population increase is sufficient evidence to show that the effects of forestry selection are almost entirely limited to one single production of trees and that the first stages of each new stand must start all over again. With good ecotypes and local races outside their natural dispersion, and on better and poorer soils, the unsatisfactory findings have shown time and time again that such selections are very closely linked with the site they are at, particularly in terms of their morphological and physiological properties.

Really good natural ecotypes and excellent locals are rather scarce because they imply an intimate genetic link between a wider variety of people who at the same time, in other areas and when used, are required to have a high degree of resistance against variable biotic and abiotic influences. Therefore, it is more exceptional to conserve the desirable characteristics and maintain the narrow populations in optimal form after they are introduced into the new site. The difference between the ecological conditions at the source site and the introductory site is greater and significant, with individual deviations and internal degradation. On the other hand, forestry selection is not able to produce long-lasting amelioration effects as long as a morphological choice is made between individuals, for which there are 8 genetic foundations, as the phenotype of trees is influenced by both accidental micro-ecological and dire conditions in a very high degree, particularly in intensely treated stands.

It thus becomes apparent that it appears that the only reliable result in the development of local races and smaller populations is greater individual and collective vitality, resulting in a more stable standard, which can ultimately contribute to the generation of a greater economic value and create favorable conditions for temporary improvement. The use of natural selections, beyond their normal range, does not guarantee positive results for growth.

A critical study of naturally occurring regeneration and its supposed benefits may soon be concluded that the bulk of positive results in practice do not in itself result from natural regeneration but are more closely linked with the technical measures taken during the period of regeneration and the characteristics of cutting and management of the forestry. Indeed, it is reported mostly in areas where intensive forestry treatments have changed and manipulated the environment that supposedly good local races with real economic value.

This also means that, following artificial regeneration, favorable conditions for the formation of local races can be created by forestry treatment at least to a similar degree. There are rather limited situations in which this has been done actively as the preference for artificial regeneration often implies a state of mind, in which the bio-ecological view of Silviculture is lost and continued treatment is neglected.

The concept of artificial regeneration is indeed given by a number of foresters to a limited content. But, as natural regeneration does not mean 'selection forest,' artificial regeneration is not always a direct preference for large-scale technical forestry or one-sided forests. Set on a large scale or expansion quickly across a large area, the natural and artificial regeneration can produce and produce critical situations.

In either case, significant costs are required for the implementation and subsequent handling of the young stands. They tend to weigh heavily on the management of forests under the social and economic conditions to be a real treat for good financial management by their concentration over a short time, because they can easily cause serious trouble in their work and funding plans.

A large number of alleged advantages of natural regenerative, especially the greater possibilities of individual selection, are completely lost as soon as early and intensive treatment for various reasons becomes impossible, particularly when there is no sufficient labor or a disadvantageous link between expenses and production.

That critical situation is already present in part of Europe where artificial regeneration of fast-growing species begins in the absence of any environmental or biological problem in areas where natural regeneration is replaced. There is an equivalent development in terms of forestation and artificial regeneration in extended areas.

The cost of afforestation has increased spectacularly; chronic labor shortages have grown so high that their counter-balsas is rather doubtless due to future economic surpluses in industrialized and densely populated countries with high standards of living, investment and expenditure for goods and services. This situation worsens, so it is quite understandable for a reversal in artificial group regeneration to allow forestry interventions to spread as well as save forest costs and expenses to treat dense natural regenerated stands on a regular basis.

This is not an unfortunate development, as it includes substantial opportunities to raise production, improve stand quality and organize stand change, which have not been sufficiently studied up to now. At the same time, artificial regeneration is compatible with the forestry bio-ecological orientation and can be easily integrated into most classical forestry systems or treatment patterns. Much has indeed arisen from the wrong formulation and role of stand regeneration and the unilateral emphasis on specific aspects of some or other regeneration methods. There is also a great deal of misunderstanding. Schaedelin believes that regeneration is only to be seen as a binding connection of generations in the forest, which is constantly under evaluation, so the attempt to minimize the importance of regeneration cannot be understood.

It really means that no superior and generally acceptable regeneration system exists, that most silvicultural systems do not have an imperative regeneration system and, in each individual case, a choice of regeneration must be made by isolation in accordance with the local conditions and with the fact that regeneration must be implemented. These general objectives can be achieved both naturally and artificially.

However, where in the first case artificial regeneration is considered quite normal, it is often considered that artificial regeneration is not compatible with the permanent forest or that the only task is to cover large areas quickly. For a more realistic approach to most silviculture and forestry problems, a better understanding of the potential of artificial regeneration is necessary.

7.5. TREE PLANTING: PROCESS AND MANAGEMENT

The renewal of forests – to change harvested trees, fire, disease, extreme weather or aging – is crucial to sustainable forestry. Regeneration could be accomplished by planting trees. The steps concerned with tree planting are talked about. Two fundamental methods may be utilized for regeneration:

7.5.1. Identifying Explanations for Planting Trees

A few preliminary actions have to be considered before you leap into buying as well as growing trees. What and also just how you intend to make use of your property will in part depend on just how. It's therefore crucial to develop a vision for the country and set goals that are applicable to your specific needs and interests. For instance, by planting trees that carry fruit or maybe species that offer a stylish nesting environment, you are able to attract bird watchers.

7.5.2. Site Evaluation

The following step in planning your forest is assessing the spot after you've decided what you would like your land to become. Each place is different, and a comprehensive analysis of the site is crucial for selecting vegetation, growing procedures, stand style and maintenance activities. This particular evaluation must aid you assess the climate, soil type, topography, existing vegetation, landscape and forest problems. You're prepared to build your plantation layout which meets your targets and it is suitable for the circumstances found on your site with your site assessment in hand. The advancement of a plantation strategy consists of deciding exactly where and where arrangements to grow trees.

Obviously, your forest has the perfect chance to grow as well as survive by selecting the proper trees for your site. Furthermore, by using the best adequate plantation system you'll best achieve your objectives. For example, plant trees in rows to optimize tree development and facilitate maintenance.

Whenever the visible effect of rows of trees will be minimized, think about alternating rows of conifers as well as hardwood trees. Create clumps of forests for protection to be able to produce wildlife habitat. In the improvement of your respective planting arrangement also think about protection elements.

Leaving room for planting as well as planting roads, firebreaks are made around and near your plantation that slow or even stop wildfires from burning the trees. The preparation of a planting location is a crucial part of the effective planting procedure, often overlooked. Implementing a couple of steps that are simple are able to mean that trees need to survive and will achieve proper planting.

Weed management is the main objective of site preparation. Controlling weeds makes it easy for trees to establish themselves with no competition for water, nutrients and sunlight. Think about the soil type, pre-existing vegetation sort as well as density, the time of yours plus monetary restrictions to determine the cost-effective and effective method of creating sites.

Mechanical site preparation as well as substance site preparation will be the two primary site preparation strategies that are used prior to planting tree plants. The preparation of mechanical websites involves the utilization of different kinds of planting tools and machines. The preparation of the substance site consists of the usage of herbicides produced commercially. Prescribed burns as well as cover crops, but particularly in conjunction with other techniques, could likewise be beneficial.

7.5.3. Careful Handling of Tree Seedlings

Careful handling of trees from nurseries to planting is crucial to a booming reforestation work. Seedlings are living organisms, and certain conditions are required by them to keep good health and vigor. Proper care and management of seedlings are able to protect your investment and ensure a great start to your forest plantation. Mechanical planting and hand planting are the primary methods for planting trees. The selection of what sort of strategy is primarily determined by the size and the field of the plantation region.

The accomplishment of the latest planting of the trees is usually based on the follow-up. Proper care for brand new tree growing contains survival monitoring, ongoing command of competing vegetation, animal safety like rodents or horses, moreover illness as well as pest monitoring. In the success of the new tree plantation, weather additionally plays a role. You cannot manage the weather conditions, though you are able to cut back countless environmental impacts by following the aspects you are able to manage. It's essential to keep track of the survival of the new plants to assess the results of your regeneration effort. The great time for evaluations is but one year after the seeding or natural regeneration.

Vegetation management could be used throughout the improvement of a forest stand and also provides among the best possibilities to make sure that the composition, as well as structure of the preferred species, develop inside an economically viable time. In reality, with no highly effective FVM, many forest regeneration tasks globally would fail or even be negatively postponed.

The handy management of vegetation needs intense silviculture designed to boost wood fiber yields per hectare. Many businesses are carried out to set the regeneration as well as consequent forest crop improvement up to the crop. This is accomplished in forestry at the different stages of development to provide its development a healthy environment.

These tasks are called tending operations:

1. **Weeding:** Weeds might be managed when you follow these techniques:
   a. Mechanical Methods.
   b. Biological Methods.

c. Chemical Methods.

2. **Cleaning:** Cleaning is done on a crop that hasn't passed the sapling stage and it is described as a cut to meet up with the very best people of an undesirable era that disrupts the development of the wanted person or perhaps is apt to hinder it. A good command of the harvest composition, particularly in crops that are mixed, will be the primary advantage provided by cleaning. Mechanical, chemical and biological cleaning techniques might be referred to as discussed in weeding.
3. **Thinning:** Thinning is described as a felling in an undeveloped stand up, to enhance the expansion as well as the condition of the other forests without constantly busting the canopy. Thinning is a cure in a crop that takes place outside of the sapling stage until the beginning of regeneration.

The thinning concepts are formulated so they just are applied to fresh as well as old plants or maybe perhaps aged trees in a place. On the foundation of the organic advancement of the stand, getting thinner concepts are designed. Thus, thinning naturally happens under the law of survival of the fittest in a heavily saved forest.

For effective forest management globally, management of the unwanted or even fighting vegetation is vital. The Pacific forests in North America's north-west, south-east and north regions are impacted by uncommon vegetation troubles, which might lessen the development as well as the survival of ideal tree species considerably when unmanaged.

Lots of techniques, which include mechanical machinery, herbicides, prescribed fire and manual cutting, are used to control forest vegetation inside each area. Interfering vegetation comprises plant life that, casting a thick shade on forest floors, prevents the germination as well as development of seedlings. The deer removes or lowers indigenous seeds, and also provides cultivating a place for much more competitive natural species, which benefit from unique lighting problems as well as selective browsing tastes. Poorly planned as well as carried through forest harvests, known as "high grading," leave low-profit trees behind. It has resulted in a shift in our woodlands towards less desirable species of trees and the poorer trees. Many overstocked stalls that could be diluted do not receive treatment due to declining markets for small-diameter trees. Thinning improves the growth of the tree and resistance to the insects and diseases.

Finally, the growing numbers of invasive plants have direct effects on the ability of forests to preserve the diversity of native wildlife. Herbicides can address all of these questions safely, effectively and cost-effectively when properly applied.

CASE STUDY

FOREST REGENERATION IN THE AMAZON RAINFOREST

The Amazon Rainforest, often referred to as the "lungs of the Earth," is one of the most biodiverse ecosystems on the planet. However, extensive deforestation for agriculture, logging, and urban development has led to significant ecological degradation. In recent years, efforts to restore this critical biome have gained momentum. A prominent initiative is the "Amazon Fund," established in 2008 with the aim of financing projects that promote the sustainable use of natural resources, combat deforestation, and foster forest regeneration. The fund, supported by both international and national partners, has facilitated the restoration of over one million hectares of degraded land by replanting native species and implementing sustainable agricultural practices.

Community involvement has been pivotal to the success of forest regeneration projects in the Amazon. Local indigenous communities have played a crucial role in these efforts, applying traditional knowledge to restore ecosystems and maintain biodiversity. Programs that integrate sustainable land use with cultural practices have not only aided in reforestation but also provided economic opportunities for these communities. For instance, agroforestry systems that combine crops with native trees have emerged as a viable solution to enhance food security while promoting forest health. These initiatives highlight the importance of including indigenous voices in conservation strategies to ensure that restoration efforts are culturally appropriate and environmentally effective.

Despite the progress made, challenges remain in the fight against deforestation and forest degradation. Economic pressures, such as land conversion for agriculture and illegal logging, continue to threaten regeneration efforts. Additionally, climate change poses a significant risk to the Amazon's ecosystems, affecting rainfall patterns and biodiversity. Continuous monitoring and adaptive management strategies are essential to address these challenges and promote long-term forest resilience. As nations around the world look to combat climate change, the Amazon Rainforest serves as a vital example of how collaborative efforts and community engagement can foster successful forest regeneration.

Discussion Questions

1. How can local communities be further empowered to take an active role in forest regeneration initiatives?
2. What policies or strategies should governments implement to balance economic development with the need for sustainable forest management and regeneration?

SUMMARY

Forest regeneration encompasses the processes through which forests naturally renew themselves or are actively restored after disturbance. This chapter explores both natural and artificial methods of regeneration, highlighting their ecological, economic, and social implications.

Natural regeneration involves the spontaneous regrowth of forests from seeds, root suckers, or surviving trees after disturbances like logging or fire. It is often preferred for its cost-effectiveness and ecological benefits, such as preserving genetic diversity and supporting wildlife habitat. However, successful natural regeneration depends on factors like seed availability, soil conditions, and disturbance history.

Artificial regeneration, on the other hand, involves human intervention through techniques like planting seeds or transplanting seedlings. This method allows for more precise control over species composition and forest structure but can be costly and labor-intensive. Vegetative methods, such as cloning or grafting, are also used to propagate specific tree species.

Factors influencing the choice between natural and artificial regeneration include ecological goals, site conditions, economic feasibility, and social preferences. Effective tree planting requires careful site evaluation, seedling selection, and post-planting management to ensure survival and growth.

Balancing natural and artificial regeneration methods is crucial for maintaining sustainable forest ecosystems, supporting biodiversity, and meeting societal needs for timber, ecosystem services, and recreational opportunities.

REVIEW QUESTIONS

1. What are the main objectives of forest regeneration? Explain.
2. Describe the process of natural regeneration and its requirements.
3. What are the advantages and disadvantages of natural regeneration?
4. Explain how artificial regeneration differs from natural regeneration.
5. What factors influence the choice between natural and artificial regeneration methods?

REFERENCES

1. Alfaro-Sánchez, R., Jump, A. S., Pino, J., Díez-Nogales, O., & Espelta, J. M. (2019). Land use legacies drive higher growth, lower wood density and enhanced climatic sensitivity in recently established forests. *Agricultural and Forest Meteorology, 276–277*, 107630. https://doi.org/10.1016/j.agrformet.2019.107630.
2. Basnou, C., Vicente, P., Espelta, J. M., & Pino, J. (2016). Of niche differentiation, dispersal ability and historical legacies: What drives woody community assembly

in recent Mediterranean forests? *Oikos, 125*(1), 107–116. https://doi.org/10.1111/oik.02534.

3. Bastin, J. F., Finegold, Y., Garcia, C., Mollicone, D., Rezende, M., Routh, D., Zohner, C. M., & Crowther, T. W. (2019). The global tree restoration potential. *Science, 365*(6448), 76. https://doi.org/10.1126/science.aax0848.
4. Benayas, J. M. R., Bullock, J. M., & Newton, A. C. (2008). Creating woodland islets to reconcile ecological restoration, conservation, and agricultural land use. *Frontiers in Ecology and the Environment, 6*(6), 329–336. https://doi.org/10.1890/070057.
5. Brancalion, P. H. S., & Holl, K. D. (2020). Guidance for successful tree planting initiatives. *Journal of Applied Ecology*. https://doi.org/10.1111/1365-2664.13725.
6. Bullock, J. M., Aronson, J., Newton, A. C., Pywell, R. F., & Rey-Benayas, J. M. (2011). Restoration of ecosystem services and biodiversity: Conflicts and opportunities. *Trends in Ecology & Evolution, 26*(10), 541–549. https://doi.org/10.1016/j.tree.2011.06.011.
7. Camargo, P. H. S. A., Pizo, M. A., Brancalion, P. H. S., & Carlo, T. A. (2020). Fruit traits of pioneer tree's structure seed dispersal across distances on tropical deforested landscapes: Implications for restoration. *Journal of Applied Ecology*. https://doi.org/10.1111/1365-2664.13697.
8. Castro-Díez, P., Vaz, A. S., Silva, J. S., van Loo, M., Alonso, Á., Aponte, C., Bayón, Á., Bellingham, P. J., Chiuffo, M. C., DiManno, N., Julian, K., Kandert, S., La Porta, N., Marchante, H., Maule, H. G., Mayfield, M. M., Metcalfe, D., Monteverdi, M. C., Núñez, M. A., & Godoy, O. (2019). Global effects of non-native tree species on multiple ecosystem services. *Biological Reviews, 94*(5), 1477–1501. https://doi.org/10.1111/brv.12511.
9. Ceccherini, G., Duveiller, G., Grassi, G., Lemoine, G., Avitabile, V., Pilli, R., & Cescatti, A. (2020). Abrupt increase in harvested forest area over Europe after 2015. *Nature, 583*(7817), 72–77. https://doi.org/10.1038/s41586-020-2438-y.
10. Chazdon, R. L. (2014). *Second Growth: The Promise of Tropical Forest Regeneration in An Age of Deforestation*. The University of Chicago Press.
11. Chazdon, R. L., Gutierrez, V., Brancalion, P. H. S., Laestadius, L., & Guariguata, M. R. (2020). Co-creating conceptual and working frameworks for implementing forest and landscape restoration based on core principles. *Forests, 11*(6), 706. https://doi.org/10.3390/f11060706.
12. Chazdon, R. L., Lindenmayer, D., Guariguata, M. R., Crouzeilles, R., Rey Benayas, J. M., & Lazos Chavero, E. (2020). Fostering natural forest regeneration on former agricultural land through economic and policy interventions. *Environmental Research Letters, 15*(4), 043002. https://doi.org/10.1088/1748-9326/ab79e6.
13. Corbera, E., & Schroeder, H. (2011). Governing and implementing REDD+. *Environmental Science & Policy, 14*(2), 89–99. https://doi.org/10.1016/j.envsci.2010.11.002.
14. Cramer, V. A., Hobbs, R. J., & Standish, R. J. (2008). What's new about old fields? Land abandonment and ecosystem assembly. *Trends in Ecology & Evolution, 23*(2), 104–112. https://doi.org/10.1016/j.tree.2007.10.005.

15. Crouzeilles, R., Beyer, H. L., Monteiro, L. M., Feltran-Barbieri, R., Pessôa, A. C. M., Barros, F. S. M., Lindenmayer, D. B., Lino, E. D. S. M., Grelle, C. E. V., Chazdon, R. L., Matsumoto, M., Rosa, M., Latawiec, A. E., & Strassburg, B. B. N. (2020). Achieving cost-effective landscape-scale forest restoration through targeted natural regeneration. *Conservation Letters, 13*(3), e12709. https://doi.org/10.1111/conl.12709.

16. Crouzeilles, R., Ferreira, M. S., Chazdon, R. L., Lindenmayer, D. B., Sansevero, J. B. B., Monteiro, L., Iribarrem, A., Latawiec, A. E., & Strassburg, B. B. N. (2017). Ecological restoration success is higher for natural regeneration than for active restoration in tropical forests. *Science Advances, 3*(11), e1701345. https://doi.org/10.1126/sciadv.1701345.

17. Cruz-Alonso, V., Ruiz-Benito, P., Villar-Salvador, P., & Rey-Benayas, J. M. (2019). Long-term recovery of multifunctionality in Mediterranean forests depends on restoration strategy and forest type. *Journal of Applied Ecology, 56*(4), 745–757. https://doi.org/10.1111/1365-2664.13340.

18. Curtis, P. G., Slay, C. M., Harris, N. L., Tyukavina, A., & Hansen, M. C. (2018). Classifying drivers of global forest loss. *Science, 361*(6407), 1108. https://doi.org/10.1126/science.aau3445.

19. Dirzo, R., Young, H. S., Galetti, M., Ceballos, G., Isaac, N. J. B., & Collen, B. (2014). Defaunation in the Anthropocene. *Science, 345*(6195), 401–406. https://doi.org/10.1126/science.1251817.

20. Elias, F., Ferreira, J., Lennox, G. D., Berenguer, E., Ferreira, S., Schwartz, G., Melo, L. D. O., Reis Júnior, D. N., Nascimento, R. O., Ferreira, F. N., Espirito-Santo, F., Smith, C. C., & Barlow, J. (2020). Assessing the growth and climate sensitivity of secondary forests in highly deforested Amazonian landscapes. *Ecology, 101*(3), e02954. https://doi.org/10.1002/ecy.2954.

21. Espelta, J. M., Cruz-Alonso, V., Alfaro-Sánchez, R., Hampe, A., Messier, C., & Pino, J. (2020). Functional diversity enhances tree growth and reduces herbivory damage in secondary broadleaf forests but does not influence resilience to drought. *Journal of Applied Ecology*, *57*(4), 781–790. https://doi.org/10.1111/1365-2664.13728.

22. Fagan, M. E., Reid, J. L., Holland, M. B., Drew, J. G., & Zahawi, R. A. (2020). How feasible are global forest restoration commitments? *Conservation Letters*, *13*(2), e12700. https://doi.org/10.1111/conl.12700.

23. Farwig, N., & Berens, D. G. (2012). Imagine a world without seed dispersers: A review of threats, consequences, and future directions. *Basic and Applied Ecology*, *13*(2), 109–115. https://doi.org/10.1016/j.baae.2012.02.006.

24. Frei, T., Derks, J., Rodríguez Fernández-Blanco, C., & Winkel, G. (2020). Narrating abandoned land: Perceptions of natural forest regrowth in Southwestern Europe. *Land Use Policy*, *99*, 105034. https://doi.org/10.1016/j.landusepol.2020.105034.

25. Gamfeldt, L., Snäll, T., Bagchi, R., Jonsson, M., Gustafsson, L., Kjellander, P., Ruiz-Jaen, M. C., Fröberg, M., Stendahl, J., Philipson, C. D., Mikusi⊠ski, G., Andersson, E., Westerlund, B., Andrén, H., Moberg, F., Moen, J., & Bengtsson, J. (2013). Higher levels of multiple ecosystem services are found in forests with more tree species. *Nature Communications*, *4*, 1340. https://doi.org/10.1038/ncomms2328.

26. García, C., Moracho, E., Díaz-Delgado, R., & Jordano, P. (2014). Long-term expansion of juniper populations in managed landscapes: Patterns in space and time. *Journal of Ecology*, *102*(6), 1562–1571. https://doi.org/10.1111/1365-2745.12297.

27. Hampe, A., Alfaro-Sánchez, R., & Martín-Fores, I. (2020). Establishment of second-growth forests in human landscapes: Ecological mechanisms and genetic consequences. *Annals of Forest Science*, *77*(1), 87. https://doi.org/10.1007/s13595-020-00993-7.

28. Hanewinkel, M., Cullmann, D. A., Schelhaas, M. J., Nabuurs, G. J., & Zimmermann, N. E. (2013). Climate change may cause severe loss in the economic value of European forest land. *Nature Climate Change*, *3*(3), 203–207. https://doi.org/10.1038/nclimate1687.

29. Holl, K. D., & Brancalion, P. H. S. (2020). Tree planting is not a simple solution. *Science*, *368*(6494), 580. https://doi.org/10.1126/science.aba8232.

30. Holl, K. D., Reid, J. L., Cole, R. J., Oviedo-Brenes, F., Rosales, J. A., & Zahawi, R. A. (2020). Applied nucleation facilitates tropical forest recovery: Lessons learned from a 15-year study. *Journal of Applied Ecology*, *57*(6), 1158–1169. https://doi.org/10.1111/1365-2664.13684.

31. Laestadius, L., Buckingham, K., Maginnis, S., & Saint-Laurent, C. (2015). Back to Bonn and beyond: A history of forest landscape restoration and an outlook for the future. *Unasylva*, *245*(66), 1–8.

32. Lamb, D., Erskine, P. D., & Parrotta, J. A. (2005). Restoration of degraded tropical forest landscapes. *Science*, *310*(5756), 1628. https://doi.org/10.1126/science.1111773.

CHAPTER 8

Forest Protection and Wildlife Biology

LEARNING OBJECTIVES

After studying this chapter, you will be able to:

- Identify important diseases and insect pests affecting forest nurseries.
- Understand the occurrence of diseases and pests in both plantations and natural forests.
- Differentiate between exotic species pests and indigenous pests.
- Explain the management strategies for controlling pests and diseases in forests.
- Assess the economic and ecological losses caused by forest diseases.

KEY TERMS FROM THIS CHAPTER

Assessment of losses
Exotic species
Forestry hazards
Managing pests and diseases
Diseases and insect pests
Forest protection
Indigenous pests

INTRODUCTION

The extent of research on various factors, such as the causes and effects of windstorms, depends on the location. While some aspects, like forest protection against pathogens, receive more attention, there are still unknown factors that require further investigation. Certain threats are diminishing or exerting minimal impact in certain countries, such as Europe, whereas they remain substantial in others, like Africa. Forest protection is a less studied discipline compared to other fields like weeds, which receive more attention from botanists. Aliens, referring to invasive organisms, are currently a prominent subject on the scientific agenda. An initial inquiry that arises in the minds of many individuals is the reason behind the lack of equilibrium in certain forest ecosystems, as well as the significance of safeguarding forests. The impact of forests is multifaceted, with human activity being the most significant factor. Human beings have exerted influence on the forest ecosystem since the beginning of human civilization. Forest conservation has had a notable influence on maintaining the stability of forests, even in earlier times when valuable metals were being extracted. However, it has encountered significant obstacles, especially since the onset of the Industrial Revolution. Forestry hazards can be categorized into specific areas of concern, leading to negative impacts on forest nurseries and stands.

Although the size of forest nurseries is decreasing, which reduces the risk of dangerous substances, the impact of forest stalls is much more diverse. Forest nurseries employ a more rigorous and potentially similar approach to agricultural intervention, while forests have a substantial influence on the frequency of interventions due to their accessibility and extended growth cycle. The analysis of the predominant techniques employed to safeguard the forest will be examined. These protections mainly consist of mechanical measures, such as movement, and chemical measures, such as the use of pesticides. Both approaches have their own set of benefits and drawbacks, but their primary justifications for implementation are their effectiveness and affordability. Biological control is presently employed and continues to be one of the most auspicious methods for safeguarding forests.

Furthermore, it is crucial to evaluate the individual and/or selected measures for all potential pests or pathogens. Contrary to popular belief, forests are not completely silent and devoid of wildlife. The wildlife ranges from insects to tigers, encompassing both small creatures like insects and large ones like tigers, as well as a diverse array of plant life including small shrubs and towering trees. This particular type is referred to as biodiversity in its biological manifestation. Forests possess a greater spectrum of life forms compared to deserts, resulting in a higher level of biodiversity. Most of the global wildlife species are widely distributed, yet their populations are rapidly diminishing. The decline of wildlife can be attributed to our technological advancements and insatiable greed. We systematically deforest areas to obtain land for agricultural or comparable activities. Consequently, the animals have no available refuge. Another alarming phenomenon is the eradication of wildlife due to the expansion of forests. This chapter will provide instruction on the identification and control of insect pests commonly encountered in forest nurseries, farm forestry, and plantations.

8.1. IMPORTANT DISEASES AND INSECT PESTS OF NURSERIES

Forests offer a variety of benefits, including the provision of wood, small forest products, fuel, and forage. They play a vital role in water and soil conservation, serve as a source of food and shelter for wildlife, and enhance people's recreational and aesthetic experiences. The main objective of forestry typically involves the establishment and preservation of forests to optimize their productivity. Biological factors such as insect pests and diseases significantly impact forest production, particularly in plantations and nurseries.

The seedlings have sustained significant damage, resulting in a decrease in both the quantity and quality of the planting stocks. High rates of mortality in the nursery, caused by diseases and pest infestations, can greatly decrease the number of seedlings in the plantation, potentially leading to significant negative consequences. They can have a substantial impact on plantations, potentially leading to a decrease in biomass production or the depletion of valuable germplasm collections (Figures 8.1 & 8.2).

Moreover, plants that are infected exhibit a state of debilitation and are incapable of enduring unfavorable circumstances in both plants and fields.

Figure 8.1. *Red pine (Pinus resinosa) nursery seedlings killed by Lophodermium seditiosum.*

Source: Courtesy US Forest Service.

Figure 8.2. *Slash pine shoot killed by pitch canker.*

Source: Courtesy US Forest Service.

Insect pests and diseases in nurseries result in substantial economic losses. To ensure both a sufficient supply of nurseries and a higher level of productivity, it is necessary to have healthy seedlings that are free from disease. This chapter explores and emphasizes different illnesses, along with the factors that impact the development, spread, and proliferation of diseases.

The chapter also provides a comprehensive analysis of the efficacy of managing these issues through the integration of control methods, including cultural practices, sanitation, and the utilization of fungicides, biocontrol agents, and biofertilizer. Tree diseases and pests can harm trees at any point in their growth, putting at risk the ability of both natural forests and plantations to achieve their management goals. Certain insects and fungi pose a threat to young tree seedlings in kindergartens, while others target fully grown trees in various ways.

Although the competition between invasive plants and seedlings, as well as the presence of harmful animals, pose ongoing challenges for forests, the focus is currently on insects and diseases. There are many instances of forest depletion resulting from various types of foreign or native diseases or pests, which require adjustments in management practices or potentially compel forestry management to accommodate different tree characteristics.

8.1.1. Occurrence of Disease and Pest in Plantations and Natural Forests

Natural forests and plantations in both tropical and temperate regions contain harmful diseases and pests. The simple idea of ecosystem dynamics is that as diversity increases, so does stability. The more animals and plants an ecosystem has, the more inspections and balances are required to keep some species from overgrowing to the point where they threaten other ecosystem components. The availability of host material will be the most significant barrier to pest and disease population growth. In an intricate rainforest, for example, where one hectare can protect up to 1000 plant species, only a small number of suitable host materials will appear in a host-specific population of phytophagous caterpillars. As a result, their numbers will remain relatively stable. The agroecosystem is on the opposite end of the diversity spectrum.

Agricultural ecosystems dominated by a single plant species frequently cover large areas and provide virtually unlimited amounts of suitable host biomass. This may result in population explosions of the organisms that host the plant. Furthermore, the lack of ecosystem variety will result in, at best, far more agroecosystems for forestry than natural forests for stability and diversity.

This leads to the conclusion that plantation forests are significantly more susceptible to disease and pesticide damage than natural forests. Even though this is generally true, it does not always imply that natural forests are immune to the damage. In mixed tropical forests, the absence of pest eruptions is commonly used to demonstrate the importance of diversity in soothing plant communities (Speights & Wainhouse, 1989).

Nonetheless, many boreal and temperate natural forests in the northern hemisphere are insect-prone and have simpler ecosystems than tropical rainforests. A variety of diseases caused by fungi, bacteria, parasites, or viruses can also harm natural forests, making them vulnerable to exotic disease and pesticide introduction.

The eastern United States saw the introduction of two Juniper insects, *Carulaspis viscí* and *Newsteadi lepidasaphis*, as well as the white pine blister rust *Cronartium ribicola* in Bermuda, the European moth *Lymantrie dis* in North America, and the *Phytophthora cinnamomi* fungus. Furthermore, *Fusarium subglutinans*, a fungus, was recently introduced to treat pitch cancer disease. Species Pini is now causing concern in *Pinus radiata* forests off the coast of California, in the United States.

The rose or hibiscus mealybug, *Maconellicoccus hirsutus*, an insect that feeds on 125–150 different plant species, was discovered in 1994–95 on the Caribbean islands of Grenada, Trinidad, and St. Kitts-Nevis. Insects pose a significant threat to the area's natural mixed rainforests, ornamental plants, and vegetable plants.

8.1.2. Exotic Species Plantations

The typical scenario for exotics in forest plantation plans is that they benefit from a "honeymoon" period with no pests, which occurs in both temperate and tropical areas. During this time, management practices frequently evolve, making plantations more vulnerable to diseases and pests as they develop (Figure 8.3).

These include:

1. Not paying attention to the correspondence between species and site.
2. Use of a narrow genetic base planting stock.
3. Inadequate storage and strength of the trees by means of intermediate cuttings.
4. Dependency in a system for a single or two species leading to a limitless source of host materials to potentially harmful organisms.

Figure 8.3. *Exotic species in a forest.*

Source: Wikimedia.

Diseases and pests are able to come from two sources when they begin to occur. An indigenous agent complex is able to familiarize to the new host or even the introduction of exotic pests. A variety of examples of native insects adapting to exotic forest plantings is known.

8.1.3. Indigenous Pests and Diseases

Indigenous insects have been commonly found in the exotic Cupressus, Pinus and Eucalypt plantations in Colombia, with caterpillars, leaf cutting ants, and stem infestation in Colombia being essentially the most commonly encountered. Rodas P. reports that in the Andes region of Colombia 30 kinds of species defoliating insects feed on exotic forest plantations.

There's a similar situation in Chile, in which a few specimens of indigenous pests now are associated with amazing plantings of forests, including *Ormiscodes cinnamomea*. Edwards and Ohmart (1991) report 96 indigenous insect feed in China, 94 in India, 223 in Brazil, 31 in New Zealand, 105 in Papua New Guinea, and 62 on Sumatra Island, Indonesia. The main harmful substances are termites and leaf cutting ants.

Indigenous organisms' diseases may also adapt and cause severe damage to new hosts. *Monochaetia unicornis*, a fungal stem cancer, has seriously damaged plantations of the *Cupressus macrocarpa* in Kenya. During the start of the 1990s numerous species of foliar and twig pathogenic agents have been affected in Vietnam by Eucalyptus plantings including *Cylindocladium* leaf blight, Coniella leaf spot, as well as Kirramyces leaf spot and rosy disease, and *Corticum salmonicolor* (Figure 8.4).

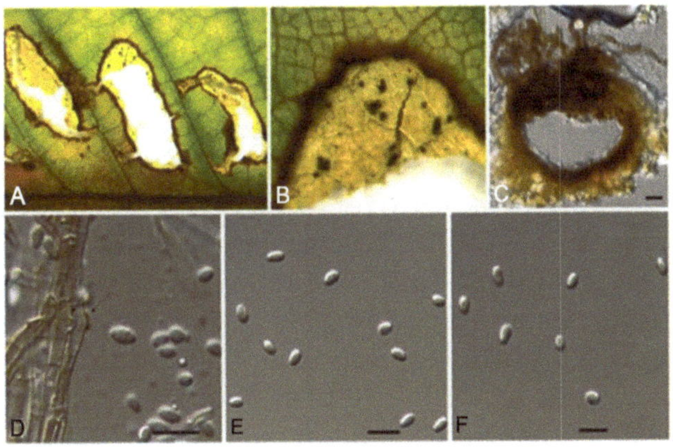

Figure 8.4. *Aureobasidium castaneae (BJFC-C007, holotype). (A) Colony on PDA; (B, D) Conidiogenous cells and conidia; (C, E, F) Conidia. Scale bars: B–F = 10 μm.*

Source: Jiang, N.; Fan, X.; Tian, C. Identification and Characterization of Leaf-Inhabiting Fungi from Castanea Plantations in China. J. Fungi 2021, 7, 64. https://doi.org/10.3390/jof7010064.

8.1.4. Exotic Pests and Disease

The primary concern regarding pest and disease damage to forest plantations will be the emergence of unfamiliar diseases and pests. In the absence of natural predators, foreign diseases and pest agents, when given ample resources in the form of suitable hosts, can proliferate rapidly and cause severe damage, with little resistance to the newly introduced disease or pest.

Pine plantations in Southern Hemisphere, particularly *Pinus radiata*, are probably the most often documented instances of exotic diseases and pests that damage exotic plantations. In Australia, and Chile, New Zealand, and South Africa *P. radiata* were initially planted at the conclusion of the nineteenth century. Another planting process was stimulated by the outstanding form and growth of the tree there (Figure 8.5).

Figure 8.5. *Wood borer and beetle damage to forest plantations in Vietnam. (a–f) damaged trees; (g–l) close-up of damage symptoms; (m–s) adults. (a, g, m) Batocera lineolata damage in Eucalyptus hybrid; (b, h, o) Euwallacea fornicatus damages in Acacia hybrid; (c, i, p) Lycaria westermanni damage in Fernandoa brilletii; (d, j, q) Tapinolachnus lacordairei damage in Chukrasia tabularis; (e, k, r) Xyleborus perforans damage in Eucalyptus urophylla; (f, l, s) Xystrocera festiva damage in Acacia mangium. Scare bar: m = 1.5 cm, o = 1.0 mm; p, q, s = 1.0 cm; r = 1.2 mm.*

Source: Thu, P.Q.; Quang, D.N.; Chi, N.M.; Hung, T.X.; Binh, L.V.; Dell, B. New and Emerging Insect Pest and Disease Threats to Forest Plantations in Vietnam. Forests 2021, 12, 1301. https://doi.org/10.3390/f12101301.

Figure 8.6. *A lateral view of Sirex noctilio Fabricius.*

Source: https://www.forestryimages.org/browse/detail.cfm?imgnum=5455076.

This was the foundation for the prosperous wood and paper industries, especially in New Zealand and Chile. In early 1900s the European wood wasp, named *Sirex noctilio*, was among the first exotic insects to show up in radiated pine plantations in New Zealand (Figure 8.6). This insect was most likely imported from Europe on unprocessed pine logs. Originally from North Africa and southern Europe, in which minimum harm is caused. In recently killed pines, noctilio usually breeds. Plantations in New Zealand stagnated between 1920 and 1930 since no market existed for tiny logs from thinning. The plantations became sensitive to *Amylostereum areolatum, Sirex noctilio*, and its associated fungi. By 1947, there was significant tree mortality, particularly in unmanaged or poorly managed plantations. (McLean, 1998).

Nowadays, however, because of natural control techniques and better stand management, Noctilio is not really a problem. It was later found in Australia in 1952 in *P. radiata* plantations (Miller & Gilbert, 1952; Taylor, 1981). An inflammation of *P. elliotii* and *P. taida* plantations in Argentina, southern Uruguay and Brazil was detected in 1980, in South America.

Noctilio was found in Cape Peninsula South Africa in *P. radiata* plantations (Tribe, 1995). The bark beetle, *Ips grandicolli*, an Australian species from the north of America (Morgan, 1989), and the European pine shoots *Rhyacionia bouliana* into Chile and Argentina are other harmful insects that must been placed on *P. radiata* plantations. This last one is today regarded as the number one insect pest in Chile.

In the Chilean radiate pine plantations during the late 1980s, European boar beetles, *Hyalstes ater* and *Hylurgus ligniperda* were introduced and could kill young pine seedlings (Ciesla, 1988). Also *P. radiata* plantations have been seriously damaged by a few fungi. In 1940 *P. radiata* plantations in Chile, parts of Eastern Africa, and New Zealand became a key concern for a needle fungus, *Dothistroma septosporum* (*D. pini*) although records of its incidence have been found in this specific region in 1940 (Manion, 1991). In some countries, the stem disease *Sphaeropsis sapinea* (*Diplodia pini*). The fungus was particularly damaging in Kenya in which it first occurred in 1973 and grew on P. radiata, a winter rainfall adapted species that's especially prone to fungal attacks during warm temperature times and rain. The *Leucaena leucocephala*, a fast growing legume originating in Central America and Mexico, is one more instance of damage caused to exotic forest plantations by the pest.

This particular tree was brought to the Philippines as soon as 1565 and quickly developed in the tropics in which it was discovered to be a great fuelwood, and later shadow for cultivation of coffee, cocoa, vanilla and pepper. In the end genetically improved varieties have been brought into the tropics, where the widest range of applications of all of the tropical legumes was considered.

In the 1980s, a relatively dark insect, *L. leucaena* psyllid was found. The natural leucocephala was found in southern Florida, USA and also was commonly found in the Hawaiian Islands in the next year. From 1985 to 1988, the insect rapidly spread throughout Asia and the Pacific.

The infestations were found one year later on in many eastern African States, including in Burundi, Kenya, Tanzania and Uganda, on the Southern Indian Ocean Islands of Mauritius and Reunion in 1991. The insect feeds Leucaena's young leaves and shoots, which cause wilting and loss of growth. Ironically speaking, attempts to bring new genetic material into a land have often led to destructive pests.

A wooly pin aphid was introduced into Kenya in 1968, *Pineus borneri* (also known as *P. pini*), on *Pinus caribaea* imported out of Australia. 20 years later, on *P. elliotii* scion material introduced from the United States, an insect scale, *Oracella acuta*, has been introduced to China eight. By 1995, over 212,500 ha *P. elliotii* were infested (Jianghua Sun et al., 1996) in Guangdong Province, Southern China.

Forests is likely to be planted with indigenous species (e.g., *Pinus massoniana* in China, *Picea abies, Pinus sylvestris* and *Populus* spp. in Europe, *Pinus taeda* and *Pinus elliotii* in South Eastern United States as wells as *Pseudotsuga menziesii* in Western USA and Canada. In most regions of North America and Asia, forest plantations are plant with indigenous species. These plantations are likely to be damaged by an exotic and indigenous disease and pest complex.

Dendrolimus punctatus, a defoliating caterpillar, is a significant pest for indigenous pine plantations in Vietnam and China. In Eastern central China, this particular insect has 2 generations in one year, defoliating *Pinus massoniana.* In the country, *D. puncto* has four generations annually, and outbreaks have happened in plantations of *P. merkusii* that usually occur in slopes that are steep where wood, fuelwood, pulpwood, and resin are a crucial source. Nun Moth is a defoliating worm in central Europe, *Pinus sylvestris* and both *Picea abie*. Originally a major natural pest of *Picea abies*, this insect was easily accommodated in the extensive pine plantations.

In the Poland, *P. sylvestris* plantations are planted on low alimentation, abandoned land or sandy land. A significant outbreak of this caterpillar required 2.5 million ha of air sprays between 1978 and 1983, that represented more than 25% of the forests in Poland (Wainhouse & Speight, 1989). The insect reached outbreaks once again in 1991, and also between 1992 and 1994 large parts of forest were treated.

Figure 8.7. *European gypsy moth.*

Source: buscandolaverdad.es.

In Scarves, *Dendrolimus pini*, *Panolis flammea*, *Bupallis piniarius*, and Europe, *diprion spp.* are also defoliating insects in Europe. *Cronartium quercuum* f.sp., a fusiform rust instigated by the fungus *P. elliottii* as wells as *P. taeda* plantations in the Southeastern US have grown to be a significant pest, fusiform.

Before the 1900s the disease had been unusual, but on account of intensive cultural practices in forestry planting, including the planting of infected plants, fire management, and fast growing trees selection and also the expansion of susceptible hosts, it became a crucial issue. The latest estimates show that $130 million in damage occurs each year and that the disease rises by 2–3% per year (Manion, 1991).

In China, Paulownia species are commonly planted in villages and intercropping systems, as ornamental products. Paulownias are often infected with the MLO, which produces big witches' brooms in tree crowns. These disease bases considerable decreases in the quantity of trees infected that cause gradual death and decrease (Tsai et al., 1988).

Borsaphelenchus xylophilus, a pine wood nematode, is a cause of the wilt disease of the pine. Since the early 1900s, this condition has caused major mortality in pine groves in Japan (Figure 8.8).

It is probably North America born and was lately introduced to China and Japan where large scale tree deaths occur on Pinus massoniana plantations. When the mature beetles feed pine shoots, then the nematode spreads from tree by longhorn wood beetles, *Monochamus* spp. (Manion, 1991).

Figure 8.8. *Dead and dying pine trees infected with the pinewood nematode, Bursaphelenchus xylophilus.*

Source: www.invasive.org.

The idea of diversity leading to healthy ecosystems suggests that mixed planting is much more resistant to disease and pest than the planting of specific species.

In Germany, Klimetzek's (1990) work shows that the inclusion in the pine forests as a mixture or perhaps in the understory of broadleaf trees will lower the incidence of *Lymantria monacha* outbreaks. The cultivation of *Picea abies* under a *P. sylvestris* plantation may, on another hand, intensify or perhaps prolong this insect's outbreaks as the two trees are hosts. Wormald (1992) states that data are at best inclusive and confusing on the relative susceptibility of pure and mixed planting species in tropical and subtropical regions. While examples are cited of diseases and pests in pure species plantations, many are caused by accidental introduction or perhaps by the adaptation of the indigenous pest type species to new conditions.

Additionally, he cites examples of mixed plantation activities which had minimum adverse effect on the reduction in losses caused by pesticides. Certain temperate areas and diseases have adapted in alternative hosts to mixed species ecosystems by developing the lives of theirs. Many rust diseases, with fusiform rust of *P. elliottii*, *Cronartium quercuum* f.sp., and *P. taeda*, fall under this category.

Quercus spp. is a fusiform with alternative oak spore stages. So it's typical, although not completely useful, for the removal of oaks from pine nurseries in Southeastern USA (Manion, 1991). Cooley spruce gall aphid, *Adelges cooleyi*, an insect native to western North America, has alternating life stages on *Pseudotsuga* and *Picea menziesii*, trees that tend to be found growing near one another in natural forests.

The number of species utilized for planting in a country is probably no more concerned with pests or perhaps diseases compared to individual planting species. Based on just one or perhaps two very closely related species may meet the wood and foreign exchange needs of a nation. But should there be a pest or perhaps a complex of pests in these plantations? Nevertheless, this dilemma has fallen to numerous places.

Examples include Kenya, where 46% of *Cupressus lusitanica* industrial plantations consist of Colombia, 66% of *Pinus patula*,

70% of Poland's forest plantations of *P. sylvestris*, and the 90% of the plantations of New Zealand and Chile consisting of *P. radiata*. The experience of New Zealand has been that planting a range of species to reduce the chance of pests and diseases hasn't worked deliberately. Rather, great place matching and forestry that comes with the research support of focus on control and disease prevention were used as a substitute strategy.

8.1.5. Managing Pests and Diseases

What is a great forest? During the mid to late 1980s, the idea of a great forest was popular.

In financial terms, a "healthy" forest are able to be identified as a "forest where pests and diseases remain at concentrations that are low and don't interfere with management goals." Environmentally, a forest is a "full functional ecosystem, in which all the components of its can interact mutually beneficially." Pests and diseases are seen as a sign of unhealthy forests instead of a situation under the general policy to protect forest health.

The aim of this section is usually to focus on the underlying causes of the disease factors including overstretching, over ripeness, poor compliance of species and sites, excess energy sources and individual species forests with very little diversity. The quest for healthy forests includes the prevention of pests and diseases based on historical records and the knowledge of the climate and the forests which in turn enhance the wealth of theirs.

This offers time for management practices to make sure that such forests are inhospitable for the advancement of harmful diseases and pesticides (Ciesla, 1998). In particular in those instances where an introduced pest causes massive damage in an easy plantation ecosystem in the lack of its normal natural enemy complex, the healthy forest perception may seem like a fitting approach to managing forest pests and diseases.

Nevertheless, even exotic diseases and pests which are much more damaging respond to stand and site conditions as well as it may be found that their activities are concentrated on' unhealthy' forests. A classic example will be *Sirex noctilium* in the Southern Hemisphere in exotic pine plantations. It was noticed in New Zealand that the very first attacks by this insect were on stressed trees on unthinned plantations.

The plantations were therefore effectively diluted through the action of its (McLean, 1998). Thinning has become part of the management of Sirex wood wasp threatened plantations. In some locations in which that insect, noctilio, was introduced. In another situation, most serious damage occurred in *Cupressus lustianica* plantations that were either overmature or perhaps found on low nutrient soils, in Southern and east Africa, where a cypress aphid originally identified as *Cinara cupressic* was present.

8.1.5.1. Integrated Pest Management Tools

A healthy forestry concept for dealing with damage caused by pests and diseases provides the framework and objective. On the contrary, Integrated Pest Management (IPM) deals with the tools to do this task. The philosophy or conception of IPM was formalized in the 1960s as a "rational" approach to pest management.

The specialists in Crop Protection have found that chemical pesticide dependency, including pesticide resistance, human health hazards, environmental damage, secondary pests, has negative effects. To be able to ensure efficient and long-term protection against harmful pests, 11 alternative approaches were also required, including cultural, genetic and biological tactics, whether in combination or isolation. In the literature there are many definitions of IPM.

Smane et al. (1976) calls IPM "an ecologically based process which integrates multi-disciplinary methodologies in the advancement of functional, effective, public and environmental health management strategies for agro-ecosystems. Pimentel (1966) calls IPM "a pesticide and nonchemical and technological pesticide control method, all of which includes wise use."

The decision making process lays the groundwork for more action, like no action. The action program may consist of one or even more ecologically, socially and economically acceptable tactics aimed at reducing pest populations to nondamaging levels (Ciesla, 1982). Decision-making is normally the complex and time-consuming aspect of IPM.

The decision making process is the pest, its host, its resource management goals and the environmental, social and economic implications of the different tactics out there must be carefully considered. There are estimated populations of projections and pesticides of anticipated loss of resource as are the costs and anticipated benefits of therapy.

If the cost of treatment exceeds losses, there might be a logical decision not to process losses and to recognize them. Another matter which should be addressed is whether natural controls will take place within sufficient time to avoid artificial controls, or even whether the consequences of the proposed treatments are extremely negative that they overweight the treatment's advantages. The damage and surveillance of forest pests and diseases is an important input to the decision making process of IPM. Pest surveillance is turning into a sophisticated technology process. Insect population levels are often monitored using other chemical regulators and pheromones. The mapping process as well as the assessment of forest damage is based on remote sensing technologies, for example aerial mapping, aerial video and aerial photography.

The places of the affected areas might be associated to key resource values, land characteristics, land ownerships and areas which are environmentally sensitive by making use of geographical info systems (GIS). The resulting injuries instigated by specific levels and the consequences of theirs can be predicted by mathematical modeling.

Pest, growth, economic models and yield in some cases are connected with projections of the impact of disease or pest. The expected results of alternative action tactics can be shown by data viewing techniques.

Two overall strategies comprise the IPM action process: prevention or direct control (suppression). Under every one of these strategies, there are certain pest management practices.

8.1.5.1.1. Prevention

Prevention involves tactics designed either to reduce the threat that a pest or perhaps disease will occur or perhaps to create inhospitable environmental conditions for its development in damaging numbers. Examples of prevention techniques are regulatory, genetic or cultural tactics. Regulatory tactics are intended to stop and prevent the introduction of exotic diseases and pests. Examples comprise the inspection of timber and wood containers at the ports during the entry to intercept pesticides, performing pest risk analyzes in new trading agreements, and creating quarantine areas when a pest species is discovered at a new place for the first time.

Cultural tactics aim at creating inhospitable conditions for the advancement of harmful diseases and pests. The tree species chosen for planting are matched to adequate sites, storage control is controlled through intermediate moves to keep the tree vigor up and plantation harvests are scheduled when mature.

The simple elimination of a species from a plantation program due to the high sensitivity it exists to certain diseases and pests is a radical but at times necessary cultural approach. An example of this is Kenya's policy to stop planting Cupressus macrocarpa because of its vulnerability to *Sphaeropsis sapinea, P. radiata*, *Monochaetia unicornosis* and stem cancer fungus.

Genetic tactics use host plant varieties that are far more understanding or perhaps less pest friendly to damage. A big investigation line following this insect's introduction in the Asia Pacific Region was testing and identification of *Leucaena leucocephala* and other Leucaena species and hybrids for tolerance or resistance to the psyllid, *Heteropsylla Cubana.*

Scan and development of fusiform rust resistant pine seedlings is an important component of the management of the disorder in the southeastern US (Manion, 1991). *Cryphonectra cubensisine* for Pinus (Pines) and *Eucalyptus radiata* breeding programs in Brazil for resistance to cancer has a rating for resistance to the *Dothistroma pini* needle fungus (McLean, 1998).

8.1.5.1.2. Suppression

Direct control or perhaps suppression tactics against the pest or perhaps disease may also be referred to. For instance, different kinds of biological, chemical or mechanical methods are included. The utilization of natural enemies of a pesticide or perhaps disease to control their numbers includes biological control.

A technique popular in agriculture, the traditional biological control involves the import of natural enemies to control the exotic pesticide. Columbia has successfully controlled the introduction and release of the egg parasitoid *Telenomus alsophilae* by the geometric *Oxydia trichiata*, an indigenous species attacking pine and cypress.

The release of a parasitoid, the *bicolor pauesia*, resulted in the collapsing of a pine aphid, a North American type, the *Cinara cronartii*, accidentally brought into pine plantations in the region of South Africa in a successful example of classical biological control. In Chile, the state of the art forest industry for the creation of the European pine shoot moth, *Orgilis obscurator*, has established mass rearing plants.

One of these simple laboratories was established in 1992, Controladora de Plagas Forestales. In 1996, 1.2 million parasitized larvae and 22,600 fully developed females parasitoids were produced in this facility for field discharge. The chance that the natural enemy introduced could also attack beneficial or harmless insects in the surroundings is a vital concern about biological control.

It's thus required, to ensure their relative host specificity before release, to carefully assess the candidate species. The risk of accidental introduction of hyperparasites, which may ultimately impact the efficacy of the agent, is one more concern. For instance, three species of hyperparasites were found on the colonies of the psyllid leucaena Olla v nigrum, released on Reunion Island of the Indian Ocean. This could significantly affect this predator's ability to act as a good biocontrol agent.

Enhanced biological control is a procedure created to improve the efficacy of existing natural enemies. A good example of this particular tactic may be the mass release of *Trichogramma dendrolimi*, the indigenous *Dendrolimus punctatus* defoliator egg parasitoid in Chinese pine plantations (Liu & Yan, 1992).

The use of biological insecticides, like a bacterial agent, *Bacillus thuringiensis* (Bt), fungal preparations or nuclear polyhedrosis for insect control is an additional biological control tactic. Bt is popular in healthy and also planting forest control of lepidoptera defoliators. The pine tuber *Dendolimus punctatus* is used in Vietnam and China as fungus *Beauvaria bassiana* (Anon., 1993).

Mechanical tactics include destruction and removal as well as rapid removal, to destroy the pest, of infected or infested trees. Examples include cutting and burning bark beetle trees or perhaps quick rescue of infested trees and destruction of the sawmill. The original response was to ruin the infested pines when pine wooly aphid, *Pineus borneri*, was initially discovered in Kenya.

Nevertheless, this insect is easily transported through the air currents and it quickly became apparent it'd spread beyond the designated area of destruction of trees (Owour, 1991). The use of chemicals from the ground or perhaps by low flying aircraft remains an important part of IPM. These materials are, however, more sparingly applied and more effectively at reduced intervals. Chemical pesticides in IPM are typically viewed as the last recourse tactic.

8.1.5.2. The Integration of New Technology

An essential element of IPM is it's usually easy to implement new technologies, regardless of how advanced, effective or sophisticated IPM system may be for a particular pest or perhaps pest complex.

These may include better pest monitoring, new pest management tactics forecasting and more effective treatments with a lot fewer undesirable side effects.

The IPM integrated pest management systems consist of a mix of pest and pest management tools for pest control and are developed at different development stages. A good illustration of a developing IPM system will be the approach taken to manage *Sirex noctilio*, a European wood wasp, in the plantations of pine in southern Brazil.

Early detection in plantations with the herbicide is done by baiting suppressed trees, an attacking wasp procedure. The inoculation of parasite nematodium, *Daladenus siricidola*, is often used for treating infestations, making the female wood wasps unable to develop eggs, or perhaps to reduce the stocking rate and sustain tree force in slimy plantations.

Damage evaluation technologies are being evaluated, for instance, aerial sketching and aerial photos. The Landsat satellite imagery is tried to determine the capability of Landsat to map the location of tiny, non-industrial private pine plantations that could be utilized to keep this damaging insect population. Another possible tactic for the control of the insect is introducing additional parasites.

One could conclude from the fundamental idea of ecological diversity leading to stability in ecosystems that mixed forest species are much less likely compared to individual species plantations to harm forest pests and diseases. Nevertheless, both natural forests and plantations are commonly recognized to be susceptible to various harmful diseases and pests.

Some temperate northern forests are quite simple ecosystems. These forests have a record of regular bark, defoliation and other insect pest's outbreaks. Additionally, an assortment of fungi, bacteria, parasites and other disease causing agents damage both natural forests and plantations. Accordingly, effective forest monitoring and protection should be an essential component of wood management, whether natural or perhaps plant species, exotic or perhaps single or indigenous species, or combined, and regardless of disease control. Forests should be successfully managed.

In forest plantations there are several records of devastating pest losses. Plants of exotic tree species can be attacked by indigenous pests adapted to the brand new host or perhaps by pests, which can form into numbers that are large, in the lack of natural enemies. Failure to manage indigenous or exotic crops will increase planting susceptibility to pest and disease damage.

8.2. ASSESSMENT OF LOSSES BECAUSE OF DISEASES

Forest disturbance is the green fluctuations and destructive events which, on a temporal or spatial level, perturb forestry health and/or structure and/or physical environment or resource (FAO, 2005). The disturbances caused by forestry ecosystems are significant. They include fire, severe weather, diseases and insect pests.

Disturbances of insect pests as well as diseases are a fundamental component of the forestry ecosystem in normal circumstances in healthy forests (Dajoz, 2000). Nevertheless, disastrous disorders are able to impact undesirably on forest ecosystems allowing it to affect environmental functions, impacting on biodiversity, impacts and livelihoods on climate change.

Figure 8.9. *Prunus persica (rust on leaves).*

Source: Wikimedia Commons.

For instance, Barber and Cochrane (2009) described the incidence of catastrophic wild fire that destroyed about 80% of principal crops in a province in Brazil between the years 1997–1998. The Chambers et al. (2007), by means of data from satellites with modest spatial resolution, have estimated that the 2006 injury to forest land caused by

Katrina hurricane in the United States led to a total annual carbon sink for all forests in the country of carbon emissions between 50 as well as 140%.

Scientific community and policymakers are growingly recognized for the several factors linked to global change affecting sustainable forest management. Forests are important to humanity since they produce a broad range of important ecosystems, but the continuing loss of forest cover means that the growing demand from a constantly decreasing resource needs to be met.

The development of global trade networks and also the amount of trade goods have been an element of population growth. The establishment of nonnative species populations in virtually every aspect of the planet has increased considerably. Although a number of these species seem to be relatively benign, several of them have significant negative effects on trees in healthy, managed and urban environments.

The interlinked and compounding of forest disturbances, deforestation caused by land use and climate change (Dale et al., 2001). Laurence and Cochrane (2008) describe just how new deforestation is able to lead to cascading effects, including increased forest flammability, increasing burning land and even further destructive fires as an outcome of changes in land use across the Amazon.

Lombardo and Ayres (2000) estimate the financial loss of over one dollars billion per year in the United States, which is apt to have wide negative effects on the distribution and severity of forest based insects and disease outbreaks. Thus, it's essential to ensure an accurate assessment of the extra size and size of forest disturbances, particularly in view of a changing global climate, for the surveillance of the Earth's natural systems.

8.3. INSECT PESTS AND MYCOFLORA OF SEEDS OF FOREST TREES AND THEIR MANAGEMENT

Seed fungi are available in all known tree species. Some fungal species, like pitch cancers and subglutenants, seem to be restricted to some pine species, as well as their amount varies with their geographical location, such as fusarium subglutinans (syn. *F. moniliforme var. subglutinans*).

8.3.1. Damage

During seed development, or germination, storage, losses as a result of seed fungi may occur. Damage is brought on by loss of seed feasibility or perhaps by seedling infection. But losses related with these fungi are generally small.

8.3.2. Diagnosis

On and in the seeds, embryos and gametophytes, seed fungi are discovered. These fungi can't be detected with some exceptions, like forms, by examining the outside of seeds. The visible occurrence of mycelium when seeds are open can detect some internal seed fungi. Nevertheless, seed samples are accurately assessed in appropriate cultivation media to evaluate the incidence of seed fungi.

On seed surface and growth medium, fungi may be found after incubation by fruiting bodies or perhaps by mycelium. Because of the great number of fungi that infect seed, it's generally required that a specialist identify them. Damage from pathogenic seed fungi can be frequently detected on infected cotyledon tissue by the presence of superficial mycelium on conifers.

8.3.3. Biology

The seeds of the forest trees contain a variety of fungi. Several of these fungal products are pathogenic and might cause spread or perhaps damping, root red or any other seedlings disease, for example Fusarium and Diplodia species. The mode of entry of internal seed fungi wasn't determined by research.

They might enter during the development of cones and seeds, or cracks, in the seed coat, especially after the cone is extracted. At any time after the seed coat is formed, external fungi are competent to create on the seed. Amounts and seed fungi types vary with the species, year and location in which they're collected. During all stages of seed development and processing, fungal populations may increase. The same fungus is usually carried by subglutinane infected orchards. The occurrence of seed fungal appears to increase, for example, by making it possible for the seeds to fall to nets or perhaps the earth before harvesting.

Cone processing processes resulting in humidity that is high and cone temperature frequently increase the fungus of seeds. Seeds can also be attached during extraction.

8.3.4. Control

Prevention - Identify seedlings with internal fungal seedlings. Don't use seed lots known to be taken over by seed borne pathogens where possible. Identify seed lots in unsound seed which are highly pathogenic and remove the unhealthy seeds. Stay away from shipping pathogenic fungal seeds.

Cultural-Monitor seed lots recognized to contain pathogenic fungi for the detection of potential disease development in a nursery.

Specific tree species have several kinds of strong seed treatment available. A few surface fungi were effectively reduced by surface drying for many species of conifer seed. A 48-hour running water reduces the seed contamination of certain pine species by bacteria and fungus. Bleach sterilization reduces the level of fungus on most seeds of the tree. Generally these compounds work better in thicker coat seeds and better damage seeds with thin seed coats.

Seed fungi can be decreased with the treatment of benomyl, thiram, or captan, but often fungicide treatments are selective since they affect just one or possibly two fungi and also may reduce germination.

8.4. BIOLOGICAL CONTROL OF INSECT PESTS AND DISEASES OF FOREST TREES

The use of harmless or beneficial organisms to control and eliminate a pest organism is biological control. You will find two aspects of natural control: classical and natural. In daily suppression of insects as well as diseases which could otherwise result in grave harm to the trees, natural biological controls are deceivingly important. Most North American native pesticides are loaded with nature's enemies that play an important part in safeguarding pesticide populations.

Figure 8.10. *Praying mantid consuming prey insect.*

Source: aroundtheworldtravels.co.uk.

Mammals, birds, insects, mites, reptiles or spiders, bacterium, fungi, nematodes and viruses are included as biological control agents and natural enemies. They act as predators, pathogens, parasites, antagonists and/or pest organism's competitors. Classical biological control, in contrast, consists of the introduction of natural enemy into specific areas because of the control of a pest that has inadvertently been in the past introduced and is an issue because native biological or perhaps environmental factors don't handle it.

Generally, the natural enemies imported come from the target pest's area of origin. The development of mass production and use of parasite microorganisms (bacteria, viruses, nematodes) which attack just one plus related insect species involves a major natural development. These microorganisms usually are good control agents and also have a little or perhaps nonapparent toxicity to humans.

In the fields and in business environments a number of caterpillars feeding on trees and on some other crops are commonly used as *Bacillus thuringiensis*, a bacterium. An excellent example of natural control of a tree's disease may be the use of competitive funguses to suppress the root activity of annus in thin pine plantations.

8.5. MOLECULAR TOOLS FOR DEVELOPING DISEASE RESISTANCE TREES

Molecular biology has alternately been hailed and condemned to be a high and costly risk venture which undermines the usual approaches. The use of molecular biology is a tool for resolving applied plant biology problems.

Recently, this particular focus has gone beyond the world of agricultural plants and they have also developed into an issue in forest biology. Long-term tree pest management requires environmentally, economical, and stable sound approaches. Pathogens are only limited in the chemical control of theirs and expensive, time consuming and potentially environmental hazard.

It appears to be very likely that the usage of pesticides in the future would be restricted to tree diseases. Cultural practices make management of diseases partly feasible, but time-consuming and costly also. There can easily be an argument for the use of resistant planting stocks as probably the most effective biological control strategy for managing disease issues.

Resistant host tissues can be used in conjunction along with other kinds of biological controls as well as offer a number of advantages in the management of diseases: host tissues may be protected during variable weather conditions all year long and not just for some time after the chemical application. The development of disease resistant trees depends to a big degree upon a good understanding of the fundamental biology and interaction between the host and the pathogen.

Various aspects of diseases biology is often identified as prerequisite to disease research, including sites, times and piercing mechanisms; nutrition, development and growth; reproductive and survival. For some disease problems, studies are initially needed to identify and characterize causal and/or vectoring agents. To be able to understand disease resistance and biological and environmental interactions, we should also conduct host response to the pathogens' challenge and host interaction with pathogen development. To identify and characterize potential mechanisms of disease resistance, an understanding of all these elements is crucial. In plants, humans and animals, numerous species of fungi are competent to result in diseases. For diagnosis, monitoring and modeling, robust and accurate exposure and quantification of fungi is crucial. Fungi may also be immediately detected, so that natural microbial communities can be much deeper, especially as many fungi are impossible or difficult to develop.

Over the last 10 years the research in fungal detection as well as identification has revolutionized efficient amplification platforms, sound development and numerous quantitative. PCR technologies probably the latest fungal detection and differentiation technologies are reviewed in this section.

8.5.1. DNA Hybridization Technology

A technique to recognize plant pathogens before introducing more sensitive, easier and speed based PCR methods using selected samples from libraries was the utilization of the Southern blot/dot blot hybridization techniques (Takamatsu et al., 1998; Levesque et al., 1998; Xu et al., 1999). Nevertheless, new and groundbreaking hybridization based methodologies were developed recently for phytopathogenic fungal diagnosis.

8.5.2. Isothermal Amplification Processes

The possibility for an isothermal amplification that requires no thermocycling appliances is a cost-effective and efficient option to PCR. Loop mediatic isothermal amplification (LAMP) uses a set of six oligonucleotide primers, which have 8 binding sites, which are specifically hybridized in various regions of the target gene. This technology is able to amplify the DNA target in under one hour using one heated block.

8.5.3. Polymerase Chain Reaction (PCR)

Most sensitive technique that is important for plant pathogens is polymerase chain reaction (PCR). The PCR enables millions of copies of specific DNA sequences to be increased by the use of specifically designed oligonucleotides (primers), deoxyribonucleotide triphosphates (dNTPs), and thermometrically supported Taq DNA polymerases (Faloona & Mullis, 1987), by repetitive cycles of denaturation, polymerization, and elongation at various temperatures).

Through an electrophorus gel stained with EtBr, SYBR Green or perhaps other safer molecule, the augmented DNA fragments are envisaged either as electrophoresis or perhaps as a colorimetric or as a fluorometric test. The existence of a specific DNA band with the anticipated size suggests that the target pathogen is present.

Progress in PCR based processes like PCR in time that is real allow for fast, accurate quantification and detection in automated reaction of plant pathogen. High sensitivity, specificity as well as reliability are key benefits of PCR technologies. Moreover, the pathogen doesn't have being isolated from infected material, therefore reducing time from week to hour for a diagnosis and allowing non-cultural pathogens to be detected and identified. In the analysis of plants with no symptoms this feature was especially helpful.

Nevertheless, the frequent existence of inhibitors of PCR in tissue or soil of plant life can significantly reduce reaction sensitivity as well as lead to false negative detection.

8.5.4. Fingerprinting

Fingerprinting approaches allow random regions of the genome to be screened to identify species related sequences when the genes preserved aren't sufficiently variable to determine the species successfully (McCartney et al., 2003).

The phylogenetic structure of fungal populations is frequently used for fingerprint analyses. Nevertheless, these methods are also helpful in identification of certain sequences for detection of fungi at minimal taxonomic level as well as also for differentiating strains with different host ranges, virulence, and compatibility group or perhaps mattress type of identical species.

8.5.5. Sequencing

Morphological properties aren't always enough to recognize a pathogen as discussed previously. Among the most direct approaches is to amplify the PCR of a target gene with universal primers, to compare and sequence the public database. Additionally, sequencing approaches have been used to refer to new fungal species.

Nevertheless, sequence databases are able to result in a number of pitfalls, including incomplete and incorrect sequences, misidentified organism sequences, inability to change or perhaps easily update data and issues associated with species boundaries, leading to erroneous interpretation of the search engine results. This disadvantage should be remedied by an attempt to create and archive quality data that are high for the research community (Kang et al., 2010).

The demand for several loci for results' robustness, and in cases in which fast results, like eradication or control of severe plant diseases outbreaks, are essential, is another constraint on sequencing as a diagnostic tool. Nevertheless, increased sequencing capacity and cost decreases allowed high fungal sequences to accumulate in openly available sequence databases and gene sequences for the identification of specific pathogenic agents and for sequence based diagnostic methods to be used carefully.

CASE STUDY

WILDLIFE BIOLOGY IN CONSERVATION EFFORTS – THE AMERICAN BISON

The American bison, once numbering in the millions across North America, faced near extinction in the late 19th century due to overhunting and habitat loss. By the early 1900s, only a few hundred bison remained, prompting significant conservation efforts. Wildlife biologists and conservationists initiated a multifaceted approach to save the species, involving habitat restoration, legal protection, and breeding programs. The establishment of national parks, such as Yellowstone National Park, provided a sanctuary for the bison and allowed for population recovery.

Through these conservation efforts, the bison population began to rebound, demonstrating the impact of targeted wildlife biology practices. By the early 21st century, the population of American bison grew to around 500,000 individuals, primarily in protected areas and private ranches. Wildlife biologists continue to monitor the genetic diversity and health of bison populations to ensure their long-term sustainability. Moreover, these efforts highlight the importance of collaboration among government agencies, non-profit organizations, and local communities in wildlife conservation.

This case exemplifies how wildlife biology not only contributes to the recovery of endangered species but also informs policies that impact ecosystems and biodiversity. The success of the American bison serves as a model for other conservation projects worldwide, showcasing the need for integrated approaches in wildlife management that consider ecological, social, and economic factors.

Discussion Questions

- What strategies can be employed to ensure the long-term sustainability of wildlife populations?
- In what ways can public awareness and community involvement enhance conservation efforts?

SUMMARY

This chapter encompasses a range of practices and studies, including habitat management, species conservation, and ecological monitoring. By understanding the interactions between wildlife and their forest habitats, practitioners can implement strategies to mitigate threats such as habitat loss, climate change impacts, and invasive species. Effective forest protection not only safeguards endangered species but also ensures the sustainability of ecosystems that provide essential services like carbon sequestration, water regulation, and recreational opportunities for communities. Through interdisciplinary approaches integrating biology, ecology, and conservation principles, forest protection and wildlife biology play a pivotal role in maintaining the balance and resilience of global forest ecosystems.

REVIEW QUESTIONS

1. What are the main types of diseases and insect pests that affect forest nurseries?
2. How do exotic species plantations contribute to the spread of pests and diseases in natural forests?
3. What are the differences between indigenous and exotic pests and diseases in forest ecosystems?
4. Discuss the strategies mentioned for managing forest pests and diseases.
5. Why is it important to assess economic and ecological losses caused by forest diseases?

REFERENCES

1. Ali, S., Wang, D., Hussain, T., Lu, X., & Nurunnabi, M. (2021). Forest resource management: An empirical study in Northern Pakistan. *Sustainability, 13*(16), 8752. https://doi.org/10.3390/su13168752.
2. Alisson, F. B., & David, L. C. (2005). Gender-specific out-migration, deforestation and urbanization in the Ecuadorian Amazon. *Global Planetary Change, 47*(1–4), 99–110. https://doi.org/10.1016/j.gloplacha.2005.02.002.
3. Arild, A., & David, K. (1999). Rethinking the causes of deforestation: Lessons from economic models. *World Bank Research Observer, 14*(1), 73–98. https://doi.org/10.1093/wbro/14.1.73.
4. Cai, J., & Li, D. (2020). How does the commercial forest redemption policy leverage social participation and economic performance? A mixed-methods research of Fujian forestry policy innovation. *Journal of Public Administration, 13*(1), 40–60, 207–208. https://doi.org/10.1108/JPA-12-2019-0090.
5. Chen, L., Chen, L., Zhu, Z., Han, L., & Cao, Y. (2021). Development potential of forestry carbon sequestration projects in natural forest protection project area: Taking Heilongjiang Natural Forest Protection Project area as an example. *Journal of Zhejiang Agricultural University, 33*(6), 944–954. https://doi.org/10.16423/jzau.2021.06.03.

6. Chen, X. (2007). *The Study on Influence Factors of Regional Forest Resources Fluctuation in China* (Master's thesis). Beijing Forestry University, Beijing, China.
7. Chen, Y., Zhang, Z., & Liao, B. (2017). Study on dynamic changes of forest ecological security of state-owned forest areas in Northeast China—Based on ecology—Industry symbiosis. *Resources Development & Market, 33*(6), 411–416. https://doi.org/10.13688/j.ydyl.2017.06.010.
8. Dai, E., Wang, X., Zhu, J., & Wang, X. (2020). Modeling the long-term impacts of harvest and artificial regeneration on forest area and aboveground biomass in Red Soil Hilly Region: A case study in Moshao forest farm of Huitong County. *Journal of Natural Resources, 35*(12), 2995–3006. https://doi.org/10.11849/zrzyxb.2020.12.003.
9. Delang, C. O., & Wang, W. (2013). Chinese forest policy reforms after 1998: The case of the Natural Forest Protection Program and the Slope Land Conversion Program. *International Forestry Review, 15*(3), 290–304. https://doi.org/10.1505/146554813807500068.
10. Deng, Y., & Deng, Y. (2019). Analysis on the factors influencing the production efficiency of northeast state-owned forestry enterprises under the background of implementing the natural forest resources protection project. *Rural Economy, 3*, 128–134. https://doi.org/10.3969/j.issn.1003-6010.2019.03.019.
11. Fan, L. (2019). Comprehensive benefit assessment of natural forests protection project in Shanxi Province. *Journal of Northwest Forestry University, 34*(2), 265–272. https://doi.org/10.3969/j.issn.1000-5382.2019.02.027.
12. Fan, S., Chen, Y., & Chen, J. (2013). Performance evaluation and comparison of ecological construction policies based on public value. *Journal of Public Management, 10*(2), 110–116, 142–143. https://doi.org/10.3969/j.issn.1672-9703.2013.02.013.
13. Farooq, M. U., Shahzad, U., Sarwar, S., & Li, Z. (2019). The impact of carbon emission and forest activities on health outcomes: Empirical evidence from China. *Environmental Science and Pollution Research, 26*(13), 12894–12906. https://doi.org/10.1007/s11356-019-04640-3.
14. Feng, Q., Chen, J., & Qin, F. (2010). Empirical analysis and policy recommendations study on Chinese state-owned forest of labor employment. *Research on Agricultural Modernization, 31*(4), 304–307. https://doi.org/10.3969/j.issn.1672-0202.2010.04.001.
15. Fortmann, L., & Bruce, J. (1991). *You've Got to Know Who Controls the Land and Trees People Use: Gender, Tenure and The Environment*. University of Zimbabwe.
16. Geist, H., & Lambin, E. (2001). *What Drives Tropical Deforestation? A Meta-Analysis of Proximate and Underlying Causes of Deforestation Based on Subnational Case Study Evidence*. LUCC International Project Office.
17. Geng, Y., & Jiang, Y. (2022). Main factors on family income of employees in state-owned forest areas after full stop of commercial logging of natural forests. *Journal of Forestry Research, 50*(1), 122–126. https://doi.org/10.1007/s11676-021-01376-5.
18. Geng, Y., Sun, S., & Youn, Y. (2021). Impact of forest logging ban on the welfare of local communities in Northeast China. *Forests, 12*(1), 3. https://doi.org/10.3390/f12010003.

19. Goytom, A., & Erwin, B. (2019). Trust, regulation and participatory forest management: Micro-level evidence on forest governance from Ethiopia. *World Development, 120*, 118–132. https://doi.org/10.1016/j.worlddev.2018.05.029.

20. Guo, Y., Zhou, W., Yu, D., Zhou, L., Gu, X., Wu, Z., Wu, S., Wang, X., & Dai, L. (2015). Research on forest vegetation carbon storage under the National Natural Forest Protection Project in the upper reaches of Yangtze River. *Resources and Environment in the Yangtze Basin, 24*(2), 221–228. https://doi.org/10.11870/cjlyzyyhj201502009.

21. He, Y., Fei, Z., Ye, X., Liu, Z., & Dong, W. (2022). Evaluation of carbon sink value of NFPP and the significance of "carbon neutralization." *Journal of Public Management, 19*, 154–163, 176. https://doi.org/10.3969/j.issn.1674-2758.2022.01.017.

22. He, Y., Xie, H., & He, Y. (2022). Problems and recommendations of conservation and restoration of natural forest in Yangtze River Economic Belt. *Forest Resources Management, 1*, 1–9. https://doi.org/10.3969/j.issn.2096-5609.2022.01.001.

23. Hu, A., & Shen, R. (2014). Pioneer of ecological civilization: Chinese forest road (1949–2013). *Journal of Tsinghua University (Philosophy and Social Sciences), 29*(1), 63–72, 171. https://doi.org/10.3969/j.issn.1004-6138.2014.01.009.

24. Huang, L., Wang, B., Niu, X., & Song, Q. (2017). Evaluation of ecological effects of the Natural Forest Protection Program in key state-owned forest districts in Northeast China and Inner Mongolia. *Science of Soil and Water Conservation, 15*(1), 89–96. https://doi.org/10.3969/j.issn.1672-8252.2017.01.014.

25. Jiang, X. (2016). *The Potential and Development Path of Sequestering Carbon in China's Forest Sector* (Ph.D. thesis). Zhejiang University, Zhejiang, China.

26. Jin, H. (2012). Interaction and integration: Sociological analysis of public goods supply behavior—Based on a case study of forest management. *Probe, 146–152*. https://doi.org/10.3969/j.issn.1005-0102.2012.12.029.

27. Ke, S., Zhu, L., Yuan, H., & Ji, P. (2018). An economic interpretation and policy enlightenment of the "Two Mountains" theory: An analysis based on the example of complete cessation of commercial logging of natural forests. *Chinese Rural Economy, 52*, 52–66. https://doi.org/10.3969/j.issn.1000-8207.2018.06.007.

28. Lan, J., Zhang, Y., Shi, Q., Li, J., Xu, Z., & Han, Y. (2018). Assessment of service functions value of the Natural Forest Protection Program in Xinjiang. *Journal of Northwest Forestry University, 33*(4), 289–296. https://doi.org/10.3969/j.issn.1000-5382.2018.04.003.

29. Li, C., Zhu, J., Zhang, F., Li, Q., Tian, Y., Xiao, W., & Chen, W. (2021). Carbon sequestration capacity of Beijing arbor forest based on NbS. *Journal of Beijing Forestry University, 43*(6), 13–22. https://doi.org/10.13332/j.1000-1522.2021.06.002.

30. Li, G., Zhao, W., Wei, Y., Fang, X., Gao, B., & Dai, L. (2015). Evaluation on the influence of Natural Forest Protection Program on forest ecosystem service function in Changbai mountain. *Acta Ecologica Sinica, 35*(4), 984–992. https://doi.org/10.5846/stxb201305101733.

31. Li, L., Deng, D., Zhang, D., & Yang, W. (2021). Analysis on socio-economic determinants of forest fragmentation in Beijing-Tianjin-Hebei region. *Forest Economics, 43*(1), 5–16. https://doi.org/10.3969/j.issn.1002-8182.2021.01.002.

32. Li, X., Cao, J., & Li, F. (2015). Influence of ecological compensation participation on farmer's income: A case study in Wuling Mountain areas. *Journal of Huazhong Agricultural University, 34*(2), 51–57. https://doi.org/10.11835/j.issn.1000-2561.2015.02.008.

33. Lin, J., Liao, W., Huang, H., & Wang, J. (2021). Will the policy of completely stopping commercial logging of natural forests affect the livelihood capital of forestry farmers? *Forest Economics, 43*(2), 5–20. https://doi.org/10.3969/j.issn.1002-8182.2021.02.001.

34. Lin, X., & Zhang, D. (2020). Construction of legal system for natural forest protection and restoration under the concept of green development. *Journal of Fujian Normal University (Philosophy and Social Sciences Edition), 71*(1), 71–79, 170. https://doi.org/10.3969/j.issn.1003-2797.2020.01.010.

35. Liu, J., & Yin, R. (2010). An estimation of the effects of China's priority forestry programs on farmers' income. *Environmental Management, 45*(3), 526–540. https://doi.org/10.1007/s00267-009-9364-8.

36. Liu, S., & Xia, J. (2021). Forest harvesting restriction and forest restoration in China. *Forest Policy and Economics, 129*, 102516. https://doi.org/10.1016/j.forpol.2021.102516.

37. Liu, S., Ma, J., & Miao, N. (2015). Achievements in natural forest protection, ecological restoration, and sustainable management in China. *Acta Ecologica Sinica, 35*(6), 212–218. https://doi.org/10.5846/stxb201305061573.

38. Liu, Y., Chen, S., Zhao, R., & Li, Q. (2021). Research on protection and restoration of natural forests in China. *World Forestry Research, 34*(4), 1–7. https://doi.org/10.13348/j.1000-4025.2021.04.001.

39. Lv, J. (2005). *A Study on Forest Public Goods Supply in China* (master's thesis). Northeast Forestry University, Harbin, China.

40. Ma, S., Tian, M., Liu, C., Wang, F., & Du, L. (2021). The function mechanism of forest certification in promoting high-quality development of natural forest. *Forest Resources Management, 30*, 19–27. https://doi.org/10.3969/j.issn.1003-2797.2021.01.003.

41. Mammo, S., Zhang, K., Wang, X., Ahmad, B., Shale, G., Kebede, G., Terfe, Y., & Liu, X. (2018). Does participatory forest management save the remnant forest in Ethiopia? *Proceedings of the National Academy of Sciences India Section B: Biological Sciences, 88*(1), 1–14. https://doi.org/10.1007/s40011-017-0905-1.

42. Narendra, C., Geoffrey, N. K., & Hugh, B. (2015). Production efficiency of community forest management in Nepal. *Forest Policy and Economics, 50*, 172–179. https://doi.org/10.1016/j.forpol.2014.09.007.

43. Ostrom, E. (2007). A diagnostic approach for going beyond panaceas. *Proceedings of the National Academy of Sciences of the United States of America, 104*(39), 15181–15187. https://doi.org/10.1073/pnas.0702288104.

44. Ostrom, E. (2008). Developing a method for analyzing institutional change. In S. Batie & N. Mercuro (Eds.), *Assessing the Evolution and Impact of Alternative Institutional Structures* (pp. 25–48). Routledge Press.

45. Ostrom, E. (2009). A general framework for analyzing sustainability of social-ecological systems. *Science, 325*(5939), 419–422. https://doi.org/10.1126/science.1172133.

46. Ostrom, E. (2009). *Understanding Institutional Diversity*. Princeton University Press.

47. Ostrom, E. (2011). Background on the institutional analysis and development framework. *Policy Studies Journal, 39*(1), 7–27. https://doi.org/10.1111/j.1541-0072.2010.00394.x.

48. Ostrom, E. (2012). Why do we need to protect institutional diversity? *European Political Science, 11*(1), 128–147. https://doi.org/10.1057/eps.2011.19.

49. Ostrom, E., & Basurto, X. (2011). Crafting analytical tools to study institutional change. *Journal of Institutional Economics,* 7(3), 317–343. https://doi.org/10.1017/S1744137411000011.

50. Pienkowski, T., Dickens, B. L., Sun, H., & Carrasco, L. R. (2017). Empirical evidence of the public health benefits of tropical forest conservation in Cambodia: A generalized linear mixed-effects model analysis. *The Lancet Planetary Health, 1*(6), e180–e187. https://doi.org/10.1016/S2542-5196(17)30069-1.

51. Prabodh, I. (2005). *Technical Efficiency in Agriculture and Dependency on Forest Resources: An Economic Analysis of Rural Households and The Conservation of Natural Forests in Sri Lanka* (Ph.D. thesis). The University of Hawaii, Honolulu, HI, USA.

52. Qin, L., & Zhang, M. (2021). An empirical analysis on the implementation performance of the second phase of the Natural Forest Protection Project. *Issues in Forest Economics, 41*, 455–463. https://doi.org/10.3969/j.issn.1003-2825.2021.41.008.

53. Reem, H., Johan, A. O., Peter, C., Peter, N., Aaron, J. M. R., & Wen, Z. (2021). A global analysis of the social and environmental outcomes of community forests. *Nature Sustainability,* 4(3), 216–224. https://doi.org/10.1038/s41893-021-00718-8.

54. Ricardo, G., Tobias, K., & Leandro, M. (2013). Beyond 'land sparing versus land sharing': Environmental heterogeneity, globalization, and the balance between agricultural production and nature conservation. *Current Opinion in Environmental Sustainability,* 5(5), 477–483. https://doi.org/10.1016/j.cosust.2013.06.007.

55. Robinson, E. (2021). Citizens, custodians, and villains: Environmentality and the politics of difference in Senegal's community forests. *Geoforum, 125*, 25–36. https://doi.org/10.1016/j.geoforum.2021.07.003.

56. Rogelj, J., Den, E., Höhne, N., Fransen, T., Fekete, H., Winkler, H., Schaeffer, R., Sha, F., Riahi, K., & Meinshausen, M. (2016). Paris Agreement climate proposals need a boost to keep warming well below 2°C. *Nature, 534*(7609), 631–639. https://doi.org/10.1038/nature18307.

57. Shen, G. (2009). Natural Forest Protection Project and sustainable forest management. *Forest Economics, 15–16*. https://doi.org/10.3969/j.issn.1003-2797.2009.06.001.

58. Sheng, C., Zhu, Z., & Chen, L. (2021). Connotation, characteristics, and development of natural forest protection public welfare undertakings in the new period. *Issues in Forest Economics, 41*, 13–19. https://doi.org/10.3969/j.issn.1003-2825.2021.41.002.

59. Shi, C. (2010). *Influence Factors Analysis on China's Forest Quality* (Ph.D. thesis). Beijing Forestry University, Beijing, China.

60. Shi, M., Yin, R., & Lv, H. (2017). An empirical analysis of the driving forces of forest cover change in Northeast China. *Forest Policy and Economics, 78*, 78–87. https://doi.org/10.1016/j.forpol.2017.01.005.

61. Song, X., Li, Q., & Li, N. (2018). Influence on reform of state-owned forest region on allocation of labor resources: An empirical analysis based on pilot project in Heilongjiang. *Resources Development and Market, 34*, 532–538. https://doi.org/10.19512/j.cnki.issn1008-9342.2018.04.015.

62. Sun, C., Zhen, L., Wang, C., Hu, J., & Du, B. (2017). Assessment of the ecological effects of the first-phase natural forest protection project in the Xiaolongshan region, Gansu Province. *Progress in Geography, 36*(6), 732–740. https://doi.org/10.11820/dlkxjz.2017.06.001.

63. Tao, J., Zhang, W., & He, Y. (2002). Natural forest conservation and its studies in China. *Forest Resources Management, 6*, 61–68. https://doi.org/10.3969/j.issn.1002-1500.2002.01.010.

64. Tian, N., Neelam, P., & Lu, F. (2021). Assessments of landowners' willingness to accept compensation for participating in forest certification in Shandong, China. *Sustainability, 13*(2), 903. https://doi.org/10.3390/su13020903.

65. Wang, H., Wang, B., Niu, X., & Song, Q. (2017). Dynamic change of natural forest ecological benefit in Changbai Forest Industry Group since implementation of Natural Forest Protection Project. *Science of Soil and Water Conservation, 15*(3), 86–93. https://doi.org/10.13869/j.cnki.stbcyj.2017.03.012.

66. Wang, Y. (2020). Ningxia forest ecological compensation policy innovation and practice. *Journal of North Minzu University*, 171–176. https://doi.org/10.13650/j.cnki.nmdx.2020.06.018.

67. Wei, Y., Zhou, W., Yu, D., Zhou, L., Fang, X., Zhao, W., Bao, Y., Meng, Y., & Dai, L. (2014). Carbon storage of forest vegetation under the Natural Forest Protection Program in Northeast China. *Acta Ecologica Sinica, 34*(20), 5696–5705. https://doi.org/10.5846/stxb201304300451.

68. Wu, H., Yin, H., & Cao, Y. (2022). Effects of the natural forest protection project on total factor productivity of forestry in China: Empirical based on provincial panel data. *Journal of Northeast Agricultural University, 53*(1), 90–96. https://doi.org/10.7668/j.issn.1000-8601.2022.01.011.

69. Yan, R., Zinda, J. A., & Ke, S. (2020). Designating tree crops as forest: Land competition and livelihood effects mediate tree crops impact on natural forest cover in South China. *Land Use Policy, 96*, 104702. https://doi.org/10.1016/j.landusepol.2020.104702.

70. Yang, H. (2017). China's Natural Forest Protection Program: Progress and impacts. *Forest Chronicle, 93*(1), 113–117. https://doi.org/10.5558/tfc2017-015.

71. Zang, L. (2014). Effects of the Natural Forest Protection Program on farmers' income in Western China. *Journal of Agro-Forestry Economics and Management, 13*(2), 201–209. https://doi.org/10.3969/j.issn.1008-1492.2014.02.014.

72. Zhang, B. (2020). *Research on Characteristic Evolution and Effects Analysis of Forestry Economic Transition Based on Forest Transition in Heilongjiang Key State-Owned Forest Areas* (Ph.D. thesis). Northeast Forestry University, Harbin, China.

73. Zhang, C., Wang, L., Song, Q., Chen, X., Gao, H., & Wang, X. (2018). Biomass carbon stocks and dynamics of forests in Heilongjiang Province from 1973 to 2013. *China Environmental Science, 38*(12), 4678–4686. https://doi.org/10.3969/j.issn.1000-6923.2018.12.002.

74. Zhang, H. (2022). *Research on the Effect of Collective Forests Tenure Reform on China's Timber Supply* (Ph.D. thesis). Nanjing Forestry University, Nanjing, China.

75. Zhang, Y., Liu, X., Gao, W., & Li, H. (2021). Dynamic changes of forest vegetation carbon storage and the characteristics of carbon sink (source) in the Natural Forest Protection Project region for the past 20 years. *Acta Ecologica Sinica, 41*(21), 5093–5105. https://doi.org/10.5846/stxb202006092452.

76. Zhao, H. (2019). *Research and Analysis on the Development Potential of Forestry Carbon Sink Projects in China* (master's thesis). Beijing Forestry University, Beijing, China.

77. Zhao, X. (2021). *Research on Benefit Accounting and Benefit Improvement of Forest Ecosystem in Yichun State-Owned Forest Region* (Ph.D. thesis). Northeast Forestry University, Harbin, China.

78. Zheng, L. (2015). *Research on the Coordinative Development of Forest Resources Management System in Heilongjiang Forest Industry Region* (Ph.D. thesis). Northeast Forestry University, Harbin, China.

79. Zheng, S., Wang, L., & Zang, S. (2021). The change of ecosystem services of Natural Forest Protection Project regions in the Da Hinggan Mountains. *Science of Geographical Sinica, 41*(9), 1295–1302. https://doi.org/10.13249/j.cnki.sgs.2021.09.003.

80. Zhi, L., Fan, S., Li, Q., & Guo, X. (2013). The empirical analysis about zoning desire of rural households toward public welfare forest: A case study of Wulong County and Yulong County in Natural Forest Protection Project area. *Forest Economics, 35*, 106–109. https://doi.org/10.3969/j.issn.1003-2825.2013.02.025.

81. Zhou, M. (2017). *Study on Endogenous Growth of Economic Transformation of State-Owned Forest Region in Heilongjiang* (master's thesis). Northeast Forestry University, Harbin, China.

82. Zhu, Y. (2015). *Study on the Supply for Forest Ecosystem Services by Government in Key State-Owned Forest Region of Heilongjiang Province* (Ph.D. thesis). Northeast Forestry University, Harbin, China.

83. Zhu, Z., Liu, X., & Cao, Y. (2018). Factor input, resource management, and process performance of forestry ecological project: An empirical analysis based on grey input-output system of the Natural Forest Protection Program. *World Forestry Research, 31*(1), 1–9. https://doi.org/10.13348/j.cnki.1000-1522.2018.01.001.

84. Zou, Y., Li, J., & Tian, G. (2020). Empirical analysis on determinants of subjective well-being in state-owned forest area based on capability approach: From the perspective of full stop commercial logging. *Forest Science, 56*(2), 154–164. https://doi.org/10.11707/j.issn.2095-0751.2020.02.013.

CHAPTER 9

Global Climate Change Impact on Forests

LEARNING OBJECTIVES

After studying this chapter, you will be able to:

- Understand the concept of climate change and its implications for global warming.
- Explore the impacts of climate change and global warming on forest ecosystems.
- Examine the relationship between climate change, warming, and biodiversity.
- Identify major and lesser greenhouse gases contributing to climate change.
- Analyze the effects and causes of modern deforestation and its implications for forests.

KEY TERMS FROM THIS CHAPTER

Biodiversity
Deforestation
Greenhouse gases
Lesser greenhouse gases
Modern deforestation
Climate change
Global warming
Hard deforestation
Major greenhouse gases

INTRODUCTION

Global warming is a phenomenon characterized by a gradual increase in the Earth's surface temperature. The basic understanding is that the atmosphere immediately surrounding the planet is getting incrementally hotter each year, leading to the warming of both land and sea surfaces (IPCC, 2014). The term "global warming" itself does not imply a cause; rather, it is a statement of fact that most in the scientific community acknowledge as true and provable (National Aeronautics and Space Administration [NASA], 2020). The primary cause of disagreement among scientists, policymakers, and the public concerns the reasons for the rising temperatures, the potential negative consequences, and what human communities can or should do to slow or reverse the trend. Most theories posit that human emissions of carbon dioxide (CO_2) gas, whether through industrial processing or vehicle emissions, are chiefly to blame (IPCC, 2014). This anthropogenic perspective is supported by a significant body of evidence showing a strong correlation between increased CO_2 levels and global temperature rise since the Industrial Revolution (Houghton, 2004). The Intergovernmental Panel on Climate Change (IPCC) states that it is "extremely likely" that more than half of the observed increase in global average surface temperature from 1951 to 2010 was caused by human activities, including greenhouse gas emissions and land use changes (IPCC, 2014). These human activities have significantly increased the concentrations of greenhouse gases in the atmosphere, enhancing the natural greenhouse effect and leading to global warming (Solomon et al., 2007). However, some argue that global warming is simply part of the Earth's natural climatic cycle. This perspective suggests that current warming trends are part of the natural variability of the Earth's climate system and will eventually stabilize without significant human intervention (Lindzen, 2009). Historical climate data indicate that the Earth has experienced periods of warming and cooling over geological timescales, which some use to argue that the current warming trend may not be solely due to human activities (Dansgaard et al., 1993). Despite these differing views, the potential negative consequences of global warming are a major concern. Rising temperatures can lead to more frequent and severe heatwaves, melting polar ice, rising sea levels, and changes in precipitation patterns, which can disrupt ecosystems, agriculture, and water supplies (IPCC, 2014; Parmesan, 2006). Additionally, extreme weather events, such as hurricanes and floods, are expected to become more intense and frequent, posing significant risks to human communities and infrastructure (Knutson et al., 2010).

To address global warming, many experts advocate for reducing greenhouse gas emissions through various means, such as transitioning to renewable energy sources, enhancing energy efficiency, and implementing policies that encourage sustainable practices (Pacala & Socolow, 2004). International agreements, like the Paris Agreement, aim to unite countries in the effort to limit global warming to well below 2° C above pre-industrial levels (UNFCCC, 2015). Therefore, there is broad agreement on the existence of global warming and its potential impacts, debates continue over the causes and appropriate responses. Most evidence supports the view that human activities are a significant driver of recent temperature increases, and mitigating these impacts will require concerted global efforts and substantial changes in energy production and consumption patterns.

9.1. CLIMATE CHANGE

Climate change is one of the greatest challenges of the 21st century, profoundly affecting the growth and productivity of forests through changes in temperature, rainfall, weather patterns, and other climatic factors. Elevated levels of carbon dioxide, for instance, have a significant impact on plant growth and, consequently, forest ecosystems (IPCC, 2014). These changes influence complex forest ecosystems in multifaceted ways (Figure 9.1).

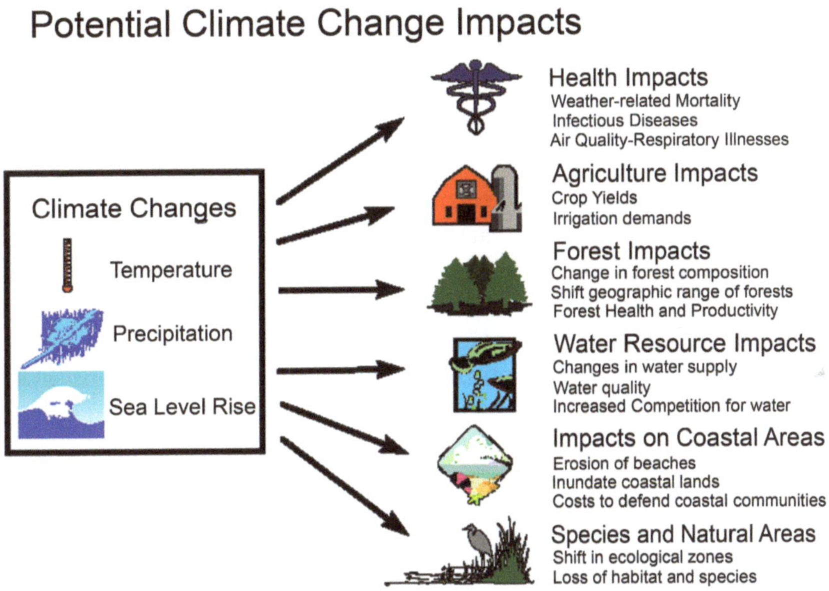

Figure 9.1. *Schematic of climate change impact on forest ecology.*

Source: http://www.atmo.arizona.edu/students/courselinks/spring07/atmo336s3/lectures/sec6/implications.html.

1. **Impact of Climate Change on Forests:** Climate is the most important determinant of vegetation patterns, significantly influencing the distribution, structure, and ecology of forests (Bonan, 2008). Climate change has already affected biodiversity, as evidenced by changes in species ranges, ecosystem boundaries, reproductive cycles, growing seasons, and species interactions, such as predation, pollination, competition, and disease (Parmesan, 2006). These effects vary between regions and ecosystems.

   Recent findings suggest that global warming is causing spatial shifts in species distributions by approximately 6.1 km per decade towards the poles or 9.05 m per decade upward, as predicted by climate change models. This shift is accompanied by the earlier arrival of spring, advancing by 2–3 days per decade in temperate latitudes (Parmesan & Yohe, 2003). Boreal and polar regions are among the ecosystems most affected by global warming (ACIA, 2004).

2. **Historical Context and Current Challenges:** Although past changes in the global climate resulted in major shifts in species ranges and significant reorganization of biological communities, landscapes, and biomes during the last 1.8 million years, these changes occurred in a landscape that was not as fragmented as it is today and with little or no pressures from human activities (Huntley, 1991). Current climate changes, coupled with anthropogenic activities, are stressing biodiversity far beyond the levels imposed by the global climatic changes that occurred in the recent evolutionary past (Sala et al., 2000).
3. **Human Influence and Biodiversity:** The human component needs to be incorporated when dealing with the impacts of climate change on biodiversity. Activities aimed at mitigating and adapting to climate change must consider biodiversity to be effective (Millennium Ecosystem Assessment, 2005). Human-induced habitat fragmentation, pollution, and overexploitation of resources exacerbate the impacts of climate change on forests, making ecosystems more vulnerable to disturbances and reducing their resilience (Brook et al., 2008).
4. **Specific Effects of Climate Change on Forests:** Climate changes directly and indirectly affect the growth and productivity of forests through changes in temperature, rainfall, weather patterns, and other factors. Elevated levels of carbon dioxide, for example, enhance photosynthesis and water-use efficiency in some plant species, potentially altering forest composition and dynamics (Norby et al., 2005). These changes can lead to shifts in species dominance, changes in forest structure, and alterations in ecosystem services provided by forests (Allen et al., 2010).
5. **Adaptation and Mitigation Strategies:** Addressing the impacts of climate change on forests requires comprehensive adaptation and mitigation strategies. These strategies should include the conservation and restoration of forest ecosystems, sustainable forest management practices, and the integration of biodiversity considerations into climate policies (Thompson et al., 2009). Enhancing the connectivity of fragmented landscapes, protecting climate refugia, and promoting genetic diversity within forest species are crucial steps in building the resilience of forest ecosystems to climate change (Heller & Zavaleta, 2009).

9.1.1. Global Climate Change and Warming

Few global issues have garnered as much attention as human-induced climate change and biodiversity conservation. The combustion of fossil fuels, primarily for energy, and deforestation have significantly increased the concentration of atmospheric carbon dioxide. Over the past three centuries, carbon dioxide levels have risen by 30%, with more than half of this increase occurring in the last 60 years—from a pre-industrial level of 280 parts per million (ppm) to the current level of around 360 ppm (IPCC, 2007). Additionally, human activities have more than doubled the concentration of methane and increased levels of other greenhouse gases, contributing to climate warming (Shindell et al., 2012).

9.1.1.1. Climate Change

Greenhouse gases reduce the efficiency with which the Earth radiates its energy into space, leading to warming of the lower atmosphere and surface of the planet. This warming potential is expressed as radiative forcing (Forster et al., 2007). The increase in atmospheric sulfur dioxide, another by-product of industrialization, causes cooling by forming aerosols that reflect solar radiation back into space and enhance cloud formation. Although aerosols have a much shorter atmospheric lifetime (days to weeks) compared to most greenhouse gases (decades to centuries), they play a significant role in radiative transfer and climate dynamics (Ramanathan et al., 2001).

The projected climate change represents the most prominent global environmental challenge faced by human society today, with far-reaching implications for food production, water supply, coastal settlements, forest ecosystems, health, and energy security (IPCC, 2014). While there is some uncertainty about the precise effects of increased carbon dioxide levels on global climate, scientific consensus indicates that a doubling of carbon dioxide concentrations could have serious environmental consequences within this century (Meehl et al., 2007). Recent climate models predict a global temperature rise of approximately 1–3.5° C by the year 2100, underscoring the urgency of addressing climate change (IPCC, 2021).

9.1.1.2. Conservation of Biodiversity

Biodiversity, the variety of life on Earth, is essential for ecosystem stability, resilience, and the provision of ecosystem services. However, human activities such as habitat destruction, pollution, overexploitation, and climate change are driving unprecedented rates of biodiversity loss (Pimm et al., 2014). The conservation of biodiversity is critical for maintaining ecosystem functions and the benefits they provide to human societies, including food, clean water, disease regulation, and cultural values (Cardinale et al., 2012).

9.1.1.3. Drivers of Biodiversity Loss

Habitat destruction, primarily through deforestation, agriculture, and urbanization, is the leading cause of biodiversity loss. Fragmentation of habitats further exacerbates the problem by isolating populations and reducing genetic diversity (Fahrig, 2003). Pollution, including chemical contaminants, plastics, and nutrient loading, also poses significant threats to terrestrial and aquatic ecosystems (Galloway et al., 2008). Overexploitation of species for food, medicine, and other resources has led to declines in many wildlife populations (Dirzo et al., 2014). Additionally, climate change is altering habitats and species distributions, often with detrimental effects on biodiversity (Bellard et al., 2012).

9.1.1.4. Conservation Strategies

Effective conservation strategies require a multifaceted approach, integrating in-situ and ex-situ conservation, sustainable management practices, and policy interventions. Protected areas play a vital role in conserving biodiversity by preserving critical habitats and providing refuges for species (Watson et al., 2014). However, the effectiveness of protected areas depends on adequate management, enforcement, and connectivity to allow for species migration and adaptation to climate change (Juffe-Bignoli et al., 2014).

Sustainable land-use practices, such as agroforestry, conservation agriculture, and sustainable forestry, can help mitigate the impacts of human activities on biodiversity while supporting livelihoods (Perfecto et al., 2009). Policy interventions, including international agreements like the Convention on Biological Diversity (CBD), national regulations, and market-based incentives, are crucial for promoting biodiversity conservation at various scales (CBD, 1992).

Addressing climate change and conserving biodiversity are intertwined challenges that require urgent and coordinated global action. Mitigating the impacts of climate change through reducing greenhouse gas emissions and enhancing carbon sequestration is critical for the long-term stability of both natural and human systems (IPCC, 2021). Simultaneously, conserving biodiversity through effective protection, sustainable management, and policy measures is essential for maintaining the resilience and functioning of ecosystems that support life on Earth (Pimm et al., 2014).

Since it is impossible to develop one or even several scenarios that capture the many possibilities for Earth's continued human development, the Intergovernmental Panel on Climate Change (IPCC) utilizes up to 40 different scenarios as evaluation tools to predict a range of possible future climates. Presented results from numerous models which were run for 35 different scenarios and concluded that globally averaged surface temperatures would increase between 1.4 and 5.8° C from AD 1990–2100 with a wider range and warmer temperatures for terrestrial surfaces especially in winter and in the Northern Hemisphere where warming is predicted to be maximal.

These changes in biodiversity alter ecosystem processes and change the resilience of ecosystems to environmental change. This has profound consequences for services that humans derive from ecosystem as this will change global biogeochemical cycles, transforming land and enhancing the mobility of biota.

9.1.2. Impact of Climate Change and Global Warming

The projected global mean temperature increase is expected to induce widespread changes in precipitation patterns, forcing the northward advance of monsoon rainfall further into Africa and Asia, thereby increasing the total rainfall in these areas. This rainfall will also be more intense in its occurrence, leading to flooding and erosion (IPCC, 2014). Climate change-affected precipitation will cause more intense storms, which will result in increased surface runoff and erosion. However, when considering warming, which causes continental drying and an associated risk of drought, increased rainfall in some areas may not translate into more mesic conditions (Huntington, 2006). Areas with increased precipitation will have excess soil moisture, causing more leaching, decreased soil fertility, and ultimately reduced productivity. Summer and winter precipitation is expected to increase at higher latitudes. Increased winter precipitation is predicted over northern mid-latitudes, tropical Africa, and Antarctica, and in the summer for southern and eastern Asia. Decreases in winter precipitation are predicted for Australia, Central America, and southern Africa (IPCC, 2013).

9.1.2.1. Water Resources

Climate change and warming are likely to worsen water availability. Under a changed climatic regime, the combined effect of lower rainfall and more evaporation would have dire consequences for any given region. Both factors would lead to less runoff, substantially changing the availability of freshwater in watersheds (Vörösmarty et al., 2000). Changes in temperature and precipitation might dramatically impact soil moisture and aridity levels of hydrological zones. With changes in flows, annual runoff, and groundwater recharge, water available for usage will further decrease (Arnell, 1999). An assessment using scenarios from Hadley Centre Model Simulations indicates that by 2050, the average annual runoff in the Brahmaputra River will decline by 14% (Gosain et al., 2006). The impact of snow melting in the high Himalayas will lead to flood disasters in Himalayan catchments, particularly in the western Himalayas, where snow contributes about 60% to the runoff of major rivers, compared to 10% on the eastern side (Immerzeel et al., 2010).

9.1.2.2. Soil Water and Water-Use Efficiency

Temperature increases will induce higher rates of evapotranspiration, causing reductions in soil moisture. The increase in atmospheric CO_2 concentration from 350–550 ppm is predicted to increase stomatal resistance by 30%, partially closing the stomata, decreasing leaf transpiration rate, and increasing water use efficiency (Lawson & Blatt, 2014). Enhanced CO_2 levels may protect crops from or increase their resilience to drought, allowing them to mature and produce a harvestable product. This could extend the distribution and boundaries into semi-desert areas for specific crops currently constrained by water supplies (Ainsworth & Rogers, 2007).

9.1.2.3. Coastal Zone

A coastal zone can be defined as an area of activity rather than an area contained by boundaries. Coastal zones, often densely populated and economically significant, are located at the interface between land and water and change frequently due to chemical, biological, and geological attributes (Nicholls & Cazenave, 2010). Coastal zones evolve due to tectonic forces and meteorological conditions. Emergent coastlines, with rugged sea cliffs and marine terraces, result from land uplift relative to the sea, while submerged coasts with gentle, sandy shorelines formed due to rising sea levels at the end of the glacial ages (Bird, 2000).

High winds and waves along the coast deposit sediment and erode rock and land continually, making coastal zones vulnerable to natural hazards such as hurricanes (Gornitz, 1995). Coastal erosion, driven by both natural processes and human activities like sand extraction, shifts and changes these zones, making defining their boundaries elusive (Small & Nicholls, 2003). Coastal zones have historically been important for human settlement, facilitating trade and communication, but overdevelopment and environmental degradation pose significant challenges. Effective management of coastal zones requires international cooperation to address issues like pollution and erosion (Klein et al., 2001).

9.1.2.4. High Latitudes

The temperature increase in temperate and polar regions will decrease the snowing period, thereby increasing the length and intensity of the growing period and growing degree units (ACIA, 2005).

9.1.2.5. Agriculture and Forestry

Agriculture and forestry will experience dual effects due to increased CO_2—first, a direct fertilizing effect from higher CO_2 concentration, and second, indirect effects from climate change on crops, livestock, pests, diseases, weeds, soils, and water supplies (Rosenzweig & Hillel, 2015). The Kyoto Protocol recognizes the role of forests in climate change, acknowledging their dual nature as both part of the problem and solution (UNFCCC, 1997). Changes in climate could alter forest ecosystems' configuration, impacting species composition, productivity, and biodiversity (IPCC, 2007).

Global warming may enhance primary production in tundra, grassland, and forest biomes by an average of 19% this century, with greater responses in colder ecosystems like the tundra (Rustad et al., 2001). Warmer temperatures and increased CO_2 may lead to greater primary production in most grasslands, though the response varies depending on the ecosystem and environmental conditions. Elevated CO_2 levels may also increase the photosynthetic rate of plants, enhancing crop yield and forest productivity (Norby et al., 2005). However, CO_2 and temperature responses are often additive, with warmer temperatures enhancing CO_2's stimulatory effect on plant growth (Ainsworth & Long, 2005). These changes in climate could profoundly impact traditional livelihoods, industry, biodiversity, soil, and water resources, and hence biological productivity (Schneider et al., 2007).

9.1.2.6. Pests and Weeds

Weeds, insects, and pathogens cause significant losses in forest and agro-ecosystems, which will be exacerbated by increased CO_2 and temperature rise. The frequency of pest and disease outbreaks will increase, with pests and disease organisms shifting poleward or to higher altitudes (Bale et al., 2002). This shift will affect plant-pest relationships, with temperature playing a dominant role in pest distribution and development. For example, spruce budworm outbreaks follow droughts and dry summers, with drought stress increasing host tree vulnerability and egg laying (Mattson & Haack, 1987). Increased CO_2 levels will also alter plant growth, leading to changes in pest feeding habits and potentially higher pest densities (Ziska & Runion, 2007).

Climate change will also impact the distribution of vector-borne diseases like malaria and dengue, as well as food- and waterborne infectious diseases like diarrhea (Githeko et al., 2000). For example, tick-borne encephalitis incidence in Sweden increased with milder winters, shifting northward from 1980 to 1994 (Lindgren et al., 2000). Weeds, particularly C_4 plants, will be less favored by rising CO_2 levels, which generally benefit C_3 crops more (Patterson, 1995). Climate change will shift agricultural and forestry potential to new regions, potentially introducing new pest species (Porter et al., 1991).

9.1.3. Climate Change, Warming, and Biodiversity

The United Nations Convention on Biodiversity (UNCBD) notes that "So far about 1.75 million species have been identified, mostly small creatures such as insects." Scientists believe that there are 13 million species, though estimates range from 3 to 100 million (UNCBD, 1992). Species have been disappearing at 50–100 times the natural rates, and this is predicted to rise dramatically. Based on current trends, an estimated 34,000 plant and 5,200 animal species face extinction (UNCBD,

1992). The impacts of climate change on biodiversity are a major concern to the Convention. At its fifth meeting in 2000, the Conference of the Parties highlighted the risks, in particular, to coral reefs and forest ecosystems, and drew attention to the serious impacts of biodiversity loss on people's livelihoods (UNCBD, 2000). More recently, the Conference of the Parties turned its attention to the potential impacts on biodiversity and ecosystems of various options for mitigating or adapting to climate change and requested the Convention's Subsidiary Body on Scientific, Technical and Technological Advice (SBSTTA) to develop scientific advice on these issues (UNCBD, 2021).

Rapid environmental change in the past has always been associated with enhanced rates of species extinction and community and biosphere restructuring. A known example of this occurred at the end of the Cretaceous period, 65 million years ago, when a significant change in global climate resulted in long-term restructuring of the vegetation and a major change in the biota of the northern hemisphere (WRI, 2005). The World Resources Institute reports that there is a link between biodiversity and climate change. Rapid global warming can affect ecosystems' chances to adapt naturally. Unfortunately, in the international policy arena, biodiversity loss and climate change have often moved in wholly unconnected domains. Climate change is also a significant driving force behind the loss of genes, species, and critical ecosystem services (WRI, 2005).

Climate is one of the major factors controlling global patterns of vegetation structure, productivity, and plant and animal species composition. Many plants can successfully reproduce and grow only within a specific range of temperatures and respond to specific amounts and seasonal patterns of precipitation. Climate change can cause plants to be displaced by competition from other species or to fail to survive (WRI, 2005). The impacts of climate change on humans and biodiversity are both direct and indirect. The Global Biodiversity Strategy identified mechanisms that affect current levels of biodiversity, nearly all of which have significant human components (WRI, 1992). The direct mechanisms following:

i. Exploitation of wild living resources.
ii. Expansion of agriculture, forestry, and aquaculture.
iii. Habitat loss and fragmentation.
iv. Indirect negative effects of species introduced by humans.
v. Indirect positive effects of species introduced by humans.
vi. Pollution of soil, water and atmosphere.
vii. Global climate change.

The indirect mechanisms identified in the Global Biodiversity Strategy and elsewhere are arguably more significant and indicate the foundations of the human impact on biodiversity.

They include:

i. Human social organization.
ii. The growth of human population.
iii. Natural resources consumption patterns.
iv. Global trade.
v. Economic system and policies that fail to value the environment and its resources.
vi. Inequity in the ownership, management and flow of benefits from both the use and conservation of biological resources.

9.1.3.1. Direct Impacts

The most important direct impacts that climate change has on biodiversity has been thoroughly assessed by the Intergovernmental Panel on Climate Change (IPCC). The IPCC has evaluated the direct effects of climate change on biological systems, noting a discernible impact of regional climate change, particularly increases in temperature, on biological systems throughout the 20th century and projecting similar impacts for the 21st century (IPCC, 2014). In many parts of the world, observed changes in these systems, whether anthropogenic or natural, are coherent across diverse localities and consistent with the expected effects of regional temperature changes. The probability that the observed changes in the expected direction, without reference to magnitude, could occur by chance alone is negligible (Parmesan & Yohe, 2003).

Such systems include the timing of reproduction or migration events, the growing season length, species distributions, and population size, all of which implicate regional climate change as a prominent contributor to observed changes (Root et al., 2003). Additionally, there have been observed changes in the types, intensity, and frequency of disturbances like fire, drought, and blowdowns, which are affected by regional climate change and land use practices. These disturbances, in turn, affect the productivity and species composition within ecosystems, particularly at high latitudes and altitudes (Walther et al., 2002).

9.1.3.2. Habitat Loss and Fragmentation

Habitat destruction is widely considered the most pervasive anthropogenic cause of the loss of biodiversity (Fahrig, 2003). Forests are the most species-rich terrestrial habitats. Deforestation affects biodiversity through habitat destruction, isolated fragments of formerly contiguous habitat, edge effects within a boundary zone between forest and deforested areas, and a variety of other mechanisms (Laurance et al., 2011).

9.1.3.3. Species Loss

Global warming is predicted to take place faster in this century than at any time for at least the last 10,000 years (IPCC, 2014). Coupled with other factors, such as continued land-clearing, this could mean the extinction of species at a rate even greater than when the dinosaurs disappeared about 65 million years ago (Thomas et al., 2004). Climate change may lead to a sharp increase in rates of extinction. Thomas et al., (2004) studied five regions of the world and predicted that if the present rate of climate change continues, 24% of species in these regions will be on their way to extinction by 2050. For many species, climate change poses a greater threat to their survival than the destruction of their natural habitat (Thomas et al., 2004). Evidence directly linking climate change and species extinction is difficult to procure, but at least one species, the golden toad of Costa Rica, may have become extinct due to climate change (Pounds et al., 2006).

A diversity of species increases the ability of ecosystems to perform functions like holding soils together, maintaining soil fertility, delivering clean water to streams and rivers, cycling nutrients, pollinating plants (including crops), and buffering against pests and diseases—these are sometimes called 'ecosystem functions' or 'ecosystem services' (Chapin et al., 2000). A loss of species could reduce this ability, particularly if environmental conditions are changing rapidly at the same time (Hooper et al., 2012). As climate changes and species are eliminated from an area, we may see changes in some ecosystem functions, which could mean more land degradation, changes in agricultural productivity, and a reduction in the quality of water delivered to human populations (Hooper et al., 2012).

9.1.3.4. Species Ranges and Distribution

Climate change is likely to have a number of impacts on biodiversity from ecosystem to species level (Parmesan & Yohe, 2003). The most obvious impact is the effect that temperature and precipitation changes have on the range or distribution of species and ecosystem boundaries. Climatically associated shifts in animal ranges and densities have been noted on most continents, especially in the polar regions and within major taxonomic groups of animals, such as insects, amphibians, birds, and mammals (Parmesan, 2006). Ecosystems will generally shift northward or upward in altitude, but in some cases, they will run out of space—as a 1° C change in temperature corresponds to a 100 km change in latitude, hence, the average shift in habitat conditions by the year 2100 will be about 140-580 km (Root et al., 2003).

Climate change may directly affect species through changes in phenology (e.g., earlier flowering of trees and egg-laying in birds), lengthening of the growing season, and changes in distribution, resulting from migration (e.g., poleward and altitudinal shifts in insect ranges) (Walther et al., 2002). Any particular ecosystem consists of an assemblage of species, some of which will be near the edge of their ranges and others which will not. Those at the edge of their ranges may need to move due to climate change. Species that are highly mobile or opportunistic are likely to benefit at the expense of those that are not (Parmesan, 2006).

The invasion of alien species may be facilitated in some parts of the world. Root et al., (2003) found that out of 1700 species reviewed globally, 81% of observed shifts in range were in the direction expected by climate change. The tree line and the level at which alpine plants are found in Europe is moving towards higher altitudes (Walther et al., 2002). Tree species may be replaced by new species that are better adapted to higher temperatures and drought stress. Plants or trees that need low temperatures in winter to trigger bud bursting in spring may be adversely affected (Walther et al., 2002).

In European seas, temperate species have migrated about 250 km northward per decade, whereas sub-Arctic and Arctic species have declined in number (Perry et al., 2005). Shifts in ecosystem boundaries could mean that protected areas no longer contain the species and habitats they were established to protect. Under existing static conservation paradigms, little emphasis is placed on changing patterns of biodiversity, and few protected area systems have been formulated with reference to climate change, even in countries where effects will probably be large (Hannah et al., 2007). The World Wide Fund for Nature (WWF) argues that protected areas offer limited defense against problems posed by rapid environmental change and that protected areas themselves will need to adapt to meet the challenges posed by global warming (WWF, 2003).

Flooding and sea level rise will affect the ranges and distribution of species and ecosystem boundaries, as well as threatening wetlands and coastal ecosystems (Nicholls & Cazenave, 2010). Manmade coastal defenses across the UK are eroding faster than ever and destroying important inter-tidal habitats for birds (Doody, 2004). Coastal areas will have a substantial increase in the risk of flooding, erosion, and wetland loss (Nicholls & Cazenave, 2010).

9.1.3.5. Phenology

Climate change is also causing changes in the timing of biological events (phenology) such as shifts in the reproductive cycles and growing seasons of many species (Root et al., 2003). This can alter the frequency of pest and disease outbreaks (Rohr et al., 2011). In Europe, phenological data shows an increase in the length of the growing season by about 10 days from 1962 to 1995 (Menzel & Fabian, 1999). Leemans and van Vliet (2004) note phenological shifts in certain mammal, reptile, fish, and bird species. For example, bird life cycle events such as migration and egg-laying appear to be occurring earlier (Crick et al., 1997). Increasingly warm winters associated with the North Atlantic Oscillation influence the development and fecundity of Red Deer (Cervus elaphus) in Norway (Post & Stenseth, 1999).

9.1.3.6. Species Interactions

The impact that climate change will have on many of the more complex interactions (predation, competition, pollination, and disease) that constitute functioning ecosystems remains largely unknown (Tylianakis et al., 2008). Changes in competitive ability may also emerge. For example, early leafing trees will get a two-week head start on their competitors in terms of growth and thus occupy an increasing proportion of woodland (Cleland et al., 2007). Numbers of long-distance migrating birds in the Lake Constance region have declined, probably due to warmer winters increasing the competitive advantage of resident bird species (Schaefer et al., 2008).

Mismatches in timing between interdependent species may occur, especially when changes in some species are cued by day length and others by temperature (Visser & Both, 2005). For example, numbers of native pests such as the green spruce aphid may increase due to milder winters, and non-native pests such as the pinewood nematode, gypsy moth, and Asian longhorn beetle are now found in southern British woods (Strauss et al., 2006).

9.1.3.7. Indirect Impacts

Important indirect impacts that climate change could have on biodiversity are also noteworthy. Classic top-down approaches to climate change often equate to large infrastructure construction projects. Projects designed to support adaptation to climate change are often associated with physical protection. For example, large sea walls may be built to protect against storm surges and floods. These are particularly important in Asia, and the Netherlands continues to invest millions in improving its sea defenses against sea level rise and storms (Nicholls & Cazenave, 2010). Such projects often negatively impact biodiversity. Alternative adaptation options such as the strategic placement of artificial wetlands (or protection of existing mangroves or coral reefs outside Europe) can benefit biodiversity (Mitsch & Gosselink, 2000). Other adaptation projects involve using pesticides and herbicides to control pests and diseases that might increase due to climate change. Use of such chemicals may damage existing plant and animal communities (Tilman et al., 2001).

Projects designed to reduce global greenhouse gas emissions and thus mitigate climate change are often associated with large renewable energy schemes. These often have poor outcomes for biodiversity. For example, large hydropower schemes can cause loss of terrestrial and aquatic biodiversity and inhibit fish migration. Locating dams near river estuaries is particularly damaging as it disrupts the whole watershed (Rosenberg et al., 2000). Dams can also be net emitters of greenhouse gases if submerged soils and vegetation decay and release carbon dioxide and methane (St. Louis et al., 2000). Europe accounts for 70% of the total installed wind energy capacity in the world (EWEA, 2012). Wind farms pose three main problems for birds: disturbance, habitat loss/damage, and collision. The effect that offshore wind farms will have on sea mammals, fish, and marine aquatic communities is largely unknown (Bailey et al., 2014).

Monocultures, however, not only reduce biodiversity but also increase the chances of pest attacks, thus challenging the permanence of carbon stocks (Tilman et al., 2001). The location of afforestation and reforestation projects is also important. Replacing native grasslands, wetlands, scrublands, or heathlands may lead to dramatic biodiversity losses and also lower the relative increase in carbon sequestered compared to implementing such projects on degraded land (Ewers & Rodrigues, 2008).

9.1.3.8. Variations Between Regions, Species, and Ecosystems

The impacts of climate change on biodiversity will vary between regions. The most rapid changes in climate are expected in the far north and south of the planet and in mountainous regions (IPCC, 2014). Unfortunately, these regions are also home to many species that have no alternative habitats to which they can migrate to survive and species that cannot easily compete with new immigrating species (Parmesan, 2006). In Europe, the greatest effects of climate change are projected for Arctic Regions, the moisture-limited ecosystems of Eastern Europe, and the Mediterranean region (IPCC, 2014). Polar warming probably should increase biological production, but different species compositions are likely on land and in the sea, with a tendency for poleward shifts in major biomes and associated animals (Parmesan & Yohe, 2003). High mountain systems such as the Alps are particularly vulnerable (IPCC, 2014). Some species and ecosystems are also more vulnerable than others, particularly small populations or those restricted to small areas (Hughes, 2000).

9.1.3.9. Ecosystems and Biodiversity

Land use changes and subsequent changes to biodiversity can both increase and decrease greenhouse gas emissions. Countries like Ireland, the Netherlands, and Denmark are significant carbon sources, but countries like Slovenia and Slovakia are significant carbon sinks (Schulp et al., 2008). Currently, some 60% of anthropogenic global greenhouse gas emissions originate from the generation and use of energy (IPCC, 2014). While the uses of renewable energy sources, such as wood, instead of fossil fuels can help mitigate climate change, this can also have negative impacts (Righelato & Spracklen, 2007). For example, using relatively undisturbed natural fuel sources (such as native forests as opposed to plantations) can lead to significant biodiversity losses (Righelato & Spracklen, 2007). Some bio-energy plantations replace sites with high biodiversity, introduce alien species, and use damaging agrochemicals (Fargione et al., 2008).

There are also many complex feedback mechanisms at work between biodiversity and climate change. For example, some species of ocean algae release dimethyl sulfide into the atmosphere. Rising ocean temperatures due to global warming can lead to algal blooms and resultant increases in dimethyl sulfide release contribute to cloud formation. This in turn may help reduce temperatures, as less heat will be able to reach the Earth's surface (Charlson et al., 1987). Rising temperatures can stimulate the growth of phytoplankton in the sea, which can increase carbon dioxide uptake (Falkowski, 1994). In high latitudes, the replacement of shrub/tundra vegetation with trees can affect the radiation balance. Unlike snow, trees are more likely to absorb sunlight rather than reflect it. This in turn can enhance climate change (Chapin et al., 2000).

9.2. GREEN HOUSE GASES

Greenhouse gas, any gas that has the property of absorbing infrared radiation (net heat energy) emitted from Earth's surface and reradiating it back to Earth's surface, thus contributing to the greenhouse effect. Carbon dioxide, methane, and water vapor are the most important greenhouse gases. (To a lesser extent, surface-level ozone, nitrous oxides, and fluorinated gases also trap infrared radiation.) Greenhouse gases have a profound effect on the energy budget of the Earth system despite making up only a fraction of all atmospheric gases. Concentrations of greenhouse gases have varied substantially during Earth's history, and these variations have driven substantial climate changes at a wide range of timescales. In general, greenhouse gas concentrations have been particularly high during warm periods and low during cold periods.

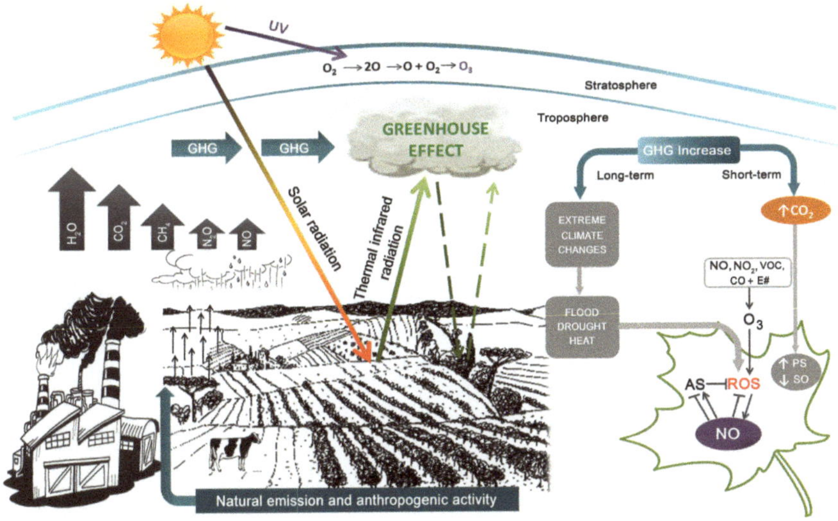

Figure 9.2. *Schematic view of greenhouse gas and their impact.*

Source: Cassia, R., Nocioni, M., Correa-Aragunde, N., and Lamattina, L. (2018) Climate Change and the Impact of Greenhouse Gasses: CO_2 and NO, Friends and Foes of Plant Oxidative Stress. Front. Plant Sci. 9:273. doi: 10.3389/fpls.2018.00273.

A number of processes influence greenhouse gas concentrations. Some, such as tectonic activities, operate at timescales of millions of years, whereas others, such as vegetation, soil, wetland, and ocean sources and sinks, operate at timescales of hundreds to thousands of years. Human activities—especially fossil-fuel combustion since the Industrial Revolution—are responsible for steady increases in atmospheric concentrations of various greenhouse gases, especially carbon dioxide, methane, ozone, and chlorofluorocarbons (CFCs). The effect of each greenhouse gas on Earth's climate depends on its chemical nature and its relative concentration in the atmosphere. Some gases have a high capacity for absorbing infrared radiation or occur in significant quantities, whereas others have considerably lower capacities for absorption or occur only in trace amounts. Radiative forcing, as defined by the Intergovernmental Panel on Climate Change (IPCC), is a measure of the influence a given greenhouse gas or other climatic factor (such as solar irradiance or albedo) has on the amount of radiant energy impinging upon Earth's surface (Figure 9.2).

9.2.1. Major Greenhouse Gases

9.2.1.1. Water Vapor

Water vapor is the most potent greenhouse gas in Earth's atmosphere, but its behavior is fundamentally different from that of the other greenhouse gases. The primary role of water vapor is not as a direct agent of radioactive forcing but rather as a climate feedback—that is, as a response within the climate system that influences the system's continued activity.

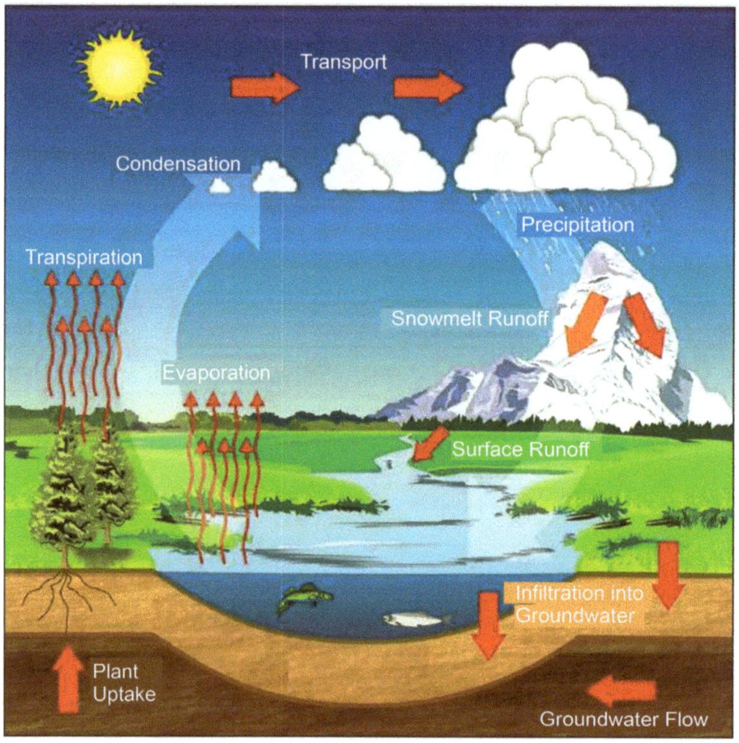

Figure 9.3. *Illustration of how atmospheric water vapor amplifies earth's greenhouse effect.*

Source: Steamy Relationships: How Atmospheric Water Vapor Amplifies Earth's Greenhouse Effect - NASA Science. (n.d.). Retrieved from https://science.nasa.gov/earth/climate-change/steamy-relationships-how-atmospheric-water-vapor-amplifies-earths-greenhouse-effect/.

This distinction arises because the amount of water vapor in the atmosphere cannot, in general, be directly modified by human behavior but is instead set by air temperatures.

The warmer the surface, the greater the evaporation rate of water from the surface. As a result, increased evaporation leads to a greater concentration of water vapor in the lower atmosphere capable of absorbing infrared radiation and emitting it back to the surface (Figure 9.3).

9.2.1.2. Carbon Dioxide

Carbon dioxide (CO_2) is the most significant greenhouse gas. Natural sources of atmospheric CO_2 include outgassing from volcanoes, the combustion and natural decay of organic matter, and respiration by aerobic (oxygen-using) organisms. These sources are balanced, on average, by a set of physical, chemical, or biological processes, called "sinks," that tend to remove CO_2 from the atmosphere. Significant natural sinks include terrestrial vegetation, which takes up CO_2 during photosynthesis.

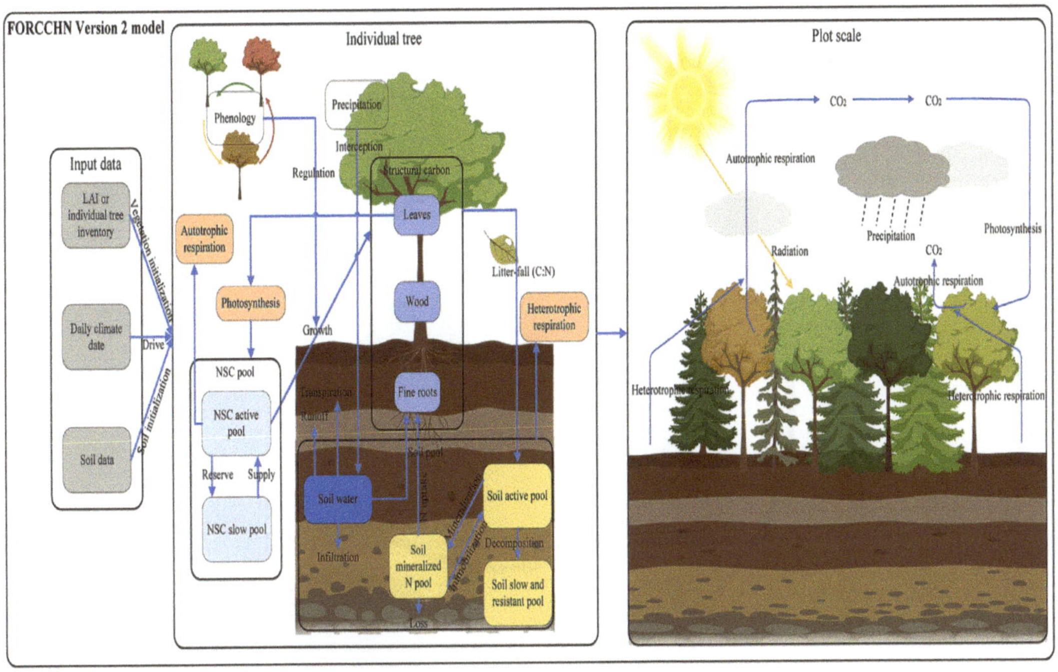

Figure 9.4. *The role of CO_2.*

Source: Fang, J., Shugart, H. H., Liu, F., Yan, X., Song, Y., and Lv, F.: FORCCHN V2.0: an individual-based model for predicting multiscale forest carbon dynamics, Geosci. Model Dev., 15, 6863–6872, https://doi.org/10.5194/gmd-15-6863-2022, 2022.

Carbon is transported in various forms through the atmosphere, the hydrosphere, and geologic formations. One of the primary pathways for the exchange of carbon dioxide (CO_2) takes place between the atmosphere and the oceans; there a fraction of the CO_2 combines with water, forming carbonic acid (H_2CO_3) that subsequently loses hydrogen ions (H^+) to form bicarbonate (HCO_3^-) and carbonate (CO_3^{2-}) ions. Mollusk shells or mineral precipitates that form by the reaction of calcium or other metal ions with carbonate may become buried in geologic strata and eventually release CO_2 through volcanic outgassing. Carbon dioxide also exchanges through photosynthesis in plants and through respiration in animals. Dead and decaying organic matter may ferment and release CO_2 or methane (CH_4) or may be incorporated into sedimentary rock, where it is converted to fossil fuels. Burning of hydrocarbon fuels returns CO_2 and water (H_2O) to the atmosphere. The biological and anthropogenic pathways are much faster than the geochemical pathways and, consequently, have a greater impact on the composition and temperature of the atmosphere (Figure 9.4).

A number of oceanic processes also act as carbon sinks. One such process, the "solubility pump," involves the descent of surface seawater containing dissolved CO_2. Another process, the "biological pump," involves the uptake of dissolved CO_2 by marine vegetation and phytoplankton (small, free-floating, photosynthetic organisms) living in the upper ocean or by other marine organisms that use CO_2 to build skeletons and other structures made of calcium carbonate ($CaCO_3$). As these organisms expire and fall to the ocean floor, their carbon is transported downward and eventually buried at depth. A long-term balance between these natural sources and sinks leads to the background, or natural, level of CO_2 in the atmosphere.

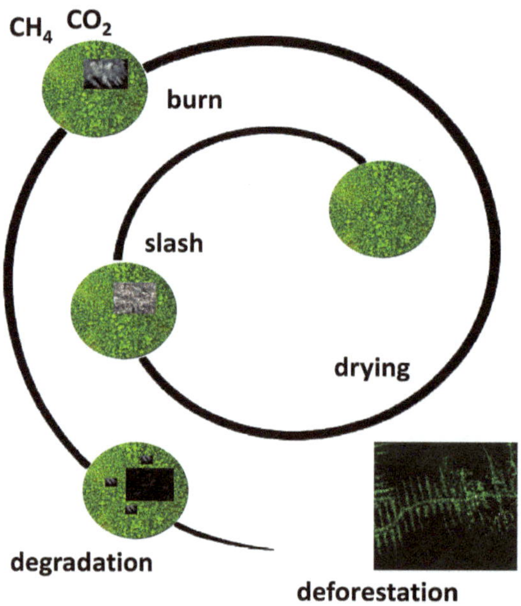

Figure 9.5. *Deforestation & greenhouse gases impact on climate change.*

Source: Silva, C.A.; Santilli, G.; Sano, E.E.; Laneve, G. Fire Occurrences and Greenhouse Gas Emissions from Deforestation in the Brazilian Amazon. Remote Sens. 2021, 13, 376. https://doi.org/10.3390/rs13030376.

On the other hand, human actions predominantly contribute to the rise of atmospheric CO_2 levels by burning fossil fuels (mainly oil and coal, and to a lesser extent natural gas) for transportation, heating, and electricity generation, as well as by producing cement. Additional human-caused sources encompass the combustion of forests and the deforestation of land (Figure 9.5). Currently, human activities are responsible for releasing approximately 7 gigatons (7 billion tons) of carbon into the atmosphere every year. Human-caused emissions account for about 3% of the total CO_2 emissions from natural sources. The additional carbon emissions from human activities greatly surpass the ability of natural sinks to absorb them, potentially by as much as 2–3 gigatons per year. The concentration of CO_2 in the atmosphere has increased steadily at an average rate of 1.4 parts per million (ppm) by volume per year from 1959 to 2006, and approximately 2.0 ppm per year from 2006 to 2018. Overall, the rate of accumulation has been consistent and constant over time, following a linear pattern. Nevertheless, specific present-day reservoirs that absorb or consume resources, such as the oceans, have the potential to transition into suppliers or contributors in the forthcoming period. This could result in a scenario where the level of atmospheric CO_2 increases exponentially, meaning that the rate of increase is also increasing over time (Figure 9.6).

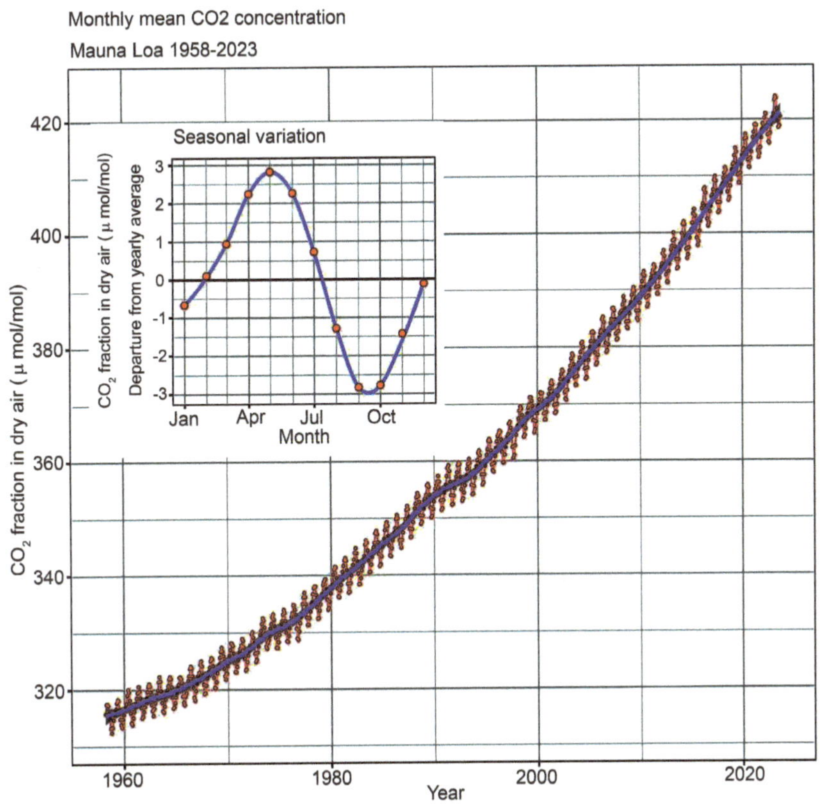

Figure 9.6. *This figure shows the history of atmospheric carbon dioxide concentrations as directly measured at Mauna Loa, Hawaii since 1958.*

Source: Wikipedia.

The Keeling Curve, named after American climate scientist Charles David Keeling, monitors fluctuations in the concentration of carbon dioxide (CO_2) in the Earth's atmosphere at a research station located on Mauna Loa in Hawaii. Despite experiencing minor seasonal variations, the concentrations of CO_2 in the atmosphere are consistently increasing over time.

The baseline concentration of carbon dioxide fluctuates over millions of years as a result of gradual changes in the release of gas through volcanic eruptions. During the Cretaceous Period, which occurred approximately 100 million years ago, it seems that the concentration of CO_2 in the atmosphere was several times greater than it is today, possibly reaching levels close to 2,000 parts per million (ppm). During the last 700,000 years, the concentrations of CO_2 have fluctuated within a relatively narrow range, approximately between 180 and 300 parts per million (ppm). These variations are connected to the Earth's orbital effects, which are also responsible for the occurrence and disappearance of ice ages in the Pleistocene epoch. During the early 21st century, the concentration of CO_2 in the atmosphere reached 384 parts per million (ppm), which is approximately 37% higher than the natural background level of around 280 ppm that was present at the start of the Industrial Revolution. The levels of atmospheric CO_2 continued to rise steadily, and by 2018, they had reached 410 parts per million (ppm). Based on ice core measurements, these levels are thought to be the highest in at least 800,000 years and, according to other evidence, possibly the highest in at least 5,000,000 years.

The radiative forcing resulting from carbon dioxide is directly proportional to the logarithm of its concentration in the atmosphere. The logarithmic relationship arises due to a saturation effect, where it becomes progressively more challenging for additional CO_2 molecules to have a further impact on the "infrared window" – a specific narrow range of wavelengths in the infrared region that is not absorbed by atmospheric gases – as CO_2 concentrations increase. According to the logarithmic relationship, the increase in surface warming potential is expected to be proportional to every doubling of CO_2 concentration. Based on the current rate of fossil fuel consumption, it is anticipated that the concentration of CO_2 will double from preindustrial levels by the middle of the 21st century, specifically when CO_2 concentrations are projected to reach 560 parts per million (ppm). An increase in CO_2 concentrations by a factor of two would result in a rise of approximately 4 watts per square meter of radiative forcing. Based on standard estimates of "climate sensitivity" without considering any counteracting factors, this increase in energy would result in a temperature rise of 2 to 5° C (3.6 to 9° F) compared to preindustrial levels. The cumulative radiative forcing resulting from human-caused CO_2 emissions since the onset of the industrial era is estimated to be around 1.66 watts per square meter.

9.2.1.3. Methane

Methane (CH_4) is the second most significant contributor to the greenhouse effect. CH_4 is more potent than CO_2 due to its higher radiative forcing per molecule. Furthermore, the infrared window exhibits lower saturation within the specific wavelength range of CH_4-absorbed radiation, allowing for a greater number of molecules to occupy this region. Nevertheless, methane (CH_4) is present in significantly lower amounts compared to carbon dioxide (CO_2) in the Earth's atmosphere. Its concentration is typically measured in parts per billion (ppb) rather than parts per million (ppm). CH_4 has a significantly shorter duration in the atmosphere compared to CO_2. The residence time of CH_4 is approximately 10 years, while CO_2 can last for hundreds of years (Figure 9.7).

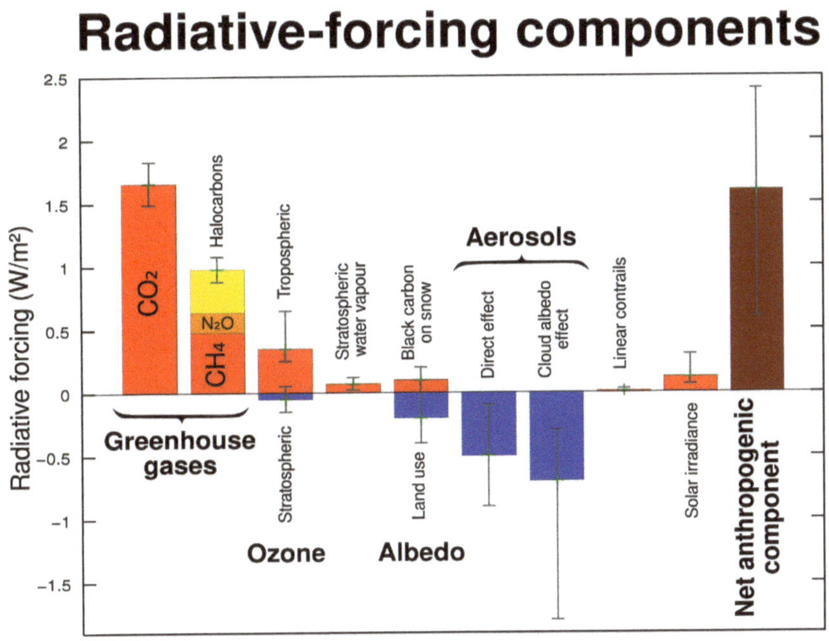

Figure 9.7. *Understanding global warming requires understanding the changes in climate forcings (radiative forcing) that have occurred since the industrial revolution.*

Source: Wikimedia Commons.

Some natural sources of methane include wetlands in tropical and northern regions, bacteria that consume organic material from termites, volcanic activity, seafloor vents in areas with organic sediment, and methane trapped in oceanic continental shelves and polar permafrost in the form of hydrates. The atmosphere serves as the main natural reservoir for methane, as methane readily undergoes a reaction with the hydroxyl radical (OH−) in the troposphere to produce CO_2 and water vapor (H_2O). CH_4 is eliminated in the stratosphere. Soil is another natural sink for methane, where bacteria oxidize it.

Similar to carbon dioxide (CO_2), human activity is causing the concentration of methane (CH_4) to increase at a rate that exceeds the capacity of natural processes to counterbalance it. Human activities currently contribute to around 70% of the total annual emissions, resulting in significant growth in concentration over time. The primary human-caused origins of atmospheric CH_4 include rice cultivation, livestock farming, the combustion of coal and natural gas, the incineration of biomass, and the decay of organic matter in landfills. Anticipating future trends is especially challenging. This is partially attributed to a limited comprehension of the climate feedback linked to CH_4 emissions. Furthermore, with the expansion of human populations, it becomes challenging to anticipate the impact of potential alterations in livestock rearing, rice farming, and energy consumption on the emission of CH_4.

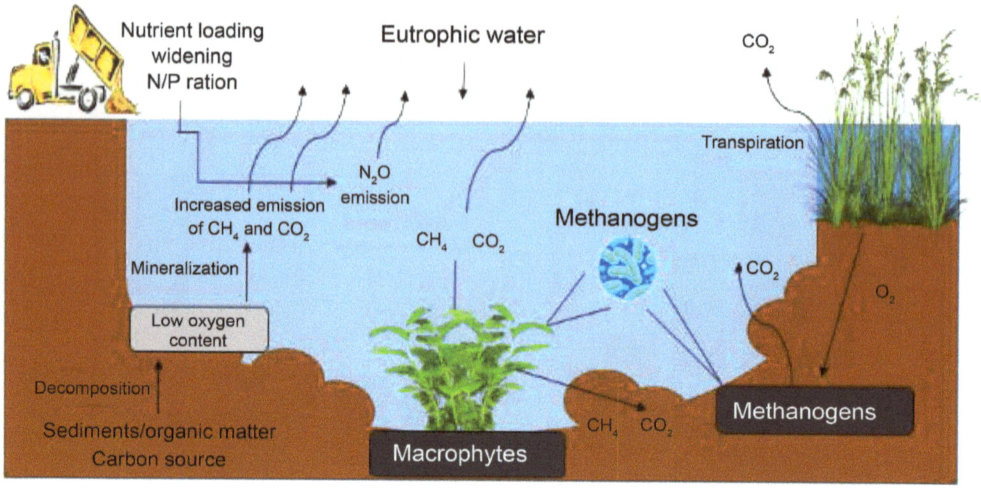

Figure 9.8. *Methane cycle in atmosphere.*

Source: *Mondal, B.; Bauddh, K.; Kumar, A.; Bordoloi, N. India's Contribution to Greenhouse Gas Emission from Freshwater Ecosystems: A Comprehensive Review. Water 2022, 14, 2965. https://doi.org/10.3390/w14192965.*

The Paleocene-Eocene Thermal Maximum (PETM) was a period characterized by a significant rise in average global temperatures, estimated to be between 4–8° C (7.2–14.4° F), over a span of a few 1,000 years. This warming event is attributed to a sudden surge in methane concentration in the atmosphere. This episode occurred approximately 55 million years ago, and the increase in CH_4 seems to have been connected to a colossal volcanic eruption that interacted with flood deposits containing methane. Consequently, significant quantities of gaseous CH_4 were released into the atmosphere. It is challenging to accurately determine the exact magnitude of these concentrations or the duration of their persistence. At extremely high concentrations, the length of time that CH_4 remains in the atmosphere can exceed the current average residence time of 10 years. However, it is probable that these concentrations reached several parts per million (ppm) during the PETM.

Methane concentrations exhibited narrower fluctuations (ranging approximately from 350 to 800 ppb) in correlation with the Pleistocene ice age cycles. The atmospheric concentration of CH_4 in preindustrial times was around 700 parts per billion (ppb), while in late 2018, it exceeded 1,867 ppb. (These concentrations correspond to the natural levels observed for at least the past 650,000 years.) The anthropogenic CH_4 emissions result in a net radiative forcing of approximately 0.5 watts per square meter, which is roughly equivalent to one-third of the radiative forcing caused by CO_2 (Figure 9.8).

9.2.2. Lesser Greenhouse Gases

9.2.2.1. Surface-Level Ozone

Surface, or low-level, ozone (O_3) is the most prominent greenhouse gas. Surface ozone is a consequence of air pollution and should be differentiated from naturally occurring stratospheric ozone, which plays a distinct role in the planetary radiation balance. The main natural origin of surface O_3 is the downward movement of stratospheric O_3 from the higher regions of the atmosphere. On the other hand, the main human-caused origin of surface O_3 is the result of photochemical reactions that involve the atmospheric pollutant carbon monoxide (CO). The most accurate assessments indicate that the typical concentration of surface O_3 in its natural state is 10 parts per billion (ppb), and the overall impact on the Earth's energy balance caused by human-generated emissions of surface O_3 is approximately 0.35 watts per square meter. In cities susceptible to photochemical smog, ozone concentrations can reach levels that are considered unhealthy. Specifically, these levels are defined as concentrations that reach or surpass 70 parts per billion (ppb) for a duration of eight hours or more.

9.2.2.2. Nitrous Oxides and Fluorinated Gases

Industrial activity generates additional trace gases with greenhouse properties, such as nitrous oxide (N_2O) and fluorinated gases (halocarbons). The halocarbons category includes CFCs, sulfur hexafluoride, hydrofluorocarbons (HFCs), and perfluorocarbons (PFCs). Nitrous oxide contributes to a radiative forcing of 0.16 watts per square meter, while fluorinated gases collectively contribute to a radiative forcing of 0.34 watt per square meter. Nitrous oxides exhibit minimal background concentrations resulting from natural biological reactions in soil and water, while fluorinated gases are predominantly derived from industrial sources.

9.3. DEFORESTATION

Deforestation, the deliberate removal or reduction of forests by human activities, is a significant problem in the global management of land (Lambin et al., 2001). Traditionally, estimates of deforestation rely on calculating the amount of forest that has been cleared for human activities, such as cutting down trees for wood products, creating croplands, and establishing grazing lands (Hansen et al., 2013). Clear-cutting involves the complete removal of all trees from a given area, resulting in the complete destruction of the forest. Occasionally, even when logging is only partially done and fires occur unintentionally, the trees can be sufficiently reduced in number to significantly alter the structure of the forest (Asner et al., 2005) (Figure 9.9).

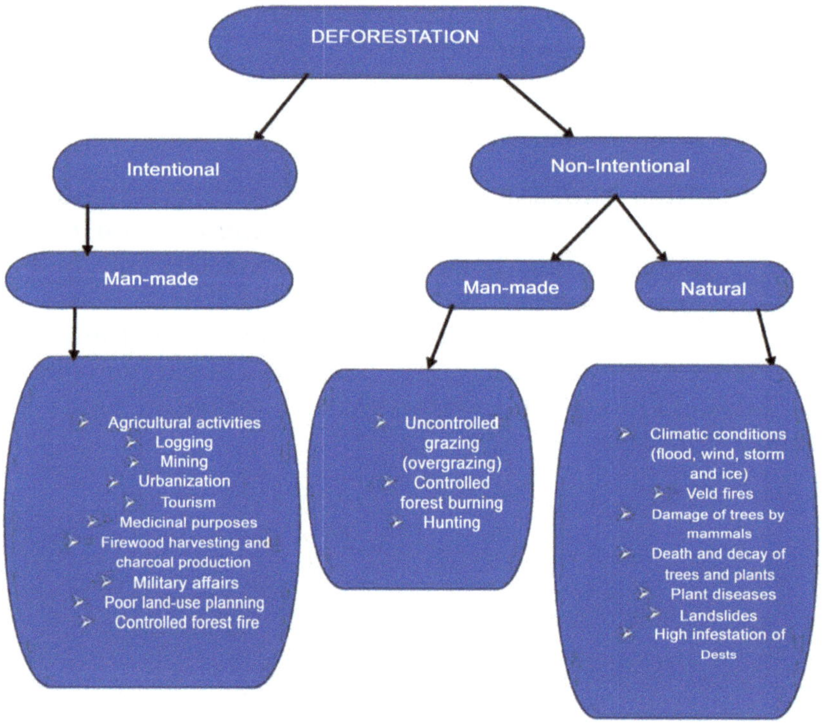

Figure 9.9. *Schematic view of deforestation.*

Source: Delineation and Dimension of Deforestation - Scientific Figure on ResearchGate. Available from: https://www.researchgate.net/figure/Schematic-illustration-of-deforestation_fig1_352375734 [accessed 4 Sept 2024].

9.3.1. Modern Deforestation

According to the United Nations Food and Agriculture Organization (FAO), the rate of deforestation is estimated to be around 1.3 million square kilometers per decade. However, in some areas, the rate has decreased in the early 21st century due to improved forest management practices and the creation of nature reserves (FAO, 2016). Significant deforestation is predominantly taking place in the tropical regions, which are home to a diverse range of forests. The forests vary in climate, ranging from consistently hot and wet rainforests to forests that are only humid and moist. Some forests experience a dry season where trees may lose their leaves, while others are dry open woodlands. Due to the arbitrary nature of the boundaries between these categories, there are varying estimates regarding the extent of deforestation in the tropics (Achard et al., 2002).

One significant cause of tropical deforestation is the implementation of slash-and-burn agriculture, also known as swidden agriculture or shifting agriculture. Small-scale farmers engage in deforestation through the method of burning forests, subsequently utilizing the nutrient-rich ashes to cultivate crops in the soil. Usually, the land yields crops for only a limited number of years before it is necessary to abandon it and burn new areas of forest. Fire is frequently employed to eradicate forests in Southeast Asia, tropical Africa, and the Americas to establish permanent oil palm plantations (Morton et al., 2006).

Other human activities that contribute to tropical deforestation include the commercial exploitation of forests through logging and the clearing of land for cattle ranching and the cultivation of economically valuable trees such as rubber trees, oil palm, and other species.

The Amazon Rainforest is the most extensive remaining expanse of moist tropical forest, with approximately two-thirds of it located in Brazil. (The remaining portion is located adjacent to the western and northern borders of that country.) Research conducted in the Amazon region indicates that approximately 5,000 square kilometers (equivalent to 1,931 square miles) undergo partial logging annually. Furthermore, on an annual basis, fires consume an area that is approximately half the size of the areas that are deliberately cleared. Even if the forest is not completely cleared, what is left is frequently a mosaic of forests and fields or, in the case of more extensive deforestation, isolated forest areas surrounded by vast deforested regions. Reforestation efforts are underway in certain regions where deforestation has occurred. Replanting is carried out for two main purposes: to restore logging areas for future use and to restore ecological balance by converting the replanted areas into protected land. Furthermore, substantial regions are cultivated as single-species plantations for the purpose of timber or paper manufacturing. These are typically cultivated areas consisting of eucalyptus or rapidly growing pine trees, which are almost always non-indigenous to their planting locations. According to the FAO, there are roughly 1.3 million square kilometers (500,000 square miles) of these plantations on Earth. The United Nations and nongovernmental organizations spearhead and finance numerous reforestation initiatives. Nevertheless, certain national governments have also embarked on ambitious reforestation initiatives. In 2017, the government of New Zealand aimed to plant over 100 million trees annually within its territory. However, the most ambitious reforestation effort occurred in India on a specific day in 2017, where citizens planted approximately 66 million trees.

9.3.2. Effects of Deforestation

Deforestation has significant global ramifications. Forests store carbon by accumulating wood and other biomass as trees grow, absorbing carbon dioxide from the atmosphere through the carbon cycle. When forests undergo combustion, the carbon they contain is released into the atmosphere in the form of carbon dioxide. This carbon dioxide is a greenhouse gas that has the ability to modify the Earth's climate by causing the greenhouse effect and contributing to global warming. Additionally, the absence of trees means that there is no longer a mechanism in place to capture and store more carbon.

Furthermore, the majority of the Earth's valuable biodiversity is concentrated in forests, especially those in tropical regions. The Amazon and other moist tropical forests harbor the highest densities of animal and plant species among all land-based ecosystems. It is estimated that approximately two-thirds of the world's species are exclusively found in these forests. As deforestation continues, it has the capacity to lead to the extinction of a growing number of these species.

At a smaller, regional level, the consequences of forest clearance, targeted logging, and fires intertwine. Selective logging enhances the combustibility of the forest by transforming a dense, moist forest into a less dense, arid one. The forest is at risk of fires spreading from nearby cleared agricultural lands and being negatively impacted by natural droughts. Due to the ongoing occurrence of wildfires, logging activities, and droughts, the forest may gradually become more exposed, ultimately resulting in the complete loss of all trees. Moreover, the incineration of tropical forests is typically a cyclical occurrence and can significantly impair air quality. The burning of oil palm plantations in Southeast Asia has led to unprecedented levels of air pollution. A significant portion of the deforested land in tropical regions is found in the shape of precipitous mountain slopes. The convergence of precipitous inclines, abundant precipitation, and the absence of arboreal roots to consolidate the soil can result in catastrophic landslides that devastate agricultural areas, residences, and human lives. Except for the forests that are destroyed for the oil palm industry, most of the humid forests that are cleared are either abandoned as croplands or used for low-density grazing due to the extremely nutrient-poor soils.

While it is possible for forests to regenerate after being cleared and subsequently abandoned, this outcome is not guaranteed, particularly if the remaining forests are extensively fragmented. Habitat fragmentation results in the isolation of plant and animal populations, hindering their ability to reproduce due to genetic bottlenecks. Additionally, the fragments may be insufficient in size to sustain large or territorial animals. In addition, lands that have been cleared of trees and replaced with commercially valuable tree species have reduced biodiversity and do not provide suitable habitats for native flora and fauna, including numerous endangered species.

1. **Local Climate Change:** Plants emit moisture into the atmosphere through cellular respiration, an intricate metabolic process through which organisms obtain energy. The presence of leaves and branches in a forest acts as a filter for sunlight, preventing direct exposure to the ground. As a result, the soil and air beneath the forest canopy generally maintain a cool and moist environment. Deforestation typically leads to soil desiccation and reduced capacity to sustain biodiversity, while also causing local temperature rise as a result of diminished shading.

Rainfall is less frequent because of the scarcity of water vapor in the atmosphere. According to certain scientists, a significant consequence of deforestation is the substantial expansion of desert landscapes. This occurs when soil, lacking regular moisture, gradually transforms into predominantly infertile sand.

2. **Agricultural and Financial Losses:** Deforestation-induced local climate change can have catastrophic consequences for farmers. Insufficient rainfall often leads to inadequate water supply for crops, resulting in reduced growth and lower crop yields, which can lead to food shortages. The reduction in supply typically leads to an increase in crop prices, resulting in far-reaching economic consequences on a global scale. The increase in expenses does not necessarily ensure that individuals engaged in agriculture can sufficiently cover their operational costs, resulting in many farmers barely reaching a breakeven point, resorting to loans or even declaring bankruptcy. Despite the presence of irrigation systems, the scarcity of rainfall frequently results in a decrease in the water table, posing a greater difficulty in utilizing these systems.

3. **Soil Erosion:** Soil erosion is an inherent phenomenon where soil is eroded by the action of flowing water, wind, or human activities like tilling. This process is considered crucial for the development of new soil. Regrettably, erosion may intensify as a result of inadequate cultivation methods, combined with windstorms and excessive precipitation. In such instances, agricultural regions and the overall ecosystem may be jeopardized. The consequences of soil erosion can encompass a decline in soil fertility at the specific location of erosion, as well as the pollution of watercourses in different areas.

Background or geological erosion, known as natural soil erosion, has been a significant factor in soil formation for a minimum of 450 million years. Geological erosion is the gradual process in which natural forces like water or ice, wind, or gravity wear down rock formations, resulting in the creation of soil. It is claimed that this gradual process removes soil at a rate equivalent to its formation.

Accelerated erosion refers to the depletion of soil at a faster rate than its creation. While geological soil erosion is commonly considered advantageous, accelerated erosion can have adverse impacts on the environment and agriculture. The primary drivers of this process are typically anthropogenic activities, including excessive farming practices, overgrazing by livestock, irrigation, deforestation, and urban development.

Soil erosion is primarily characterized by a decline in soil fertility and an escalation in water pollution. The topmost layers of soil are regarded as the most fertile and are believed to aid in water retention. When these layers undergo erosion, vegetation may experience a lack of vital nutrients and moisture. Soil erosion can result in the displacement of soil to different areas, leading to environmental complexities. Transferring this soil to watercourses can cause ecological disturbances in lakes, sedimentation in dams, and contamination of drinking water with pollutants.

Given the adverse effects of soil erosion on agriculture and the environment, soil conservation is regarded as indispensable. Several techniques can be employed to mitigate erosion. Contour plowing, crop rotation, and terracing along slopes are among the commonly employed techniques. Planting trees or flowers is a prevalent method for controlling erosion in residential areas. Additional techniques involve applying nutrient-dense mulch or fertilizer to the area, as well as constructing miniature retaining walls around garden beds.

4. **Global Warming:** Photosynthetic organisms, such as trees and other plants, undergo a process called photosynthesis where they transform carbon dioxide into oxygen that can be breathed in. Deforestation reduces the number of plants available to carry out this process. In addition, when plants undergo combustion, whether it is a deliberate act to clear land or a natural occurrence like a lightning strike, they emit carbon dioxide into the atmosphere. The presence of natural gas impedes the dissipation of heat from Earth into space, potentially exacerbating the greenhouse effect and intensifying global warming.

   The global increase in temperatures over time is a highly controversial subject, with certain individuals even claiming that the entire notion of global warming is merely a hoax perpetuated by radical environmentalists. The scientific community, for the most part, acknowledges it as a reality, although there is some disagreement regarding the underlying causes. Scientists frequently caution about the profound and enduring consequences of climate change, such as the depletion of suitable habitats for species like polar bears due to the melting of arctic ice.

5. **Cultural Displacement:** Oftentimes, individuals residing in a wooded region lack legal entitlements to the land. Their presence in that location is primarily due to the fact that their predecessors settled there, and the land has remained sufficiently habitable to support successive generations over an extended period. Companies aiming to develop these regions occasionally endeavor to facilitate the transition of indigenous populations into a more industrialized way of life. However, frequently, these groups are merely displaced to alternative territories without proper consideration. These individuals typically need to rapidly adjust to ensure their survival. However, the changes they embrace often expose them to the possibility of losing their traditional lifestyles.

6. **Addressing the Problem:** Increasing the number of trees through reforestation is a straightforward method to combat the adverse consequences of deforestation. However, the time it takes for the forest to reach maturity depends on the specific species chosen, often requiring many years of waiting. Other organisms require a longer period of time to acclimate to the environment and establish their presence within a unique ecosystem. Consequently, the significant level of devastation necessitates individuals to promptly take additional measures to address the issue.

   Certain individuals and businesses are advocating for the environment by adopting a paperless approach, utilizing technology for communication, tasks, and record-keeping. Recycling also aids in diminishing the quantity of trees that are needlessly felled. One can consider supporting organizations like the Forest Stewardship Council as an alternative, along with advocating for stricter regulations concerning environmental preservation and the influence of logging, construction, road, mining, and hydroelectric dam corporations.

9.3.3. Severe Deforestation

Severe deforestation refers to the complete removal of forested areas, followed by actions that prevent the regrowth of trees. In contrast, reversible deforestation occurs when trees are cleared, but they have the ability to regenerate. Biologists are deeply concerned about the consequences of extensive deforestation, as it involves the irreversible destruction of forested areas worldwide, which is a rapidly increasing phenomenon. Scientists in multiple countries, including the United States and Great Britain, are conducting thorough research on deforestation and developing strategies to prevent it.

Figure 9.10. *Schematic view of clearcutting of forest for the purpose of urbanization.*

Source: flickr.com.

New England highlights notable instances of both gradual and abrupt deforestation. Upon their arrival in New England, the early settlers of the United States engaged in extensive deforestation to facilitate agricultural development. Over time, agricultural production relocated to different regions of the country, resulting in the regrowth of these trees. This serves as an instance of soft deforestation, where trees are initially removed but eventually replenished. Europe also experiences the widespread occurrence of gradual deforestation. Biologists have observed that the regrowth of vegetation in previously deforested areas can be remarkably robust, and these areas, which were once completely cleared of trees, now offer a wide range of advantages to the surrounding communities.

Biologists often highlight the regeneration of forested land in New England as an example of nature's ability to rebound from detrimental human actions. The Harvard

Forest in Massachusetts has identified a concerning and escalating pattern: the deliberate removal of trees through clearcutting for the purpose of urbanization, leading to severe deforestation (Figure 9.10). Hard deforestation involves the conversion of forested areas into infrastructure such as roads, parking lots, malls, and urban developments, which hinders the natural recovery of the environment. Intense deforestation leads to a permanent overall reduction in forested areas.

Woodlands are highly valuable due to multiple reasons. From an environmental perspective, trees play a crucial role in capturing and storing carbon emissions, offering shelter and living space for animals, and functioning as regulators of temperature. Parks and wooded areas in regions can mitigate temperature fluctuations, resulting in cooler summers and warmer winters. Additionally, there are individuals who derive pleasure from woodlands, appreciating their beauty and serenity. These individuals are also concerned about the detrimental impact on their overall well-being resulting from extensive deforestation.

Organizations striving to combat severe deforestation aim to promote the appreciation of forests in their natural state, without the necessity for development. Additionally, certain groups advocate for the practice of selective timber harvesting, which enhances the economic worth of forested areas. Alternatively, some propose the idea of populating woodlands with game for the purpose of hunting, or transforming them into recreational parks for people's enjoyment, thereby enhancing the inherent worth of wooded areas.

9.3.4. Causes of Deforestation

- **Mining**: The increase of mining on tropical forests is furthering damage due to the rising demand and high mineral prices. These projects are often accompanied by large infrastructure construction, such as roads, railways, and power systems. This contributor to deforestation is putting additional pressure on our forests and freshwater ecosystems.
- **Paper:** America, China, Japan, and Canada make up more than of the world's paper production—400 million tons a year. Approximately 640 million trees represent the paper that's thrown away each year, according to the Environment Paper Network. If we recycled, we could save 27.5 million tons of carbon dioxide from going into the atmosphere. By using recycled paper, we allow the forests to remain as an ecosystem and wildlife habitat.
- **Overpopulation**: Due to overpopulation, more land is needed to establish housing and settlements. As well as many more roads and highways are being built to accommodate a larger sum of people driving. With more people that come with a large need for food and farmland to grow on and raise livestock—resulting in deforestation. Logging industries will cut down trees for furniture, paper, building materials, and many more products. These are a direct result of growing human population and is why it's important to purchase from sustainable companies which actively work against deforestation.

- **Logging:** Wood-based industries such as paper, matchsticks, and furniture need a substantial quantity of wood. Lumber and charcoal are common examples of trees being used as fuel. Cooking and heating all around the world use these resources, and half of the illegal removal from forests is thought to be used as fuel wood. Large areas are also cleared to construct roads for large trucks to have entry to logging sites. Selective logging is where only the most valuable trees are felled, however, this doesn't help our problem as one large tree may bring down surrounding trees and thin the forest canopy. The forest canopy is extremely important to the ecosystem as it houses animals, protects plants and insect population, and protects the forest floor.
- **Agriculture Expansion & Livestock Ranching:** A major cause of deforestation is agriculture plantations. An increasing supply-demand for products such as palm oil and soybeans are driving producers to clear forests at an unnerving rate. Farmers often clear the land for cattle by using slash and burn techniques (cutting down trees and burning them). Unfortunately, they will then use the property until the soil is completely degraded and repeat the process on a new patch of woodland. Eventually, it'll reforest, but it will take many years to return to its original condition.

  Cattle ranching and deforestation are strongest in Latin America. Over the past 40 years, forest area has reduced my almost an astounding 40%. During the same period, 40 years, pasture regions and cattle population have grown significantly and rapidly.
- **Climate Change:** Forests are essentially the lungs of our planet. All plants take in carbon dioxide and release oxygen. Trees are able to convert more carbon dioxide than a regular plant, though. Forest loss is often caused by climate change. Tropical rainforests are extremely humid due to the water vapor released along with the oxygen. But when a forest is cut down, the humidity levels decrease and causes the remaining plants to dry out. For example, drying out our tropical rainforests increases fire damage. Fires can be both accidental and intentional but destroy forests quickly (Juffe-Bignoli et al., 2014) (Figure 9.11).

Figure 9.11. *Causes of deforestation.*

Source: www.slideserve.com.

SUMMARY

Climate change directly and indirectly impacts forest growth and productivity through shifts in temperature, rainfall patterns, and other environmental factors. This phenomenon, exacerbated by human-induced alterations to Earth's climate and efforts to conserve biodiversity, poses significant challenges. Rising temperatures and altered precipitation patterns are likely to reduce water availability, exacerbating soil moisture depletion due to increased evapotranspiration rates. Coastal zones, defined by their economic significance and interface between land and water, are particularly vulnerable to these changes. The greenhouse effect, driven by gases like water vapor, exacerbates these effects by trapping heat in the atmosphere. Deforestation further amplifies these issues by reducing carbon sequestration capacity, affecting global carbon cycles. Concurrently, soil erosion, driven by natural processes and human activities like deforestation and agriculture, exacerbates environmental degradation. Addressing these interconnected challenges requires integrated approaches to mitigate climate impacts, conserve biodiversity, and sustainably manage land and water resources.

REVIEW QUESTIONS

1. What is climate change and how does it contribute to global warming?
2. How does climate change affect forest ecosystems?
3. What is the relationship between climate change and biodiversity?
4. What are the major and lesser greenhouse gases that contribute to climate change?
5. What are the causes of modern deforestation and how does it impact forests?

REFERENCES

1. ACIA (Arctic Climate Impact Assessment). (2004). *Impacts of a Warming Arctic: Arctic Climate Impact Assessment*. Cambridge University Press.
2. ACIA (Arctic Climate Impact Assessment). (2005). *Arctic Climate Impact Assessment*. Cambridge University Press.
3. Ainsworth, E. A., & Long, S. P. (2005). What have we learned from 15 years of free-air CO2 enrichment (FACE)? A meta-analytic review of the responses of photosynthesis, canopy properties, and plant production to rising CO2. *New Phytologist, 165*(2), 351–371. https://doi.org/10.1111/j.1469-8137.2004.01224.x.
4. Ainsworth, E. A., & Rogers, A. (2007). The response of photosynthesis and stomatal conductance to rising CO2: Mechanisms and environmental interactions. *Plant, Cell & Environment, 30*(3), 258–270. https://doi.org/10.1111/j.1365-3040.2007.01641.x.
5. Albritton, D., Cubasch, U., & Noguer, M. (2001). Summary for policy makers. In J. T. Houghton, Y. Ding, D. J. Griggs, M. Noguer, & P. J. van der Linden (Eds.), *Climate Change 2001: The Scientific Basis* (pp. 1–20). Cambridge University Press.

6. Allen, C. D., Macalady, A. K., Chenchouni, H., Bachelet, D., McDowell, N., Vennetier, M., & Cobb, N. (2010). A global overview of drought and heat-induced tree mortality reveals emerging climate change risks for forests. *Forest Ecology and Management, 259*(4), 660–684. https://doi.org/10.1016/j.foreco.2009.09.001.

7. Arnell, N. W. (1999). Climate change and global water resources. *Global Environmental Change, 9*(S1), S31–S49. https://doi.org/10.1016/S0959-3780(99)00017-5.

8. Bale, J. S., Masters, G. J., Hodkinson, I. D., Awmack, C., Bezemer, T. M., Brown, V. K., Butterfield, J., Buse, A., Coulson, J. C., Farrar, J., Good, J. E. G., Harrington, R., Hartley, S., Jones, T. H., Lindroth, R. L., Press, M. C., Symrnioudis, I., Watt, A. D., & Whittaker, J. B. (2002). Herbivory in global climate change research: Direct effects of rising temperature on insect herbivores. *Global Change Biology, 8*(1), 1–16. https://doi.org/10.1046/j.1365-2486.2002.00451.x.

9. Bellard, C., Bertelsmeier, C., Leadley, P., Thuiller, W., & Courchamp, F. (2012). Impacts of climate change on the future of biodiversity. *Ecology Letters, 15*(4), 365–377. https://doi.org/10.1111/j.1461-0248.2011.01736.x.

10. Bird, E. (2000). *Coastal Geomorphology: An Introduction*. John Wiley & Sons.

11. Bonan, G. B. (2008). Forests and climate change: Forcings, feedbacks, and the climate benefits of forests. *Science, 320*(5882), 1444–1449. https://doi.org/10.1126/science.1155121.

12. Brook, B. W., Sodhi, N. S., & Bradshaw, C. J. A. (2008). Synergies among extinction drivers under global change. *Trends in Ecology & Evolution, 23*(8), 453–460. https://doi.org/10.1016/j.tree.2008.03.011.

13. Brown, D., & Williams, A. (2003). *Int Forestry Rev, 5*, 148–155.

14. Calosi, P., Bilton, D. T., Spicer, J. I., & Atfield, A. (2008). Thermal tolerance and geographical range size in the Agabus brunneus group of European diving beetles (Coleoptera: Dytiscidae). *Journal of Biogeography, 35*(2), 295–305. https://doi.org/10.1111/j.1365-2699.2007.01793.x.

15. Cardinale, B. J., Duffy, J. E., Gonzalez, A., Hooper, D. U., Perrings, C., Venail, P., & Naeem, S. (2012). Biodiversity loss and its impact on humanity. *Nature, 486*(7401), 59–67. https://doi.org/10.1038/nature11148.

16. CBD (Convention on Biological Diversity). (1992). *Convention on Biological Diversity*. United Nations.

17. Coulson, S. J., Hodkinson, I. D., Webb, N. R., Mikkola, K., Harrison, J. A., & Pedgley, D. E. (2002). Aerial colonization of high Arctic islands by invertebrates: The diamondback moth *Plutella xylostella* (Lepidoptera: Yponomeutidae) as a potential indicator species. *Diversity and Distributions, 8*(5), 327–334. https://doi.org/10.1046/j.1472-4642.2002.00155.x.

18. Cramer, W., Bondeau, A., Woodward, F. I., Prentice, I. C., Betts, R. A., Brovkin, V., Cox, P. M., Fisher, V., Foley, J., Friend, A. D., et al. (2001). Global response of terrestrial ecosystem structure and function to CO_2 and climate change. *Global Change Biology, 7*(4), 357–373. https://doi.org/10.1046/j.1365-2486.2001.00383.x.

19. Dale, V. H., Joyce, L. A., McNulty, S., Neilson, R. P., Ayres, M. P., Flannigan, M. D.,

Hanson, P. J., Irland, L. C., Lugo, A. E., Peterson, C. J., Simberloff, D., Swanson, F. J., Stocks, B. J., & Wotton, B. M. (2001). Climate change and forest disturbances. *Bioscience, 51*(9), 723–734. https://doi.org/10.1641/0006-3568(2001)051[0723:CCAFD]2.0.CO;2.

20. Dansgaard, W., Johnsen, S. J., Clausen, H. B., Dahl-Jensen, D., Gundestrup, N. S., Hammer, C. U., & Oeschger, H. (1993). Evidence for general instability of past climate from a 250-kyr ice-core record. *Nature, 364*(6434), 218–220. https://doi.org/10.1038/364218a0.
21. Das, H. P. (2004). Adaptation strategies required to reduce vulnerability in agriculture and forestry to climate change, climate variability and climate extremes. In *Management Strategies in Agriculture and Forestry for Mitigation of Greenhouse Gas Emissions and Adaptation to Climate Variability and Climate Change* (pp. 41–92). World Meteorological Organization (WMO). Technical Note No. 202, WMO No. 969, Geneva, WMO.
22. Dirzo, R., Young, H. S., Galetti, M., Ceballos, G., Isaac, N. J., & Collen, B. (2014). Defaunation in the Anthropocene. *Science, 345*(6195), 401–406. https://doi.org/10.1126/science.1251817.
23. Fahrig, L. (2003). Effects of habitat fragmentation on biodiversity. *Annual Review of Ecology, Evolution, and Systematics, 34*(1), 487–515. https://doi.org/10.1146/annurev.ecolsys.34.011802.132419.
24. Forster, P., Ramaswamy, V., Artaxo, P., Berntsen, T., Betts, R., Fahey, D. W., & Zhang, H. (2007). Changes in atmospheric constituents and in radiative forcing. In S. Solomon, D. Qin, M. Manning, Z. Chen, M. Marquis, K. B. Averyt, M. Tignor, & H. L. Miller (Eds.), *Climate Change 2007: The Physical Science Basis. Contribution of Working Group I to the Fourth Assessment Report of the Intergovernmental Panel on Climate Change* (pp. 129–234). Cambridge University Press.
25. Galloway, J. N., Townsend, A. R., Erisman, J. W., Bekunda, M., Cai, Z., Freney, J. R., & Sutton, M. A. (2008). Transformation of the nitrogen cycle: Recent trends, questions, and potential solutions. *Science, 320*(5878), 889–892. https://doi.org/10.1126/science.1136674.
26. Gibbard, S., Caldeira, K., Bala, G., Phillips, T. J., & Wickett, M. (2005). *Geophysical Research Letters, 32*, L23705. https://doi.org/10.1029/2005GL023814.
27. Githeko, A. K., Lindsay, S. W., Confalonieri, U. E., & Patz, J. A. (2000). Climate change and vector-borne diseases: A regional analysis. *Bulletin of the World Health Organization, 78*(9), 1136–1147. https://www.who.int/bulletin/archives/78(9)1136.pdf.
28. Gornitz, V. (1995). Sea-level rise: A review of recent past and near-future trends. *Earth Surface Processes and Landforms, 20*(1), 7–20. https://doi.org/10.1002/esp.3290200103.
29. Gosain, A. K., Rao, S., & Basuray, D. (2006). Climate change impact assessment on hydrology of Indian river basins. *Current Science, 90*(3), 346–353. https://www.currentscience.ac.in/Volumes/90/03/0346.pdf (accessed on 29 August 2024).
30. Hagler, R. (1998). *The Global Timber Supply/Demand Balance to 2030: Has the Equation Changed?* A multi-client study by Wood Resources International. Wood Resources International, Reston, VA.

31. Heller, N. E., & Zavaleta, E. S. (2009). Biodiversity management in the face of climate change: A review of 22 years of recommendations. *Biological Conservation, 142*(1), 14–32. https://doi.org/10.1016/j.biocon.2008.10.008.

32. Houghton, J. (2004). *Global Warming: The Complete Briefing*. Cambridge University Press.

33. Huntington, T. G. (2006). Evidence for intensification of the global water cycle: Review and synthesis. *Journal of Hydrology, 319*(1–4), 83–95. https://doi.org/10.1016/j.jhydrol.2005.07.003.

34. Huntley, B. (1991). How plants respond to climate change: Migration rates, individualism and the consequences for plant communities. *Annals of Botany, 67*(Supplement_1), 15–22. https://doi.org/10.1006/anbo.1991.1014.

35. Immerzeel, W. W., van Beek, L. P. H., & Bierkens, M. F. P. (2010). Climate change will affect the Asian water towers. *Science, 328*(5984), 1382–1385. https://doi.org/10.1126/science.1183188.

36. IPCC (Intergovernmental Panel on Climate Change). (2007). *Climate Change 2007: Synthesis Report*. Contribution of Working Groups I, II and III to the Fourth Assessment Report of the Intergovernmental Panel on Climate Change. IPCC.

37. IPCC (Intergovernmental Panel on Climate Change). (2014). *Climate Change 2014: Impacts, Adaptation, and Vulnerability. Part A: Global and Sectoral Aspects*. Contribution of Working Group II to the Fifth Assessment Report of the Intergovernmental Panel on Climate Change. Cambridge University Press.

38. IPCC (Intergovernmental Panel on Climate Change). (2014). *Climate Change 2014: Synthesis Report*. Contribution of Working Groups I, II, and III to the Fifth Assessment Report of the Intergovernmental Panel on Climate Change. IPCC.

39. IPCC (Intergovernmental Panel on Climate Change). (2021). *Climate Change 2021: The Physical Science Basis*. Contribution of Working Group I to the Sixth Assessment Report of the Intergovernmental Panel on Climate Change. Cambridge University Press.

40. Irland, L. C., Adams, D., Alig, R., Betz, C. J., Chen, C. C., Hutchins, M., McCarl, B. A., Skog, K., & Sohngen, B. L. (2001). *BioScience, 51*, 753–764. https://doi.org/10.1641/0006-3568(2001)051[0753:BCVAAD]2.0.CO;2.

41. Juffe-Bignoli, D., Burgess, N. D., Bingham, H., Belle, E. M., de Lima, M. G., Deguignet, M., & Kingston, N. (2014). *Protected Planet Report 2014*. UNEP-WCMC.

42. Karnosky, D. F., Percy, K. E., Xiang, B., Callan, B., Noormets, A., Mankovska, B., Hopkin, A., Sober, J., Jones, W., Dickson, R. E., & Isebrands, J. G. (2002). Interacting elevated CO2 and tropospheric O3 predisposes aspen (Populus tremuloides Michx.) to infection by rust (Melampsora medusae f.sp. tremuloidae). *Global Change Biology, 8*(4), 329–338. https://doi.org/10.1046/j.1365-2486.2002.00492.x.

43. Kauppi, P. E., Assubel, J. H., Fang, J., Mather, A., Sedjo, R. A., & Waggoner, P. E. (2006). Changes in the carbon sink of forests. *Proceedings of the National Academy of Sciences USA, 103*(11), 17574–17579. https://doi.org/10.1073/pnas.0608355103.

44. Knutson, T. R., McBride, J. L., Chan, J., Emanuel, K., Holland, G., Landsea, C., &

Sugi, M. (2010). Tropical cyclones and climate change. *Nature Geoscience, 3*(3), 157–163. https://doi.org/10.1038/ngeo779.

45. Kozlov, M. V. (2008). Losses of birch foliage due to herbivory along geographical gradients in Europe: A climate-driven pattern? *Climatic Change, 87*(1), 107–117. https://doi.org/10.1007/s10584-007-9277-0.
46. Lindzen, R. S. (2009). Climategate and the calls for action. *Wall Street Journal*. Retrieved from: https://www.wsj.com (accessed on 29 August 2024).
47. Meehl, G. A., Stocker, T. F., Collins, W. D., Friedlingstein, P., Gaye, A. T., Gregory, J. M., & Zhao, Z. C. (2007). Global climate projections. In S. Solomon, D. Qin, M. Manning, Z. Chen, M. Marquis, K. B. Averyt, M. Tignor, & H. L. Miller (Eds.), *Climate Change 2007: The Physical Science Basis. Contribution of Working Group I to the Fourth Assessment Report of the Intergovernmental Panel on Climate Change* (pp. 747–846). Cambridge University Press.
48. Millennium Ecosystem Assessment. (2005). *Ecosystems and Human Well-Being: Biodiversity Synthesis*. World Resources Institute.
49. NASA. (2020). *Global Climate Change: Vital Signs of the Planet*. Retrieved from: https://climate.nasa.gov/ (accessed on 29 August 2024).
50. Norby, R. J., DeLucia, E. H., Gielen, B., Calfapietra, C., Giardina, C. P., King, J. S., & Karnosky, D. F. (2005). Forest response to elevated CO2 is conserved across a broad range of productivity. *Proceedings of the National Academy of Sciences, 102*(50), 18052–18056. https://doi.org/10.1073/pnas.0509478102.
51. Pacala, S., & Socolow, R. (2004). Stabilization wedges: Solving the climate problem for the next 50 years with current technologies. *Science, 305*(5686), 968–972. https://doi.org/10.1126/science.1100103.
52. Parmesan, C. (2006). Ecological and evolutionary responses to recent climate change. *Annual Review of Ecology, Evolution, and Systematics, 37*, 637–669. https://doi.org/10.1146/annurev.ecolsys.37.091305.110100.
53. Parmesan, C., & Yohe, G. (2003). A globally coherent fingerprint of climate change impacts across natural systems. *Nature, 421*(6918), 37–42. https://doi.org/10.1038/nature01286.
54. Perfecto, I., Vandermeer, J., & Wright, A. (2009). *Nature's Matrix: Linking Agriculture, Conservation and Food Sovereignty*. Earthscan.
55. Pimm, S. L., Jenkins, C. N., Abell, R., Brooks, T. M., Gittleman, J. L., Joppa, L. N., & Sexton, J. O. (2014). The biodiversity of species and their rates of extinction, distribution, and protection. *Science, 344*(6187), 1246752. https://doi.org/10.1126/science.1246752.
56. Ramanathan, V., Crutzen, P. J., Kiehl, J. T., & Rosenfeld, D. (2001). Aerosols, climate, and the hydrological cycle. *Science, 294*(5549), 2119–2124. https://doi.org/10.1126/science.1064034.
57. Sala, O. E., Chapin III, F. S., Armesto, J. J., Berlow, E., Bloomfield, J., Dirzo, R., & Wall, D. H. (2000). Global biodiversity scenarios for the year 2100. *Science, 287*(5459), 1770–1774. https://doi.org/10.1126/science.287.5459.1770.

58. Salama, N. K. G., Knowler, J. T., & Adams, C. E. (2007). Increasing abundance and diversity in the moth assemblage of east Loch Lomondside, Scotland over a 35-year period. *Journal of Insect Conservation, 11*(2), 151–156. https://doi.org/10.1007/s10841-006-9010-3.

59. Shindell, D., Kuylenstierna, J. C., Vignati, E., Van Dingenen, R., Amann, M., Klimont, Z., & Fowler, D. (2012). Simultaneously mitigating near-term climate change and improving human health and food security. *Science, 335*(6065), 183–189. https://doi.org/10.1126/science.121002

60. Solomon, S., Qin, D., Manning, M., Marquis, M., Averyt, K., Tignor, M. M. B., & Chen, Z. (2007). *Climate Change 2007: The Physical Science Basis*. Contribution of Working Group I to the Fourth Assessment Report of the Intergovernmental Panel on Climate Change. Cambridge University Press.

61. Stireman, J. O., III, Dyer, L. A., Janzen, D. H., Singer, M. S., Lill, J. T., Marquis, J. R., Ricklefs, R. E., Gentry, G. L., Hallwachs, W., Coley, P. D., Barone, J. A., Greemey, H. F., Connahs, H., Barbosa, P., Morais, H. C., & Diniz, I. R. (2005). Climatic unpredictability and parasitism of caterpillars: Implications of global warming. *Proceedings of the National Academy of Sciences, 102*(49), 17384–17387. https://doi.org/10.1073/pnas.0508694102.

62. Thomas, J. A. (2005). Monitoring change in the abundance and distribution of insects using butterflies and other indicator groups. *Philosophical Transactions of the Royal Society B, 360*(1454), 339–357. https://doi.org/10.1098/rstb.2004.1597.

63. Thompson, I. D., Mackey, B., McNulty, S., & Mosseler, A. (2009). *Forest Resilience, Biodiversity, and Climate Change*. Secretariat of the Convention on Biological Diversity.

64. UNFCCC (United Nations Framework Convention on Climate Change). (2015). *Adoption of the Paris Agreement*. United Nations Framework Convention on Climate Change. Retrieved from: https://unfccc.int (accessed on 29 August 2024).

65. Vanhanen, H., Veteli, T. O., Päivinen, S., Kellomäki, S., & Niemelä, P. (2007). Climate change and range shifts in two insect defoliators: Gypsy moth and nun moth–A model study. *Silva Fennica, 41*(4), 621–638. https://doi.org/10.14214/sf.309.

66. Veteli, T. O., Lahtinen, A., Repo, T., Niemelä, P., & Varama, M. (2005). Geographic variation in winter freezing susceptibility in the eggs of the European pine sawfly (*Neodiprion sertifer*). *Agricultural and Forest Entomology, 7*(2), 115–120. https://doi.org/10.1111/j.1461-9563.2005.00259.x.

67. Watson, J. E., Dudley, N., Segan, D. B., & Hockings, M. (2014). The performance and potential of protected areas. *Nature, 515*(7525), 67–73. https://doi.org/10.1038/nature13947.

Index